Modern Statistics with R

The past decades have transformed the world of statistical data analysis, with new methods, new types of data, and new computational tools. *Modern Statistics with R* introduces you to key parts of this modern statistical toolkit. It teaches you:

- **Data wrangling** – importing, formatting, reshaping, merging, and filtering data in R.
- **Exploratory data analysis** – using visualisations and multivariate techniques to explore datasets.
- **Statistical inference** – modern methods for testing hypotheses and computing confidence intervals.
- **Predictive modelling** – regression models and machine learning methods for prediction, classification, and forecasting.
- **Simulation** – using simulation techniques for sample size computations and evaluations of statistical methods.
- **Ethics in statistics** – ethical issues and good statistical practice.
- **R programming** – writing code that is fast, readable, and (hopefully!) free from bugs.

No prior programming experience is necessary. Clear explanations and examples are provided to accommodate readers at all levels of familiarity with statistical principles and coding practices. A basic understanding of probability theory can enhance comprehension of certain concepts discussed within this book.

In addition to plenty of examples, the book includes more than 200 exercises, with fully worked solutions available at: www.modernstatisticswithr.com.

Måns Thulin is a consultant, researcher, and teacher in Statistics. He started teaching Statistics with R at Uppsala University in 2007, while still an undergraduate student. Since then, he has used Statistics and R to tackle problems in diverse fields, ranging from how to identify nuclear fuel and how to improve milking robots, to wine tastings and designing music videos. His award-winning research in Statistical Methodology is concerned with modern approaches to classical statistical methods.

Modern Statistics with R

From Wrangling and Exploring Data to Inference and Predictive Modelling

Second Edition

Måns Thulin

CRC Press
Taylor & Francis Group
Boca Raton London New York

CRC Press is an imprint of the
Taylor & Francis Group, an **informa** business

A CHAPMAN & HALL BOOK

Designed cover image: © Måns Thulin

MATLAB® and Simulink® are trademarks of The MathWorks, Inc. and are used with permission. The MathWorks does not warrant the accuracy of the text or exercises in this book. This book's use or discussion of MATLAB® or Simulink® software or related products does not constitute endorsement or sponsorship by The MathWorks of a particular pedagogical approach or particular use of the MATLAB® and Simulink® software.

Second edition published 2025
by CRC Press
2385 NW Executive Center Drive, Suite 320, Boca Raton FL 33431

and by CRC Press
4 Park Square, Milton Park, Abingdon, Oxon, OX14 4RN

CRC Press is an imprint of Taylor & Francis Group, LLC

© 2025 Måns Thulin

First edition published by Eos Chasma Press 2021

Library of Congress Cataloging-in-Publication Data

Names: Thulin, Måns, author.
Title: Modern statistics with R : from wrangling and exploring data to inference and predictive modelling / Måns Thulin.
Description: Second edition. | Boca Raton, FL : CRC Press, [2024] | Includes bibliographical references and index.
Identifiers: LCCN 2024006535 (print) | LCCN 2024006536 (ebook) | ISBN 9781032497457 (hbk) | ISBN 9781032512440 (pbk) | ISBN 9781003401339 (ebk)
Subjects: LCSH: R (Computer program language) | Statistics.
Classification: LCC QA276.45.R3 T555 2024 (print) | LCC QA276.45.R3 (ebook) | DDC 519.50285/5133--dc23/eng/20240509
LC record available at https://lccn.loc.gov/2024006535
LC ebook record available at https://lccn.loc.gov/2024006536

ISBN: 978-1-032-49745-7 (hbk)
ISBN: 978-1-032-51244-0 (pbk)
ISBN: 978-1-003-40133-9 (ebk)

DOI: 10.1201/9781003401339

Typeset in LM Roman
by KnowledgeWorks Global Ltd.

Publisher's note: This book has been prepared from camera-ready copy provided by the authors.

Access the Instructor and Student Resources: www.routledge.com/9781032512440.

To Lisa, Alvar, and Irma

Contents

List of Figures

1

Introduction

1.1 Welcome to R

Welcome to the wonderful world of R!

R is not like other statistical software packages. It is free, versatile, fast, and modern. It has a large and friendly community of users that help answer questions and develop new R tools. With more than 17,000 add-on packages available, R offers more functions for data analysis than any other statistical software. This includes specialised tools for disciplines as varied as political science, environmental chemistry, and astronomy, and new methods come to R long before they come to other programs. R makes it easy to construct reproducible analyses and workflows that allow you to easily repeat the same analysis more than once.

R is not like other programming languages. It was developed by statisticians as a tool for data analysis and not by software engineers as a tool for other programming tasks. It is designed from the ground up to handle data, which is evident. But it is also flexible enough to be used to create interactive web pages, automated reports, and application programming interfaces (APIs).

Simply put, R is currently the best tool there is for data analysis.

1.2 About this book

This book was born out of lecture notes and materials that I created for courses at the University of Edinburgh, Uppsala University, Dalarna University, the Swedish University of Agricultural Sciences, and Karolinska Institutet. It can be used as a textbook, for self-study, or as a reference manual for R. No background in programming is assumed.

This is not a book that has been written with the intention that you should read it back-to-back. Rather, it is intended to serve as a guide to what to do next as you explore R. Think of it as a conversation, where you and I discuss different topics related to data analysis and data wrangling. At times I'll do the talking, introduce concepts and pose questions. At times you'll do the talking, working with exercises and discovering all that R has to offer. The best way to learn R is to use R. You should strive for active learning, meaning that you should spend more time with R and less time stuck with your nose in a book. Together we will strive for an exploratory approach, where the text guides you to discoveries and the exercises challenge you to go further. This is how I've been teaching R since 2008, and I hope that it's a way that you will find works well for you. The book contains more than 200 exercises. Apart from a number of open-ended questions about ethical issues, all exercises involve R code. These exercises all have worked solutions available on the book's webpage, http://www.modernstatisticswithr.com. It is highly recommended that you actually work with all the exercises, as they are central to the approach to learning that this book seeks to support: using R to solve problems is a much better way to learn the language than to just read about how to use R to solve problems. Once you have finished an exercise (or attempted but failed to finish it) read the proposed solution; it may differ from what you came up with and will sometimes contain comments that you may find interesting. Treat the proposed solutions as a part of our conversation. As you work with the exercises and compare your solutions to those in the back of the book, you will gain more and more experience working with R and build your own library of examples of how problems can be solved.

Some books on R focus entirely on data science – data wrangling and exploratory data analysis – ignoring the many great tools R has to offer for deeper data analyses. Others focus on predictive modelling or classical statistics but ignore data-handling, which is a vital part of modern statistical work. Many introductory books on statistical methods put too little focus on recent advances in computational statistics and advocate methods that have become obsolete. Far too few books contain discussions of ethical issues in statistical practice. This book aims to cover all of these topics and show you the state-of-the-art tools for all these tasks. It covers data science and (modern!) classical statistics as well as predictive modelling and machine learning, and deals with important topics that rarely appear in other introductory texts, such as simulation. It is written for R 4.3 or later and will teach you powerful add-on packages like `data.table`, `dplyr`, `ggplot2`, and `caret`.

The expanded second edition includes new and updated examples throughout the book, and new material on, among other things, fundamental statistical concepts, survival analysis, and structural equation models.

The book is organised as follows:

Chapter 2 covers basic concepts and shows how to use R to import and handle data, compute descriptive statistics, create nice-looking plots, and use pipes.

Chapter 3 is concerned with concepts like hypothesis testing and confidence intervals, and how to use common statistical methods like the t-test and χ^2-tests in R. It also covers the basics of computer-intensive methods like permutation tests and the bootstrap.

Chapter 4 covers exploratory data analysis using statistical graphics, as well as unsupervised learning techniques like principal components analysis and clustering. It also contains an introduction to R Markdown, a powerful markup language that can be used, e.g., to create reports.

Chapter 5 describes how to deal with messy data – including filtering, rearranging and merging datasets – and different data types.

Chapter 6 deals with programming in R and covers concepts such as iteration, conditional statements and functions.

Chapters 4-6 can be read in any order.

Chapter 7 is concerned with simulation and its role in modern statistics. This includes sample size computations, the bootstrap, and evaluation of statistical methods. It is not required reading for understanding most of the material in subsequent chapters, but the principles presented in this chapter are used in some of the more advanced examples.

Chapter 8 deals with various regression models, including linear, generalised linear and mixed models. It also includes materials on multiple imputation of missing values, along with methods for creating matched samples.

Chapter 9 is about survival analysis and methods for analysing different kinds of censored data, such as data with detection limits.

Chapter 10 describes exploratory factor analysis and confirmatory analyses using structural equation models, as well as mediation analysis.

Chapter 11 covers predictive modelling, including regularised regression, machine learning techniques, and an introduction to forecasting using time series models. Much focus is given to cross-validation and ways to evaluate the performance of predictive models.

Chapter 12 gives an overview of more advanced topics, including parallel computing, matrix computations, and integration with other programming languages.

Chapter 13 covers debugging, i.e., how to spot and fix errors in your code. It includes a list of more than 25 common error and warning messages, and advice on how to resolve them.

Chapter 14 covers some mathematical aspects of methods used in Chapters 7-11.

Finally, Chapter 15, available online, contains fully worked solutions to all exercises in the book.

The datasets that are used for the examples and exercises can be downloaded from:

http://www.modernstatisticswithr.com/data.zip

I have opted not to put the datasets in an R package, because I want you to practice loading data from files, as this is what you'll be doing whenever you use R for real work.

This book is available both in print and as an open access online book at: http://www.mo dernstatisticswithr.com.

I am indebted to team at Chapman & Hall for their work on this book, and to the numerous readers who have provided feedback on various drafts. My sincerest thanks go out to all of you. Any remaining errors are, obviously, entirely my own fault.

Finally, there are countless packages and statistical methods that deserve a mention but aren't included in the book. Like any author, I've had to draw the line somewhere. If you feel that something is missing, feel free to get in touch, and I'll gladly consider it for future revisions.

2

The basics

Let's start from the very beginning. This chapter acts as an introduction to R. It will show you how to install and work with R and RStudio.

After working with the material in this chapter, you will be able to:

- Create reusable R scripts,
- Store data in R,
- Use functions in R to analyse data,
- Install add-on packages adding more features to R,
- Compute descriptive statistics like the mean and the median, including for subgroups,
- Do mathematical calculations,
- Create nice-looking plots, including scatterplots, boxplots, histograms and bar charts,
- Distinguish between different data types,
- Import data from Excel spreadsheets and csv text files,
- Add new variables to your data,
- Modify variables in your data,
- Remove variables from your data,
- Save and export your data,
- Work with RStudio projects,
- Use |> pipes to chain functions together, and
- Find errors in your code.

2.1   Installing R and RStudio

To download R, go to the R Project website:

https://cran.r-project.org/mirrors.html

Choose a *download mirror*, i.e., a server to download the software from. I recommend choosing a mirror close to you. You can then choose to download R for either Linux[1], Mac or Windows by following the corresponding links (Figure 2.1).

The version of R that you should download is called the (base) binary. Download and run it to install R. You may see mentions of 64-bit and 32-bit versions of R; if you have a modern computer (which in this case means a computer from 2010 or later), you should go with the 64-bit version.

You have now installed the R programming language. Working with it is easier with an *integrated development environment* (IDE) which allows you to easily write, run and debug

---

[1]For many Linux distributions, R is also available from the package management system.

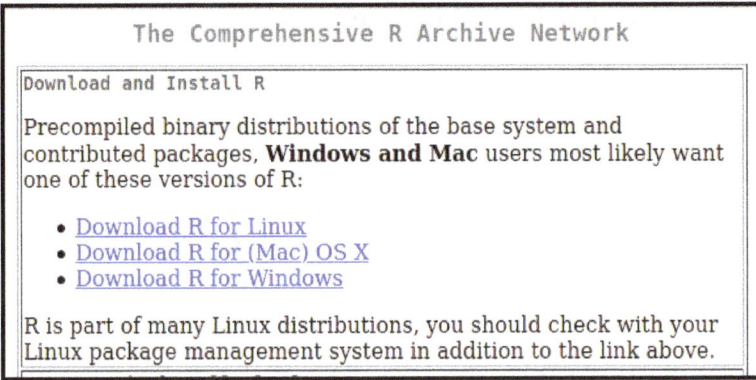

FIGURE 2.1 Screenshot from the R download page at https://ftp.acc.umu.se/mirror/ CRAN/

your code. This book is written for use with the RStudio IDE, but 99.9% of it will work equally well with other IDEs, like Emacs with ESS or Jupyter notebooks.

To download RStudio, go to the RStudio download page:

https://rstudio.com/products/rstudio/download/#download

Click on the link to download the installer for your operating system, and then run it.

2.2 A first look at RStudio

When you launch RStudio, you will see three or four panes, as seen in Figure 2.2.

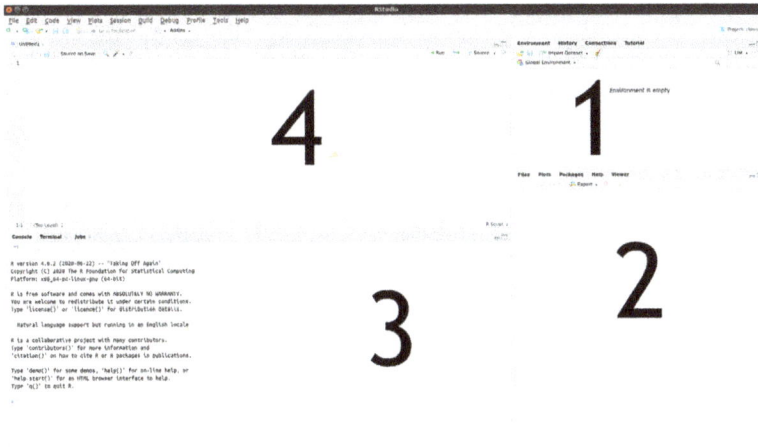

FIGURE 2.2 The four RStudio panes.

1. The *Environment* pane, where a list of the data you have imported and created can be found.
2. The *Files*, *Plots* and *Help* pane, where you can see a list of available files, will be

able to view graphs that you produce, and can find help documents for different parts of R.

3. The *Console* pane, used for running code. This is where we'll start with the first few examples.

4. The *Script* pane, used for writing code. This is where you'll spend most of your time working.

If you launch RStudio by opening a file with R code, the *Script* pane will appear; otherwise, it won't. Don't worry if you don't see it at this point; you'll learn how to open it soon enough.

The *Console* pane will contain R's startup message, which shows information about which version of R you're running[2]:

```
R version 4.4.0 (2024-04-24) -- "Puppy Cup"
Copyright (C) 2024 The R Foundation for Statistical Computing
Platform: aarch64-apple-darwin20 (64-bit)

R is free software and comes with ABSOLUTELY NO WARRANTY.
You are welcome to redistribute it under certain conditions.
Type 'license()' or 'licence()' for distribution details.

  Natural language support but running in an English locale

R is a collaborative project with many contributors.
Type 'contributors()' for more information and
'citation()' on how to cite R or R packages in publications.

Type 'demo()' for some demos, 'help()' for on-line help, or
'help.start()' for an HTML browser interface to help.
Type 'q()' to quit R.
```

You can resize the panes as you like, either by clicking and dragging their borders or using the minimise/maximise buttons in the upper right corner of each pane.

When you exit RStudio, you will be asked if you wish to *save your workspace*, meaning that the data that you've worked with will be stored so that it is available the next time you run R. That might sound like a good idea, but in general, I recommend that you don't save your workspace, as that often turns out to cause problems down the line. It is almost invariably a much better idea to simply rerun the code you worked with in your next R session.

2.3 Running R code

Everything that we do in R revolves around *code*. The code will contain instructions for how the computer should treat, analyse and manipulate[3] data. Thus each line of code tells R to do something: compute a mean value, create a plot, sort a dataset, or something else.

---

[2]In addition to the version number, each release of R has a nickname referencing a Peanuts comic by Charles Schulz.

[3]The word "manipulate" has different meanings. Just to be perfectly clear: whenever I speak of *manipulating data* in this book, I will mean *handling and transforming the data*, not tampering with it.

Throughout the text, there will be code chunks that you can paste into the Console pane. Here is the first example of such a code chunk. Type or copy the code into the Console and press Enter on your keyboard:

```
1+1
```

Code chunks will frequently contain multiple lines. You can select and copy both lines from the digital version of this book and simultaneously paste them directly into the Console:

```
2*2
1+2*3-5
```

As you can see, when you type the code into the Console pane and press Enter, R *runs* (or *executes*) the code and returns an answer. To get you started, the first exercise will have you write a line of code to perform a computation. You can find a solution to this and other exercises at the book's webpage: http://www.modernstatisticswithr.com.

~

Exercise 2.1. Use R to compute the product of the first 10 integers: $1 \cdot 2 \cdot 3 \cdot 4 \cdot 5 \cdot 6 \cdot 7 \cdot 8 \cdot 9 \cdot 10$.

2.3.1   R scripts

When working in the Console pane (i.e., when the Console pane is active and you see a blinking text cursor in it), you can use the up arrow ↑ on your keyboard to retrieve lines of code that you've previously used. There is however a much better way of working with R code: to put it in *script files*. These are files containing R code that you can save and then run again whenever you like.

To create a new script file in RStudio, press Ctrl+Shift+N on your keyboard, or select *File > New File > R Script* in the menu. This will open a new Script pane (or a new tab in the Script pane, in case it was already open). You can then start writing your code in the Script pane. For instance, try the following:

```
1+1
2*2
1+2*3-5
(1+2)*3-5
```

In the Script pane, when you press Enter, you insert a new line instead of running the code. That's because the Script pane is used for *writing* code rather than *running* it. To actually run the code, you must send it to the Console pane. This can be done in several ways. Let's give them a try to see which you prefer.

To run the entire script, do one of the following:

- Press the Source button in the upper right corner of the Script pane.
- Press Ctrl+Shift+Enter on your keyboard.
- Press Ctrl+Alt+Enter on your keyboard to run the code without printing the code and its output in the Console.

To run a part of the script, first select the lines you wish to run, e.g., by highlighting them using your mouse. Then do one of the following:

- Press the Run button in the upper right corner of the Script pane.
- Press Ctrl+Enter on your keyboard (this is how I usually do it).

To save your script, click the Save icon, choose *File > Save* in the menu or press Ctrl+S. R script files should have the file extension .R, e.g. My first R script.R. Remember to save your work often, and to save your code for all the examples and exercises in this book. You will likely want to revisit old examples in the future to see how something was done.

2.4   Variables and functions

Of course, R is so much more than just a fancy calculator. To unlock its full potential, we need to discuss two key concepts: *variables* (used for storing data) and *functions* (used for doing things with the data).

2.4.1   Storing data

Without data, there is no data analytics. So how can we store and read data in R? The answer is that we use *variables*. A variable is a name used to store data, so that we can refer to a dataset when we write code. As the name *variable* implies, what is stored can change over time[4].

The code

```
x <- 4
```

is used to *assign* the value 4 to the *variable* x. It is read as "assign 4 to x". The <- part is made by writing a less than sign (<) and a hyphen (-) with no space between them[5].

If we now type x in the Console, R will return the answer 4. Well, almost. In fact, R returns the following rather cryptic output:

```
[1] 4
```

The meaning of the 4 is clear – it's a 4. We'll return to what the [1] part means soon.

Now that we've created a variable, called x, and assigned a value (4) to it, x will have the value 4 whenever we use it again. This works just like a mathematical formula, where we for instance can insert the value $x = 4$ into the formula $x + 1$. The following two lines of code will compute $x + 1 = 4 + 1 = 5$ and $x + x = 4 + 4 = 8$:

```
x + 1
x + x
```

---

[4]If you are used to programming languages like C or Java, you should note that R is *dynamically typed*, meaning that the data type of an R variable also can change over time. This also means that there is no need to declare variable types in R (which is either liberating or terrifying, depending on what type of programmer you are).

[5]In RStudio, you can also create the assignment operator <- by using the keyboard shortcut Alt+- (i.e., press Alt and the - button at the same time).

Once we have assigned a value to x, it will appear in the Environment pane in RStudio, where you can see both the variable's name and its value.

The left-hand side of the assignment x <- 4 is always the name of a variable, but the right-hand side can be any piece of code that creates some sort of object to be stored in the variable. For instance, we could perform a computation on the right-hand side and then store the result in the variable:

```
x <- 1 + 2 + 3 + 4
```

R first evaluates the entire right-hand side, which in this case amounts to computing 1+2+3+4, and then assigns the result (10) to x. Note that the value previously assigned to x (i.e., 4) now has been replaced by 10. After a piece of code has been run, the values of the variables affected by it will have changed. There is no way to revert the run and get that 4 back, save to rerun the code that generated it in the first place.

You'll notice that in the code above, I've added some spaces, for instance between the numbers and the plus signs. This is simply to improve readability. The code works just as well without spaces:

```
x<-1+2+3+4
```

or with spaces in some places, but not in others:

```
x<- 1+2+3 + 4
```

However, you cannot place a space in the middle of the <- arrow. The following will *not* assign a value to x:

```
x < - 1 + 2 + 3 + 4
```

Running that piece of code rendered the output FALSE. This is because < - with a space has a different meaning than <- in R, one that we shall return to later in this chapter.

In some cases, you may want to switch the direction of the arrow, so that the variable name is on the right-hand side. This is called right-assignment and works just fine too:

```
2 + 2 -> y
```

Later on, we'll see plenty of examples where right-assignment comes in handy.

∼

Exercise 2.2. Do the following using R:

1. Compute the sum $924 + 124$ and assign the result to a variable named a.

2. Compute $a \cdot a$.

2.4.2   What's in a name?

You now know how to assign values to variables. But what should you call your variables? Of course, you can follow the examples in the previous section and give your variables names like x, y, a and b. However, you don't have to use single-letter names, and for the sake of readability, it is often preferable to give your variables more informative names. Compare the following two code chunks:

```
y <- 100
z <- 20
x <- y - z
```

and

```
income <- 100
taxes <- 20
net_income <- income - taxes
```

Both chunks will run without any errors and yield the same results, and yet there is a huge difference between them. The first chunk is opaque – in no way does the code help us conceive *what it actually does*. On the other hand, it is perfectly clear that the second chunk is used to compute a net income by subtracting taxes from income. You don't want to be a chunk-one type R user, who produces impenetrable code with no clear purpose. You want to be a chunk-two type R user, who writes clear and readable code where the intent of each line is clear. Take it from me; for years I was a chunk-one guy. I managed to write a lot of useful code, but whenever I had to return to my old code to reuse it or fix some bug, I had difficulties understanding what each line was supposed to do. My new life as a chunk-two guy is better in every way.

So, what's in a name? Shakespeare's balcony-bound Juliet would have us believe that that which we call a rose by any other name would smell as sweet. Translated to R practice, this means that your code will run just fine no matter what names you choose for your variables. But when you or somebody else reads your code, it will help greatly if you call a rose, a rose and not x or my_new_variable_5.

You should note that R is case-sensitive, meaning that my_variable, MY_VARIABLE, My_Variable, and mY_VariABle are treated as different variables. To access the data stored in a variable, you must use its exact name – including lower- and uppercase letters in the right places. Writing the wrong variable name is one of the most common errors in R programming.

You'll frequently find yourself wanting to compose variable names out of multiple words, as we did with net_income. However, R does not allow spaces in variable names, and so net income would not be a valid variable name. There are a few different naming conventions that can be used to name your variables:

- snake_case, where words are separated by an underscore (_). Example: household_net_income.
- camelCase or CamelCase, where each new word starts with a capital letter. Example: householdNetIncome or HouseholdNetIncome.
- period.case, where each word is separated by a period (.). You'll find this used a lot in R,

but I'd advise that you don't use it for naming variables, as a period in the middle of a name can have a different meaning in more advanced cases[6]. Example: `household.net.income`.

- `concatenatedwordscase`, where the words are concatenated using only lowercase letters. Adownsidetothisconventionisthatitcanmakevariablenamesverydifficultoreadsousethis atyourownrisk. Example: `householdnetincome`
- `SCREAMING_SNAKE_CASE`, which mainly is used in Unix shell scripts these days. You can use it in R if you like, although you will run the risk of making others think that you are either angry, super excited or stark staring mad[7]. Example: `HOUSEHOLD_NET_INCOME`.

Some characters, including spaces, `-`, `+`, `*`, `:`, `=`, `!` and `$` are not allowed in variable names, as these all have other uses in R. The plus sign `+`, for instance, is used for addition (as you would expect), and allowing it to be used in variable names would therefore cause all sorts of confusion. In addition, variable names can't start with numbers. Other than that, it is up to you how you name your variables and which convention you use. Remember, your variable will "smell as sweet" regardless of what name you give it, but using a good naming convention will improve readability[8].

Another great way to improve the readability of your code is to use *comments*. A comment is a piece of text, marked by `#`, that is ignored by R. As such, it can be used to explain what is going on to people who read your code (including future you) and to add instructions for how to use the code. Comments can be placed on separate lines or at the end of a line of code. Here is an example:

```
###############################################################
#   This lovely little code snippet can be used to compute   #
#                    your net income.                        #
###############################################################

# Set income and taxes:
income <- 100   # Replace 100 with your income
taxes <- 20      # Replace 20 with how much taxes you pay

# Compute your net income:
net_income <- income - taxes
# Voilà!
```

In the Script pane in RStudio, you can comment and uncomment (i.e., remove the `#` symbol) a row by pressing Ctrl+Shift+C on your keyboard. This is particularly useful if you wish to comment or uncomment several lines – simply select the lines and press Ctrl+Shift+C.

∼

Exercise 2.3. Answer the following questions:

1. What happens if you use an invalid character in a variable name? Try e.g., the following:

---

[6]Specifically, the period is used to separate methods and classes in object-oriented programming, which is hugely important in R (although you can use R for several years without realising this).

[7]I find myself using screaming snake case on occasion. Make of that what you will.

[8]I recommend `snake_case` or `camelCase`, just in case that wasn't already clear.

```
net income <- income - taxes
net-income <- income - taxes
ca$h <- income - taxes
```

2. What happens if you put R code as a comment? For example,

```
income <- 100
taxes <- 20
net_income <- income - taxes
# gross_income <- net_income + taxes
```

3. What happens if you remove a line break and replace it by a semicolon ; ? For example,

```
income <- 200; taxes <- 30
```

4. What happens if you do two assignments on the same line? For example,

```
income2 <- taxes2 <- 100
```

2.4.3 Vectors and data frames

Almost invariably, you'll deal with more than one figure at a time in your analyses. For instance, we may have a list of the ages of customers at a bookstore:

$$28, 48, 47, 71, 22, 80, 48, 30, 31$$

Of course, we could store each observation in a separate variable:

```
age_person_1 <- 28
age_person_2 <- 48
age_person_3 <- 47
# ...and so on
```

...but this quickly becomes awkward. A much better solution is to store the entire list in just one variable. In R, such a list is called a *vector*. We can create a vector using the following code, where c stands for *combine*:

```
age <- c(28, 48, 47, 71, 22, 80, 48, 30, 31)
```

The numbers in the vector are called *elements*. We can treat the vector variable age just as we treated variables containing a single number. The difference is that the operations will apply to all elements in the list. So, for instance, if we wish to express the ages in months rather than years, we can convert all ages to months using:

```
age_months <- age * 12
```

Most of the time, data will contain measurements of more than one quantity. In the case of our bookstore customers, we also have information about the amount of money they spent on their last purchase:

$$20, 59, 2, 12, 22, 160, 34, 34, 29$$

First, let's store this data in a vector:

```
purchase <- c(20, 59, 2, 12, 22, 160, 34, 34, 29)
```

It would be nice to combine these two vectors into a table, like we would do in a spreadsheet software such as Excel. That would allow us to look at relationships between the two vectors – perhaps we could find some interesting patterns? In R, tables of vectors are called *data frames*. We can combine the two vectors into a data frame as follows:

```
bookstore <- data.frame(age, purchase)
```

If you type bookstore into the Console, it will show a simply formatted table with the values of the two vectors (and row numbers):

```
> bookstore
  age purchase
1  28       20
2  48       59
3  47        2
4  71       12
5  22       22
6  80      160
7  48       34
8  30       34
9  31       29
```

A better way to look at the table may be to click on the variable name bookstore in the Environment pane, which will open the data frame in a spreadsheet format.

You will have noticed that R tends to print a [1] at the beginning of the line when we ask it to print the value of a variable:

```
> age
[1] 28 48 47 71 22 80 48 30 31
```

Why? Well, let's see what happens if we print a longer vector:

```
# When we enter data into a vector, we can put line breaks between
# the commas:
distances <- c(687, 5076, 7270, 967, 6364, 1683, 9394, 5712, 5206,
```

```
          4317, 9411, 5625, 9725, 4977, 2730, 5648, 3818, 8241,
          5547, 1637, 4428, 8584, 2962, 5729, 5325, 4370, 5989,
          9030, 5532, 9623)
distances
```

Depending on the size of your Console pane, R will require a different number of rows to display the data in distances. The output will look something like this:

```
> distances
 [1]  687 5076 7270  967 6364 1683 9394 5712 5206 4317 9411 5625 9725
[14] 4977 2730 5648 3818 8241 5547 1637 4428 8584 2962 5729 5325 4370
[27] 5989 9030 5532 9623
```

or, if you have a narrower pane,

```
> distances
 [1]  687 5076 7270  967 6364 1683 9394
 [8] 5712 5206 4317 9411 5625 9725 4977
[15] 2730 5648 3818 8241 5547 1637 4428
[22] 8584 2962 5729 5325 4370 5989 9030
[29] 5532 9623
```

The numbers within the square brackets – [1], [8], [15], and so on – tell us which *elements* of the vector are printed first on each row. So in the latter example, the first element in the vector is 687, the 8th element is 5712, the 15th element is 2730, and so forth. Those numbers, called the *indices* of the elements, aren't exactly part of your data, but as we'll see later they are useful for keeping track of it.

This also tells you something about the inner workings of R. The fact that

```
x <- 4
x
```

renders the output

```
> x
[1] 4
```

tells us that x in fact is a vector, albeit with a single element. Almost everything in R is a vector, in one way or another.

Being able to put data on multiple lines when creating vectors is hugely useful, but can also cause problems if you forget to include the closing bracket). Try running the following code, where the final bracket is missing, in your Console pane:

```
distances <- c(687, 5076, 7270, 967, 6364, 1683, 9394, 5712, 5206,
          4317, 9411, 5625, 9725, 4977, 2730, 5648, 3818, 8241,
          5547, 1637, 4428, 8584, 2962, 5729, 5325, 4370, 5989,
          9030, 5532, 9623
```

When you hit Enter, a new line starting with a + sign appears. This indicates that R doesn't think that your statement has finished. You can either cancel running the code by pressing Escape on your keyboard, or finish it by typing) in the Console and then pressing Enter.

Vectors and data frames are hugely important when working with data in R. Chapter 5 is devoted to how to work with these objects.

~

Exercise 2.4. Do the following:

1. Create two vectors, height and weight, containing the heights and weights of five fictional people (i.e., just make up some numbers!).

2. Combine your two vectors into a data frame.

You will use these vectors in Exercise 2.6.

Exercise 2.5. Try creating a vector using x <- 1:5. What happens? What happens if you use 5:1 instead? How can you use this notation to create the vector $(1, 2, 3, 4, 5, 4, 3, 2, 1)$?

2.4.4 Functions

You have some data. Great. But simply having data is not enough – you want to *do* something with it. Perhaps you want to draw a graph, compute a mean value or apply some advanced statistical model to it. To do so, you will use a *function*.

A function is a ready-made set of instructions, code, that tells R to do something. There are thousands of functions in R. Typically, you insert a variable into the function, and it returns an answer. The code for doing this follows the pattern function_name(variable_name). As a first example, consider the function mean, which computes the mean of a variable:

```
# Compute the mean age of bookstore customers:
age <- c(28, 48, 47, 71, 22, 80, 48, 30, 31)
mean(age)
```

Note that the code follows the pattern function_name(variable_name): the function's name is mean and the variable's name is age.

Some functions take more than one variable as input, and may also have additional *arguments* (or *parameters*) that you can use to control the behaviour of the function. One such example is cor, which computes the correlation, a measure of (a certain kind of) dependence between two variables ranging from -1 (strong negative dependence) to 1 (strong positive dependence) via 0 (no dependence). You'll learn more about correlations in Section 3.8.

```
# Compute the correlation between the variables age and purchase:
age <- c(28, 48, 47, 71, 22, 80, 48, 30, 31)
purchase <- c(20, 59, 2, 12, 22, 160, 34, 34, 29)
cor(age, purchase)
```

The answer, 0.59, means that there appears to be a fairly strong positive correlation between age and the purchase size, which implies that older customers tend to spend more. On the other hand, just by looking at the data we can see that the oldest customer, aged 80, spent much more than anybody else – 160 monetary units. It can happen that such *outliers* strongly influence the computation of the correlation. By default, cor uses the Pearson correlation formula, which is known to be sensitive to outliers. It is therefore of interest to also perform the computation using a formula that is more robust to outliers, such as the Spearman correlation. This can be done by passing an additional *argument* to cor, telling it which method to use for the computation:

```
cor(age, purchase, method = "spearman")
```

The resulting correlation, 0.35, is substantially lower than the previous result. Perhaps the correlation isn't all that strong after all.

So, how can we know what arguments to pass to a function? Luckily, we don't have to memorise all possible arguments for all functions. Instead, we can look at the *documentation*, i.e., help file, for a function that we are interested in. This is done by typing ?function_name in the Console pane, or doing a web search for R function_name. To view the documentation for the cor function, type:

```
?cor
```

The documentation for R functions all follow the same pattern:

- *Description*: a short (and sometimes quite technical) description of what the function does.
- *Usage*: an abstract example of how the function is used in R code.
- *Arguments*: a list and description of the input arguments for the function.
- *Details*: further details about how the function works.
- *Value*: information about the output from the function.
- *Note*: additional comments from the function's author (not always included).
- *References*: references to papers or books related to the function (not always included).
- *See Also*: a list of related functions.
- *Examples*: practical (and sometimes less practical) examples of how to use the function.

The first time that you look at the documentation for an R function, all this information can be a bit overwhelming. Perhaps even more so for cor, which is a bit unusual in that it shares its documentation page with three other (heavily related) functions: var, cov and cov2cor. Let the section headlines guide you when you look at the documentation. What information are you looking for? If you're just looking for an example of how the function is used, scroll down to Examples. If you want to know what arguments are available, have a look at Usage and Arguments.

Finally, there are a few functions that don't require any input at all, because they don't do anything with your variables. One such example is Sys.time() which prints the current time on your system:

```
Sys.time()
```

Note that even though Sys.time doesn't require any input, you still have to write the parentheses (), which tells R that you want to run a function.

~

Exercise 2.6. Using the data you created in Exercise 2.4, do the following:

1. Compute the mean height of the people.

2. Compute the correlation between height and weight.

Exercise 2.7. Do the following:

1. Read the documentation for the function `length`. What does it do? Apply it to your `height` vector.

2. Read the documentation for the function `sort`. What does it do? What does the argument `decreasing` (the values of which can be either `FALSE` or `TRUE`) do? Apply the function to your `weight` vector.

2.4.5   Mathematical operations

To perform addition, subtraction, multiplication and division in R, we can use the standard symbols `+`, `-`, `*`, `/`. As in mathematics, expressions within parentheses are evaluated first, and multiplication is performed before addition. So `1 + 2*(8/2)` is $1+2\cdot(8/2) = 1+2\cdot4 = 1+8 = 9$.

In addition to these basic arithmetic operators, R has a number of mathematical functions that you can apply to your variables, including square roots, logarithms and trigonometric functions. Below is an incomplete list, showing the syntax for using the functions on a variable x. Throughout, a is supposed to be a number.

- `abs(x)`: computes the absolute value $|x|$.
- `sqrt(x)`: computes $\sqrt{x}$.
- `log(x)`: computes the logarithm of x with the natural number e as the base.
- `log(x, base = a)`: computes the logarithm of x with the number a as the base.
- `a^x`: computes a^x.
- `exp(x)`: computes e^x.
- `sin(x)`: computes $\sin(x)$.
- `sum(x)`: when x is a vector $x = (x_1, x_2, x_3, \ldots, x_n)$, computes the sum of the elements of x: $\sum_{i=1}^{n} x_i$.
- `prod(x)`: when x is a vector $x = (x_1, x_2, x_3, \ldots, x_n)$, computes the product of the elements of x: $\prod_{i=1}^{n} x_i$.
- `pi`: a built-in variable with value π, the ratio of the circumference of a circle to its diameter.
- `x %% a`: computes x modulo a.
- `factorial(x)`: computes $x!$.
- `choose(n,k)`: computes $\binom{n}{k}$.

~

Exercise 2.8. Compute the following:

1. $\sqrt{\pi}$

2. $e^2 \cdot log(4)$

Exercise 2.9. R will return non-numerical answers if you try to perform computations where the answer is infinite or undefined. Try the following to see some possible results:

1. Compute $1/0$.

2. Compute $0/0$.

3. Compute $\sqrt{-1}$.

2.5 Packages

R comes with a ton of functions, but of course these cannot cover all possible things that you may want to do with your data. That's where *packages* come in. Packages are collections of functions and datasets that add new features to R. Do you want to apply some obscure statistical test to your data? Plot your data on a map? Run C++ code in R? Speed up some part of your data handling process? There are R packages for that. In fact, with more than 17,000 packages and counting, there are R packages for just about anything that you could possibly want to do. All packages have been contributed by the R community, i.e., by users like you and me.

Most R packages are available from CRAN, the official R repository – a network of servers (so-called *mirrors*) around the world. Packages on CRAN are checked before they are published, to make sure that they do what they are supposed to do and don't contain malicious components. Downloading packages from CRAN is therefore generally considered to be safe.

In the rest of this chapter, we'll make use of a package called `ggplot2`, which adds additional graphical features to R. To install the package from CRAN, you can either select *Tools > Install packages* in the RStudio menu and then write `ggplot2` in the text box in the pop-up window that appears, or use the following line of code:

```
install.packages("ggplot2")
```

A menu may appear where you are asked to select the location of the CRAN mirror to download from. Pick the one the closest to you, or just use the default option – your choice can affect the download speed, but will in most cases not make much difference. There may also be a message asking whether to create a folder for your packages, which you should agree to do.

As R downloads and installs the packages, a number of technical messages are printed in the Console pane. `ggplot2` depends on a number of packages that R will install for you, so expect this to take a few minutes. If the installation finishes successfully, it will finish with a message saying:

```
package 'ggplot2' successfully unpacked and MD5 sums checked
```

Or, on some systems,

```
⋆ DONE (ggplot2)
```

If the installation fails for some reason, there will usually be a (sometimes cryptic) error message. You can read more about troubleshooting errors in Section 2.18. There is also a list of common problems when installing packages available on the RStudio support page at: https://support.rstudio.com/hc/en-us/articles/200554786-Problem-Installing-Packages.

After you've installed the package, you're still not finished quite yet. The package may have been installed, but its functions and datasets won't be available until you *load* it. This is something you need to do each time that you start a new R session. Luckily, it is done with a single short line of code using the `library` function[9], that I recommend putting at the top of your script file:

```
library(ggplot2)
```

We'll discuss more details about installing and updating R packages in Section 12.1.

∼

Exercise 2.10. Install the `palmerpenguins` package from CRAN. We'll use it for many of the exercises and examples that follow.

2.6   Descriptive statistics

In the next few sections, we will study a dataset that is shipped with the `ggplot2` package, `msleep`, which describes the sleep times of 83 mammals. In addition, in the exercises, we'll work with a dataset from the `palmerpenguins` package, which you installed in Exercise 2.10.

`msleep`, as well as some other datasets, is automatically loaded as a data frame when you load `ggplot2`:

```
library(ggplot2)
```

To begin with, let's explore the `msleep` dataset. To have a first look at it, type the following in the Console pane:

```
msleep
```

That shows you the first 10 rows of the data, and some of its columns. It also gives another important piece of information: `83 x 11`, meaning that the dataset has 83 rows (i.e., 83 observations) and 11 columns (with each column corresponding to a variable in the dataset).

There are, however, better methods for looking at the data. To view all 83 rows and all 11 variables, use:

---

[9]The use of `library` causes people to erroneously refer to R packages as *libraries*. Think of the library as the place where you store your packages, and calling `library` means that you go to your library to fetch the package.

```
View(msleep)
```

You'll notice that some cells have the value NA instead of a proper value. NA stands for Not Available, and is a placeholder used by R to point out *missing data*. In this case, it means that the value is unknown for the animal.

To find information about the data frame containing the data, some useful functions are:

```
head(msleep)
tail(msleep)
dim(msleep)
str(msleep)
names(msleep)
```

dim returns the numbers of rows and columns of the data frame, whereas str returns information about the 11 variables. Of particular importance are the *data types* of the variables (chr and num, in this instance), which tell us what kind of data we are dealing with (numerical, categorical, dates, or something else). We'll delve deeper into data types in Chapter 5. Finally, names returns a vector containing the names of the variables.

Like functions, datasets that come with packages have documentation describing them. The documentation for msleep gives a short description of the data and its variables. Read it to learn a bit more about the variables:

```
?msleep
```

Finally, you'll notice that msleep isn't listed among the variables in the Environment pane in RStudio. To include it there, you can run:

```
data(msleep)
```

2.6.1 Numerical data

Now that we know what each variable represents, it's time to compute some statistics. A convenient way to get some descriptive statistics giving a summary of each variable is to use the summary function:

```
summary(msleep)
```

For the text variables, this doesn't provide any information at the moment. But for the numerical variables, it provides a lot of useful information. For the variable sleep_rem, for instance, we have the following:

```
   sleep_rem
Min.    :0.100
1st Qu.:0.900
Median :1.500
Mean    :1.875
```

```
3rd Qu.:2.400
Max.   :6.600
NA's   :22
```

This tells us that the mean of `sleep_rem` is 1.875, that smallest value is 0.100 and that the largest is 6.600. The first quartile[10] is 0.900, the median is 1.500 and the third quartile is 2.400. Finally, there are 22 animals for which there are no values (missing data is represented by `NA`).

Sometimes we want to compute just one of these, and other times we may want to compute summary statistics not included in `summary`. Let's say that we want to compute some descriptive statistics for the `sleep_total` variable. To access a vector inside a data frame, we use a dollar sign: `data_frame_name$vector_name`. So, to access the `sleep_total` vector in the `msleep` data frame, we write:

```
msleep$sleep_total
```

Some examples of functions that can be used to compute descriptive statistics for this vector are:

```
mean(msleep$sleep_total)      # Mean
median(msleep$sleep_total)    # Median
max(msleep$sleep_total)       # Max
min(msleep$sleep_total)       # Min
sd(msleep$sleep_total)        # Standard deviation
var(msleep$sleep_total)       # Variance
quantile(msleep$sleep_total)  # Various quantiles
```

To see how many animals sleep for more than 8 hours a day, we can use the following:

```
sum(msleep$sleep_total > 8)   # Frequency (count)
mean(msleep$sleep_total > 8)  # Relative frequency (proportion)
```

`msleep$sleep_total > 8` checks whether the total sleep time of each animal is greater than 8. We'll return to expressions like this in Section 2.11.

Now, let's try to compute the mean value for the length of REM sleep for the animals:

```
mean(msleep$sleep_rem)
```

The above call returns the answer `NA`. The reason is that there are `NA` values in the `sleep_rem` vector (22 of them, as we saw before). What we actually wanted was the mean value among the animals for which we know the REM sleep. We can have a look at the documentation for `mean` to see if there is some way we can get this:

---

[10]The first quartile is a value such that 25% of the observations are smaller than it; the third quartile is a value such that 25% of the observations are larger than it.

```
?mean
```

The argument `na.rm` looks promising; it is "a logical value indicating whether NA values should be stripped before the computation proceeds". In other words, it tells R whether or not to ignore the `NA` values when computing the mean. In order to ignore `NA`:s in the computation, we set `na.rm = TRUE` in the function call:

```
mean(msleep$sleep_rem, na.rm = TRUE)
```

Note that the `NA` values have not been removed from `msleep`. Setting `na.rm = TRUE` simply tells R to ignore them in a particular computation, not to delete them.

We run into the same problem if we try to compute the correlation between `sleep_total` and `sleep_rem`:

```
cor(msleep$sleep_total, msleep$sleep_rem)
```

A quick look at the documentation (`?cor`), tells us that the argument used to ignore `NA` values has a different name for `cor` - it's not `na.rm` but `use`. The reason will become evident later on, when we study more than two variables at a time. For now, we set `use = "complete.obs"` to compute the correlation using only observations with complete data (i.e., no missing values):

```
cor(msleep$sleep_total, msleep$sleep_rem, use = "complete.obs")
```

2.6.2   Categorical data

Some of the variables, like `vore` (feeding behaviour) and `conservation` (conservation status) are *categorical* rather than *numerical*. It therefore makes no sense to compute means or largest values. For categorical variables (often called *factors* in R), we can instead create a table showing the frequencies of different categories using `table`:

```
table(msleep$vore)
```

To instead show the proportion of different categories, we can apply `proportions` to the table that we just created:

```
proportions(table(msleep$vore))
```

The `table` function can also be used to construct a cross-table that shows the counts for different combinations of two categorical variables:

```
# Counts:
table(msleep$vore, msleep$conservation)

# Proportions, per row:
proportions(table(msleep$vore, msleep$conservation),
            margin = 1)
```

```
# Proportions, per column:
proportions(table(msleep$vore, msleep$conservation),
            margin = 2)

# Proportions, out of total:
proportions(table(msleep$vore, msleep$conservation))
```

You'll learn much more about how to create (better-looking!) tables in Section 3.2.

~

Exercise 2.11. Load the `palmerpenguins` package that you installed in Exercise 2.10, using `library(palmerpenguins)`. In this exercise, we'll study the `penguins` dataset contained in said package.

    1. View the documentation for the `penguins` data and read about its variables.

    2. Check the data structures: how many observations and variables and what type of variables (numeric, categorical, etc.) are there?

    3. Compute summary statistics (means, median, min, max, counts for categorical variables). Are there any missing values?

2.7 Plotting numerical data

There are several different approaches to creating plots with R. In this book, we will mainly focus on creating plots using the `ggplot2` package, which allows us to create good-looking plots using the so-called *grammar of graphics*. The grammar of graphics is a set of structural rules that helps us establish a language for graphics. The beauty of this is that (almost) all plots will be created with functions that all follow the same logic, or grammar. That way, we don't have to learn new arguments for each new plot. You can compare this to the problems we encountered when we wanted to ignore NA values when computing descriptive statistics – `mean` required the argument `na.rm`, whereas `cor` required the argument `use`. By using a common grammar for all plots, we reduce the number of arguments that we need to learn.

The three key components to grammar of graphics plots are:

* **Data**: observations in your dataset,
* **Aesthetics**: mappings from the data to visual properties (like axes and sizes of geometric objects), and
* **Geoms**: geometric objects representing what you see in the plot, e.g., lines or points.

When we create plots using `ggplot2`, we must define what data, aesthetics and geoms to use. If that sounds a bit strange, it will hopefully become a lot clearer once we have a look at some examples. To begin with, we will illustrate how this works by visualising some continuous variables in the `msleep` data.

2.7.1 Our first plot

As a first example, let's make a scatterplot by plotting the total sleep time of an animal against the REM sleep time of an animal.

Using base R, we simply do a call to the `plot` function in a way that is analogous to how we'd use, e.g., `cor`:

```
plot(msleep$sleep_total, msleep$sleep_rem)
```

The code for doing this using `ggplot2` is more verbose:

```
library(ggplot2)
ggplot(msleep, aes(x = sleep_total, y = sleep_rem)) + geom_point()
```

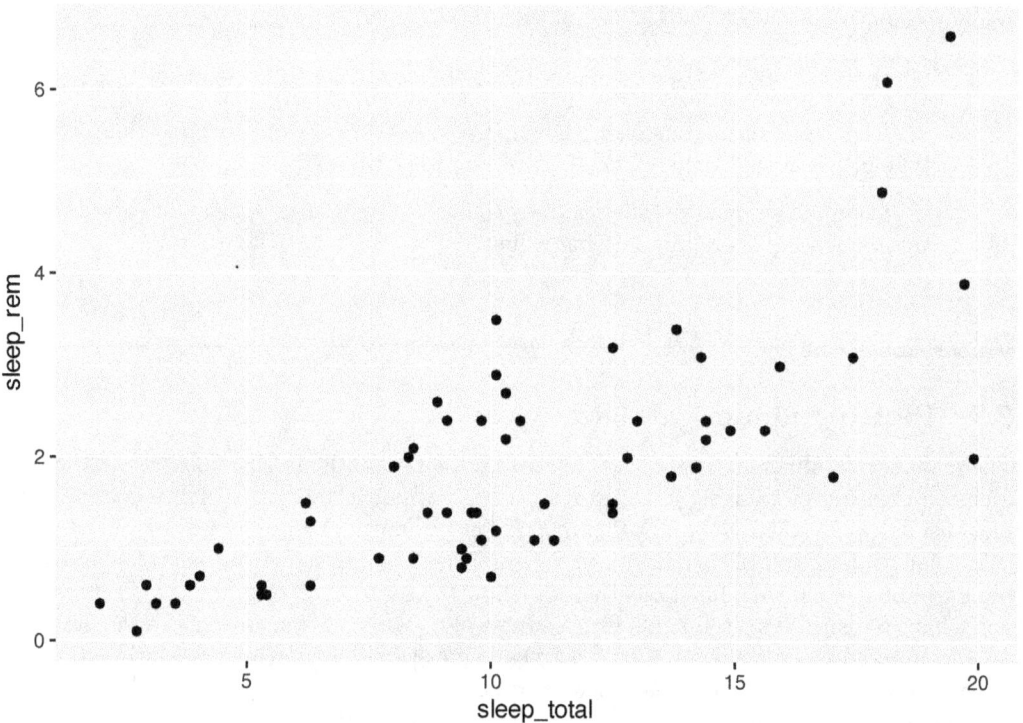

FIGURE 2.3 Scatterplot of mammal sleep times.

The code consists of three parts:

- **Data**: given by the first argument in the call to `ggplot`: `msleep`.
- **Aesthetics**: given by the second argument in the `ggplot` call: `aes`, where we map `sleep_total` to the x-axis and `sleep_rem` to the y-axis.
- **Geoms**: given by `geom_point`, meaning that the observations will be represented by points.

At this point you may ask why on earth anyone would ever want to use `ggplot2` code for creating plots. It's a valid question. The base R code looks simpler and is consistent with other functions that we've seen. The `ggplot2` code looks... different. This is because it uses the *grammar of graphics*, which in many ways is a language of its own, different from how we otherwise work with R.

But, the plot created using `ggplot2` also looked different. It used filled circles instead of empty circles for plotting the points and had a grid in the background. In both base R graphics and `ggplot2` we can change these settings, and many others. We can create something similar to the `ggplot2` plot using base R as follows, using the `pch` argument and the `grid` function:

```
plot(msleep$sleep_total, msleep$sleep_rem, pch = 16)
grid()
```

Some people prefer the look and syntax of base R plots, while others argue that `ggplot2` graphics has a prettier default look. I can sympathise with both groups. Some types of plots are easier to create using base R, and some are easier to create using ggplot2. I like base R graphics for their simplicity and prefer them for quick-and-dirty visualisations as well as for more elaborate graphs where I want to combine many different components. For everything in between, including exploratory data analysis where graphics are used to explore and understand datasets, I prefer `ggplot2`. In this book, we'll occasionally use base graphics for some quick-and-dirty plots, but put more emphasis on `ggplot2` and how it can be used to explore data.

The syntax used to create the `ggplot2` scatterplot was in essence `ggplot(data, aes) + geom`. All plots created using `ggplot2` follow this pattern, regardless of whether they are scatterplots, bar charts or something else. The plus sign in `ggplot(data, aes) + geom` is important, as it implies that we can add more geoms to the plot, for instance a trend line, and perhaps other things as well. We will return to that shortly.

Unless the user specifies otherwise, the first two arguments to `aes` will always be mapped to the x and y axes, meaning that we can simplify the code above by removing the x = and y = bits (at the cost of a slight reduction in readability). Moreover, it is considered good style to insert a line break after the + sign. The resulting code is:

```
ggplot(msleep, aes(sleep_total, sleep_rem)) +
      geom_point()
```

Note that this does not change the plot in any way; the difference is merely in the style of the code.

~

Exercise 2.12. Create a scatterplot with total sleep time along the x-axis and time awake along the y-axis (using the `msleep` data). What pattern do you see? Can you explain it?

2.7.2 Colours, shapes and axis labels

You now know how to make scatterplots, but if you plan to show your plot to someone else, there are probably a few changes that you'd like to make. For instance, it's usually a good idea to change the label for the x-axis from the variable name "sleep_total" to something like "Total sleep time (h)". This is done by using the + sign again, adding a call to `labs` to the plot:

```
ggplot(msleep, aes(sleep_total, sleep_rem)) +
    geom_point() +
    labs(x = "Total sleep time (h)")
```

Note that the plus signs must be placed at the end of a row rather than at the beginning. To change the y-axis label, add y = instead.

To change the colour of the points, you can set the colour in geom_point:

```
ggplot(msleep, aes(sleep_total, sleep_rem)) +
    geom_point(colour = "red") +
    labs(x = "Total sleep time (h)")
```

In addition to "red", there are a few more colours that you can choose from. You can run colours() in the Console to see a list of the 657 colours that have names in R (examples of which include "papayawhip", "blanchedalmond", and "cornsilk4"), or use colour hex codes like "#FF5733".

Alternatively, you may want to use the colours of the point to separate different categories. This is done by adding a colour argument to aes, since you are now mapping a data variable to a visual property. For instance, we can use the variable vore to show differences between herbivores, carnivores and omnivores:

```
ggplot(msleep, aes(sleep_total, sleep_rem, colour = vore)) +
    geom_point() +
    labs(x = "Total sleep time (h)")
```

You can change the legend label in labs:

```
ggplot(msleep, aes(sleep_total, sleep_rem, colour = vore)) +
    geom_point() +
    labs(x = "Total sleep time (h)",
        colour = "Feeding behaviour")
```

What happens if we use a continuous variable, such as the sleep cycle length sleep_cycle to set the colour?

```
ggplot(msleep, aes(sleep_total, sleep_rem, colour = sleep_cycle)) +
    geom_point() +
    labs(x = "Total sleep time (h)")
```

You'll learn more about customising colours (and other parts) of your plots in Section 4.2.

~

Exercise 2.13. Using the penguins data, do the following:

1. Create a scatterplot with bill length along the x-axis and flipper length along the y-axis. Change the x-axis label to read "Bill length (mm)" and the y-axis label to "Flipper length (mm)". Use species to set the colour of the points.

2. Try adding the argument `alpha = 1` to `geom_point`, i.e., `geom_point(alpha = 1)`. Does anything happen? Try changing the `1` to `0.75` and `0.25` and see how that affects the plot.

Exercise 2.14. Similar to how you changed the colour of the points, you can also change their size and shape. The arguments for this are called `size` and `shape`.

1. Change the scatterplot from Exercise 2.13 so that animals from different islands are represented by different shapes.

2. Then change it so that the size of each point is determined by the body mass, i.e., the variable `body_mass_g`.

2.7.3 Axis limits and scales

Next, assume that we wish to study the relationship between animals' brain sizes and their total sleep time. We create a scatterplot using:

```
ggplot(msleep, aes(brainwt, sleep_total, colour = vore)) +
    geom_point() +
    labs(x = "Brain weight",
         y = "Total sleep time")
```

There are two animals with brains that are much heavier than the rest (African elephant and Asian elephant). These outliers distort the plot, making it difficult to spot any patterns. We can try changing the x-axis to only go from 0 to 1.5 by adding `xlim` to the plot, to see if that improves it:

```
ggplot(msleep, aes(brainwt, sleep_total, colour = vore)) +
    geom_point() +
    labs(x = "Brain weight",
         y = "Total sleep time") +
    xlim(0, 1.5)
```

This is slightly better, but we still have a lot of points clustered near the y-axis, and some animals are now missing from the plot. If instead we wished to change the limits of the y-axis, we would have used `ylim` in the same fashion.

Another option is to rescale the x-axis by applying a log transform to the brain weights, which we can do directly in `aes`:

```
ggplot(msleep, aes(log(brainwt), sleep_total, colour = vore)) +
    geom_point() +
    labs(x = "log(Brain weight)",
         y = "Total sleep time")
```

This is a better-looking scatterplot, with a weak declining trend. We didn't have to remove the outliers (elephants) to create it, which is good. The downside is that the x-axis now has become difficult to interpret. A third option that mitigates this is to add `scale_x_log10` to the plot, which changes the scale of the x-axis to a $\log_{10}$ scale (which increases interpretability because the values shown at the ticks still are on the original x-scale).

```
ggplot(msleep, aes(brainwt, sleep_total, colour = vore)) +
    geom_point() +
    labs(x = "Brain weight (logarithmic scale)",
        y = "Total sleep time") +
    scale_x_log10()
```

The numbers on the x-axis are displayed with *scientific notation,* where for instance `1e-04` means 10^{-4}. You can change the settings in R to use decimals, instead, using the `options` function as follows:

```
options(scipen = 1000)
ggplot(msleep, aes(brainwt, sleep_total, colour = vore)) +
    geom_point() +
    labs(x = "Brain weight (logarithmic scale)",
        y = "Total sleep time") +
    scale_x_log10()
```

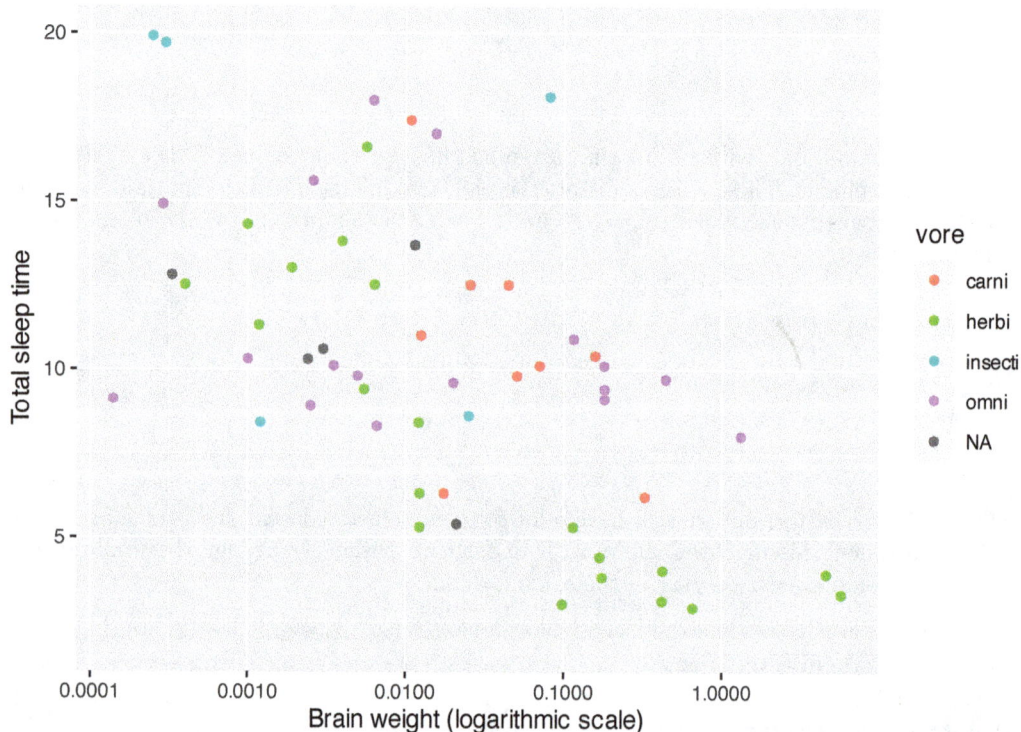

FIGURE 2.4 Example of a logarithmic x-axis.

~

Exercise 2.15. Using the `msleep` data, create a plot of log-transformed body weight versus log-transformed brain weight. Use total sleep time to set the colours of the points. Change the text on the axes to something informative.

2.7.4 Comparing groups

We frequently wish to make visual comparisons of different groups. One way to display differences between groups in plots is to use *facetting*, i.e., to create a grid of plots corresponding to the different groups. For instance, in our plot of animal brain weight versus total sleep time, we may wish to separate the different feeding behaviours (omnivores, carnivores, etc.) in the `msleep` data using facetting instead of different coloured points. In `ggplot2` we do this by adding a call to `facet_wrap` to the plot:

```
ggplot(msleep, aes(brainwt, sleep_total)) +
    geom_point() +
    labs(x = "Brain weight (logarithmic scale)",
        y = "Total sleep time") +
    scale_x_log10() +
    facet_wrap(~ vore)
```

Note that the x-axes and y-axes of the different plots in the grid all have the same scale and limits.

~

Exercise 2.16. Using the `penguins` data, do the following:

    1. Create a scatterplot with `bill_length_mm` along the x-axis and `flipper_length_mm` along the y-axis, facetted by `species`.

    2. Read the documentation for `facet_wrap` (`?facet_wrap`). How can you change the number of rows in the plot grid? Create the same plot as in part 1, but with 2 rows.

2.7.5 Boxplots

Another option for comparing groups is boxplots (also called box-and-whiskers plots). Using `ggplot2`, we create boxplots for animal sleep times, grouped by feeding behaviour, with `geom_boxplot`. Using base R, we use the `boxplot` function instead:

```
# Base R:
boxplot(sleep_total ~ vore, data = msleep)

# ggplot2:
ggplot(msleep, aes(vore, sleep_total)) +
    geom_boxplot()
```

The boxes visualise important descriptive statistics for the different groups, similar to what we got using `summary`:

- *Median*: thick black line inside the box.
- *First quartile*: bottom of the box.
- *Third quartile*: top of the box.
- *Minimum*: end of the line ("whisker") that extends from the bottom of the box.

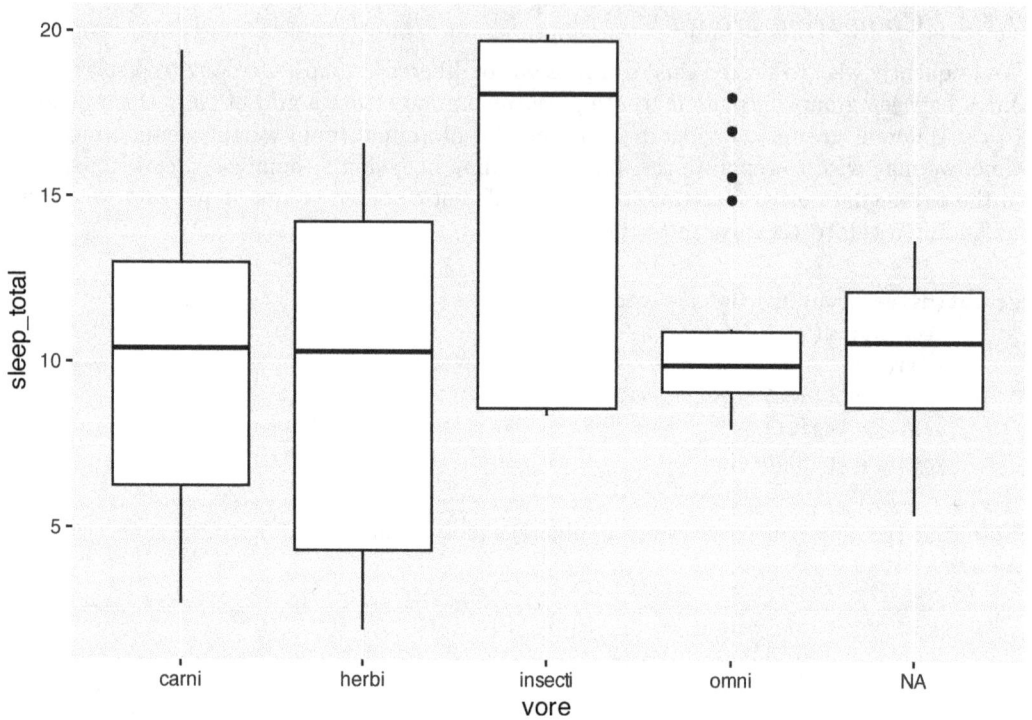

FIGURE 2.5 Boxplots showing mammal sleep times.

- *Maximum*: end of the line that extends from the top of the box.
- *Outliers*: observations that deviate too much[11] from the rest are shown as separate points. These outliers are not included in the computation of the median, quartiles and the extremes.

Note that just as for a scatterplot, the code consists of three parts:

- **Data**: given by the first argument in the call to `ggplot`: `msleep`
- **Aesthetics**: given by the second argument in the `ggplot` call: `aes`, where we map the group variable `vore` to the x-axis and the numerical variable `sleep_total` to the y-axis.
- **Geoms**: given by `geom_boxplot`, meaning that the data will be visualised with boxplots.

~

Exercise 2.17. Using the `penguins` data, do the following:

1. Create boxplots of bill lengths, grouped by `species`.

2. Read the documentation for `geom_boxplot`. How can you change the colours of the boxes and their outlines?

3. Add `geom_jitter(size = 0.5, alpha = 0.25)` to the plot. What happens?

---

[11]In this case, *too much* means that they are more than 1.5 times the height of the box away from the edges of the box.

2.7.6 Histograms

To show the distribution of a continuous variable, we can use a histogram, in which the data is split into a number of bins, and the number of observations in each bin is shown by a bar. The `ggplot2` code for histograms follows the same pattern as other plots, while the base R code uses the `hist` function:

```
# Base R:
hist(msleep$sleep_total)

# ggplot2:
ggplot(msleep, aes(sleep_total)) +
    geom_histogram()
```

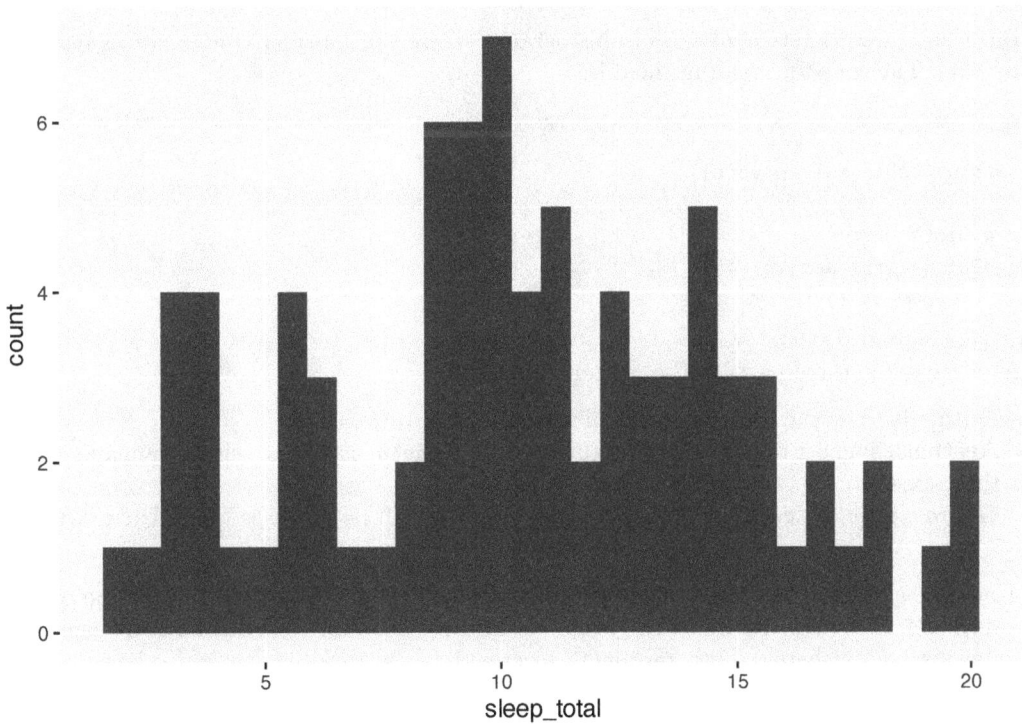

FIGURE 2.6 Histogram for mammal sleep times.

As before, the three parts in the `ggplot2` code are:

- **Data**: given by the first argument in the call to `ggplot`: `msleep`.
- **Aesthetics**: given by the second argument in the `ggplot` call: `aes`, where we map `sleep_total` to the x-axis.
- **Geoms**: given by `geom_histogram`, meaning that the data will be visualised by a histogram.

~

Exercise 2.18. Using the `penguins` data, do the following:

1. Create a histogram of bill lengths.

2. Create histograms of bill lengths for different species, using facetting.

3. Add a suitable argument to `geom_histogram` to add black outlines around the bars[12].

2.8 Plotting categorical data

When visualising categorical data, we typically try to show the counts, i.e., the number of observations, for each category. The most common plot for this type of data is the bar chart.

2.8.1 Bar charts

Bar charts are discrete analogues to histograms, where the category counts are represented by bars. The code for creating them is:

```
# Base R
barplot(table(msleep$vore))


# ggplot2
ggplot(msleep, aes(vore)) +
    geom_bar()
```

As always, the three parts in the `ggplot2` code are:

- **Data**: given by the first argument in the call to `ggplot`: `msleep`
- **Aesthetics**: given by the second argument in the `ggplot` call: `aes`, where we map `vore` to the x-axis.
- **Geoms**: given by `geom_bar`, meaning that the data will be visualised by a bar chart.

To create a stacked bar chart using `ggplot2`, we map all groups to the same value on the x-axis and then map the different groups to different colours. This can be done using the following hack, where we use `factor(1)` to create a new variable that's displayed on the x-axis:

```
ggplot(msleep, aes(factor(1), fill = vore)) +
    geom_bar()
```

~

Exercise 2.19. Using the `penguins` data, do the following:

1. Create a bar chart of species.

2. Add different colours to the bars by adding a `fill` argument to `geom_bar`.

---

[12]Personally, I don't understand why anyone would ever plot histograms without outlines!

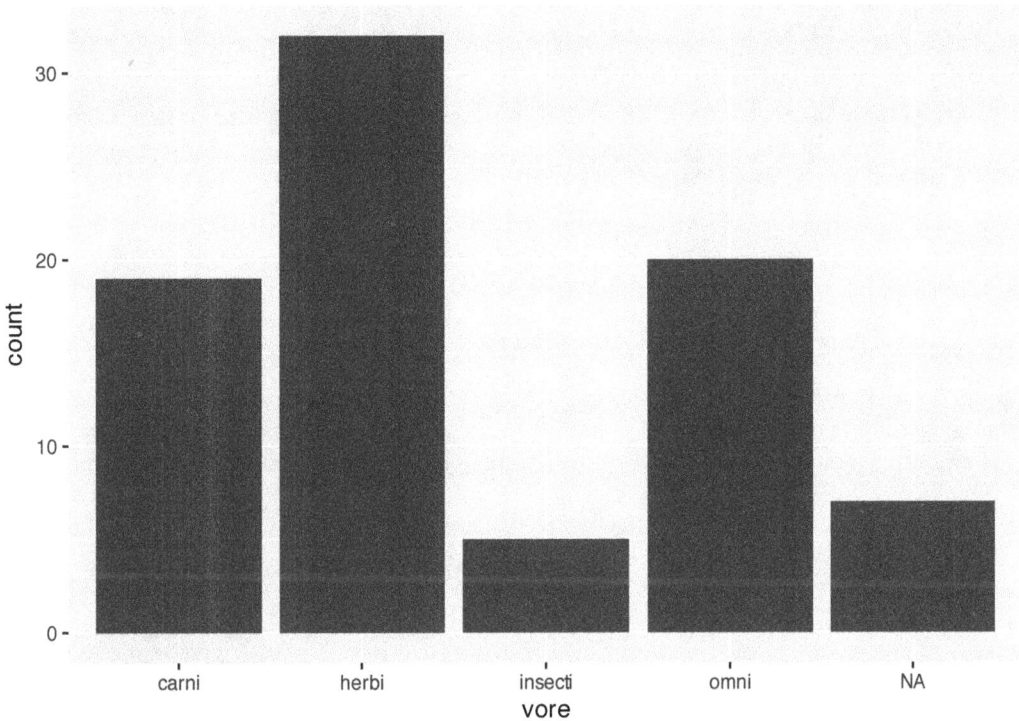

FIGURE 2.7 Bar chart for the mammal sleep data.

3. Check the documentation for `geom_bar`. How can you decrease the width of the bars?

4. Return to the code you used for part 1. Add `fill = sex` to the `aes`. What happens?

5. Next, add `position = "dodge"` to `geom_bar`. What happens?

6. Add `coord_flip()` to the plot. What happens?

2.9 Saving your plot

When you create a `ggplot2` plot, you can save it as a plot object in R:

```
library(ggplot2)
myPlot <- ggplot(msleep, aes(sleep_total, sleep_rem)) +
    geom_point()
```

To plot a saved plot object, just write its name:

```
myPlot
```

If you like, you can add things to the plot, just as before:

```
myPlot + labs(x = "I forgot to add a label!")
```

To save your plot object as an image file, use `ggsave`. The `width` and `height` arguments allows us to control the size of the figure (in inches, unless you specify otherwise using the `units` argument).

```
ggsave("filename.pdf", myPlot, width = 5, height = 5)
```

If you don't supply the name of a plot object, `ggsave` will save the last `ggplot2` plot you created.

In addition to pdf, you can save images, e.g., as jpg, tif, eps, svg, and png files, simply by changing the file extension in the filename. Alternatively, graphics from both base R and `ggplot2` can be saved using the `pdf` and `png` functions, using `dev.off` to mark the end of the file:

```
pdf("filename.pdf", width = 5, height = 5) # Creates an empty pdf file
myPlot # Adds the plot to the pdf
dev.off() # Saves and closes the pdf


png("filename.png", width = 500, height = 500)
plot(msleep$sleep_total, msleep$sleep_rem)
dev.off()
```

Note that you also can save graphics by clicking on the Export button in the Plots pane in RStudio. Using code to save your plot is usually a better idea, because of reproducibility. At some point, you'll want to go back and make changes to an old figure, and that will be much easier if you already have the code to export the graphic.

~

Exercise 2.20. Do the following:

1. Create a plot object and save it as a 4-by-4 inch png file.

2. When preparing images for print, you may want to increase their resolution. Check the documentation for `ggsave`. How can you increase the resolution of your png file to 600 dpi?

You've now had a first taste of graphics using R. We have, however, only scratched the surface and will return to the many uses of statistical graphics in Chapter 4.

2.10 Data frames and data types

2.10.1 Types and structures

We have already seen that different kinds of data require different kinds of statistical methods. For numeric data, we create boxplots and compute means, but for categorical data we don't.

Instead we produce bar charts and display the data in tables. It is no surprise then that R also treats different kinds of data differently.

In programming, a variable's *data type* describes what kind of object is assigned to it. We can assign many different types of objects to the variable a: it could for instance contain a number, text, or a data frame. In order to treat a correctly, R needs to know what data type its assigned object has. In some programming languages, you have to explicitly state what data type a variable has, but not in R. This makes programming R simpler and faster, but can cause problems if a variable turns out to have a different data type than what you thought[13].

R has six basic data types. For most people, it suffices to know about the first three in the list below:

- `numeric`: numbers like 1 and 16.823 (sometimes also called `double`).
- `logical`: true/false values (Boolean): either TRUE or FALSE.
- `character`: text, e.g., "a", "Hello! I'm Ada." and "name@domain.com".
- `integer`: integer numbers, denoted in R by the letter L: 1L, 55L.
- `complex`: complex numbers, like 2+3i. Rarely used in statistical work.
- `raw`: used to hold raw bytes. Don't fret if you don't know what that means. You can have a long and meaningful career in statistics, data science, or pretty much any other field without ever having to worry about raw bytes. We won't discuss `raw` objects again in this book.

In addition, these can be combined into special data types sometimes called *data structures*, examples of which include vectors and data frames. Important data structures include `factor`, which is used to store categorical data, and the awkwardly named `POSIXct` which is used to store date and time data.

To check what type of object a variable is, you can use the `class` function:

```
x <- 6
y <- "Scotland"
z <- TRUE

class(x)
class(y)
class(z)
```

What happens if we use `class` on a vector?

```
numbers <- c(6, 9, 12)
class(numbers)
```

class returns the data type of the elements of the vector. So what happens if we put objects of different types together in a vector?

```
all_together <- c(x, y, z)
all_together
class(all_together)
```

---

[13]And the subsequent troubleshooting makes programming R more difficult and slower.

In this case, R has coerced the objects in the vector to all be of the same type. Sometimes that is desirable, and sometimes it is not. The lesson here is to be careful when you create a vector from different objects. We'll learn more about coercion and how to change data types in Section 5.1.

2.10.2   Types of tables

The basis for most data analyses in R are data frames: spreadsheet-like tables with rows and columns containing data. You encountered some data frames in previous examples. Have a quick look at them to remind yourself of what they look like:

```
# Bookstore example
age <- c(28, 48, 47, 71, 22, 80, 48, 30, 31)
purchase <- c(20, 59, 2, 12, 22, 160, 34, 34, 29)
bookstore <- data.frame(age, purchase)
View(bookstore)

# Animal sleep data
library(ggplot2)
View(msleep)
```

Notice that all three data frames follow the same format: each column represents a *variable* (e.g., age) and each row represents an *observation* (e.g., an individual). This is the standard way to store data in R (as well as the standard format in statistics in general). In what follows, we will use the terms column and variable interchangeably, to describe the columns/variables in a data frame.

This kind of table can be stored in R as different types of objects, i.e., in several different ways. As you'd expect, the different types of objects have different properties and can be used with different functions. Here's the rundown of four common types:

- `matrix`: a table where all columns must contain objects of the same type (e.g., all `numeric` or all `character`). Uses less memory than other types and allows for much faster computations, but it is difficult to use for certain types of data manipulation, plotting and analyses.
- `data.frame`: the most common type, where different columns can contain different types (e.g., one `numeric` column, one `character` column).
- `data.table`: an enhanced version of `data.frame`.
- `tbl_df` ("tibble"): another enhanced version of `data.frame`.

First of all, in most cases it doesn't matter which of these four you use to store your data. In fact, they all look similar to the user. Have a look at the following datasets (`WorldPhones` and `airquality` come with base R):

```
# First, an example of data stored in a matrix:
?WorldPhones
class(WorldPhones)
View(WorldPhones)

# Next, an example of data stored in a data frame:
?airquality
class(airquality)
```

```
View(airquality)

# Finally, an example of data stored in a tibble:
library(ggplot2)
?msleep
class(msleep)
View(msleep)
```

That being said, in some cases, it *really* matters which one you use. Some functions require that you input a matrix, while others may break or work differently from what was intended if you input a tibble instead of an ordinary data frame. Luckily, you can convert objects into other types:

```
WorldPhonesDF <- as.data.frame(WorldPhones)
class(WorldPhonesDF)

airqualityMatrix <- as.matrix(airquality)
class(airqualityMatrix)
```

~

Exercise 2.21. The following tasks are all related to data types and data structures:

1. Create a text variable using, e.g., a <- "A rainy day in Edinburgh". Check that it gets the correct type. What happens if you use single quotes marks instead of double quotes when you create the variable?

2. What data types are the sums 1 + 2, 1L + 2 and 1L + 2L?

3. What happens if you add a numeric to a character, e.g., "Hello" + 1?

4. What happens if you perform mathematical operations involving a numeric and a logical, e.g., FALSE * 2 or TRUE + 1?

Exercise 2.22. What do the functions ncol, nrow, dim, names, and row.names return when applied to a data frame?

Exercise 2.23. matrix tables can be created from vectors using the function of the same name. Using the vector x <- 1:6 use matrix to create the following matrices:

$$\begin{pmatrix} 1 & 2 & 3 \\ 4 & 5 & 6 \end{pmatrix}$$

and

$$\begin{pmatrix} 1 & 4 \\ 2 & 5 \\ 3 & 6 \end{pmatrix}.$$

Remember to check ?matrix to find out how to set the dimensions of the matrix, and how it is filled with the numbers from the vector!

2.11 Vectors in data frames

In the next few sections, we will explore the `airquality` dataset. It contains daily air quality measurements from New York during a period of 5 months:

- `Ozone`: mean ozone concentration (ppb).
- `Solar.R`: solar radiation (Langley).
- `Wind`: average wind speed (mph).
- `Temp`: maximum daily temperature in degrees Fahrenheit.
- `Month`: numeric month (May=5, June=6, and so on).
- `Day`: numeric day of the month (1-31).

There are lots of things that would be interesting to look at in this dataset. What was the mean temperature during the period? Which day was the hottest? Which day was the windiest? What days was the temperature more than 90 degrees Fahrenheit? To answer these questions, we need to be able to access the vectors inside the data frame. We also need to be able to quickly and automatically screen the data in order to find interesting observations (e.g., the hottest day)

2.11.1 Accessing vectors and elements

In Section 2.6, we learned how to compute the mean of a vector. We also learned that to compute the mean of a vector *that is stored inside a data frame*[14] we could use a dollar sign: `data_frame_name$vector_name`. Here is an example with the `airquality` data:

```
# Extract the Temp vector:
airquality$Temp
```

```
# Compute the mean temperature:
mean(airquality$Temp)
```

If we want to grab a particular element from a vector, we must use its *index* within square brackets: `[index]`. The first element in the vector has index 1, the second has index 2, the third has index 3, and so on. To access the fifth element in the `Temp` vector in the `airquality` data frame, we can use:

```
airquality$Temp[5]
```

The square brackets can also be applied directly to the data frame. The syntax for this follows that used for matrices in mathematics: `airquality[i, j]` means the element at the i:th row and j:th column of `airquality`. We can also leave out either `i` or `j` to extract an entire row or column from the data frame. Here are some examples:

```
# First, we check the order of the columns:
names(airquality)
# We see that Temp is the 4th column.
```

---

[14]This works regardless of whether this is a regular `data.frame`, a `data.table` or a tibble.

```
airquality[5, 4]     # The 5th element from the 4th column,
                     # i.e. the same as airquality$Temp[5]
airquality[5,]       # The 5th row of the data
airquality[, 4]      # The 4th column of the data, like airquality$Temp
airquality[[4]]      # The 4th column of the data, like airquality$Temp
airquality[, c(2, 4, 6)] # The 2nd, 4th and 6th columns of the data
airquality[, -2]     # All columns except the 2nd one
airquality[, c("Temp", "Wind")] # The Temp and Wind columns
```

~

Exercise 2.24. The following tasks all involve using the `[i, j]` notation for extracting data from data frames:

1. Why does `airquality[, 3]` not return the third row of `airquality`?

2. Extract the first five rows from `airquality`. *Hint:* a fast way of creating the vector c(1, 2, 3, 4, 5) is to write 1:5.

3. Compute the correlation between the `Temp` and `Wind` vectors of `airquality` without referring to them using `$`.

4. Extract all columns from `airquality` *except* `Temp` and `Wind`.

2.11.2   Adding and changing data using dollar signs

The `$` operator can be used not just to extract data from a data frame, but also to manipulate it. Let's return to our `bookstore` data frame and see how we can make changes to it using the dollar sign.

```
age <- c(28, 48, 47, 71, 22, 80, 48, 30, 31)
purchase <- c(20, 59, 2, 12, 22, 160, 34, 34, 29)
bookstore <- data.frame(age, purchase)
```

Perhaps there was a data entry error – the second customer was actually 18 years old and not 48. We can assign a new value to that element by referring to it in either of two ways:

```
bookstore$age[2] <- 18
# or
bookstore[2, 1] <- 18
```

We could also change an entire column if we like. For instance, if we wish to change the `age` vector to months instead of years, we could use

```
bookstore$age <- bookstore$age * 12
```

What if we want to add another variable to the data, for instance the length of the customers' visits in minutes? There are several ways to accomplish this, one of which involves the dollar sign:

```
bookstore$visit_length <- c(5, 2, 20, 22, 12, 31, 9, 10, 11)
bookstore
```

As you see, the new data has now been added to a new column in the data frame.

∼

Exercise 2.25. Use the `bookstore` data frame to do the following:

> 1. Add a new variable `rev_per_minute` which is the ratio between purchase and the visit length.

> 2. Oh no, there's been an error in the data entry! Replace the purchase amount for the 80-year-old customer with `16`.

2.11.3 Filtering using conditions

A few paragraphs ago, we were asking which was the hottest day in the `airquality` data. Let's find out! We already know how to find the maximum value in the `Temp` vector:

```
max(airquality$Temp)
```

But can we find out which day this corresponds to? We could of course manually go through all 153 days, e.g., by using `View(airquality)`, but that seems tiresome and wouldn't even be possible in the first place if we'd had more observations. A better option is therefore to use the function `which.max`:

```
which.max(airquality$Temp)
```

`which.max` returns the index of the observation with the maximum value. If there is more than one observation attaining this value, it only returns the first of these.

We've just used `which.max` to find out that day `120` was the hottest during the period. If we want to have a look at the entire row for that day, we can use

```
airquality[120,]
```

Alternatively, we could place the call to `which.max` inside the brackets. Because `which.max(airquality$Temp)` returns the number `120`, this yields the same result as the previous line:

```
airquality[which.max(airquality$Temp),]
```

Were we looking for the day with the lowest temperature, we'd use `which.min` analogously. In fact, we could use any function or computation that returns an index in the same way, placing it inside the brackets to get the corresponding rows or columns. This is extremely useful if we want to extract observations with certain properties, for instance all days where the temperature was above 90 degrees. We do this using *conditions*, i.e., by giving statements that we wish to be fulfilled.

As a first example of a condition, we use the following, which checks if the temperature exceeds 90 degrees:

```
airquality$Temp > 90
```

For each element in `airquality$Temp` this returns either TRUE (if the condition is fulfilled, i.e., when the temperature is greater than 90) or FALSE (if the condition isn't fulfilled, i.e., when the temperature is 90 or lower). If we place the condition inside brackets following the name of the data frame, we will extract only the rows corresponding to those elements which were marked with TRUE:

```
airquality[airquality$Temp > 90, ]
```

If you prefer, you can also store the TRUE or FALSE values in a new variable. This creates a new variable indicating whether the condition is true or false:

```
airquality$Hot <- airquality$Temp > 90
```

If you don't like the bracket notation that we used when we filtered our data using `airquality[airquality$Temp > 90, ]`, you can try using the function `subset` instead. It achieves the same filtering with different syntax:

```
subset(airquality, Temp > 90)
```

There are several logical operators and functions which are useful when stating conditions in R. Here are some examples:

```
a <- 3
b <- 8
```

```
a == b      # Check if a equals b
a > b       # Check if a is greater than b
a < b       # Check if a is less than b
a >= b      # Check if a is equal to or greater than b
a <= b      # Check if a is equal to or less than b
a != b      # Check if a is not equal to b
is.na(a)    # Check if a is NA
a %in% c(1, 4, 9) # Check if a equals at least one of 1, 4, 9
```

When checking a condition for all elements in a vector, we can use `which` to get the indices of the elements that fulfill the condition:

```
which(airquality$Temp > 90)
```

If we want to know if all elements in a vector fulfill the condition, we can use `all`:

```
all(airquality$Temp > 90)
```

In this case, it returns FALSE, meaning that not all days had a temperature above 90 (phew!). Similarly, if we wish to know whether *at least one* day had a temperature above 90, we can use any:

```
any(airquality$Temp > 90)
```

To find how many elements fulfill a condition, we can use sum:

```
sum(airquality$Temp > 90)
```

Why does this work? Remember that sum computes the sum of the elements in a vector, and that when logical values are used in computations, they are treated as 0 (FALSE) or 1 (TRUE). Because the condition returns a vector of logical values, the sum of them becomes the number of 1s – the number of TRUE values – i.e., the number of elements that fulfill the condition.

To find the proportion of elements that fulfill a condition, we can count how many elements fulfill it and then divide by how many elements are in the vector. This is exactly what happens if we use mean:

```
mean(airquality$Temp > 90)
```

Finally, we can combine conditions by using the logical operators & (AND), | (OR), and, less frequently, xor (exclusive or, XOR). Here are some examples:

```
a <- 3
b <- 8

# Is a less than b and greater than 1?
a < b & a > 1

# Is a less than b and equal to 4?
a < b & a == 4

# Is a less than b and/or equal to 4?
a < b | a == 4

# Is a equal to 4 and/or equal to 5?
a == 4 | a == 5

# Is a less than b XOR equal to 4?
# I.e. is one and only one of these satisfied?
xor(a < b, a == 4)
```

~

Exercise 2.26. The following tasks all involve checking conditions for the airquality data:

1. Which was the coldest day during the period?

2. How many days was the wind speed greater than 17 mph?

3. How many missing values are there in the `Ozone` vector?

4. How many days are there where the temperature was below 70 and the wind speed was above 10?

Exercise 2.27. The function `cut` can be used to create a categorical variable from a numerical variable, by dividing it into categories corresponding to different intervals. Read its documentation and then create a new categorical variable in the `airquality` data, `TempCat`, which divides `Temp` into three intervals `(50, 70]`, `(70, 90]`, `(90, 110]`[15].

2.12   Grouped summaries

Being able to compute the mean temperature for the `airquality` data during the entire period is great, but it would be even better if we also had a way to compute it for each month. The `aggregate` function can be used to create that kind of *grouped summary*.

To begin with, let's compute the mean temperature for each month. Using `aggregate`, we do this as follows:

```
aggregate(Temp ~ Month, data = airquality, FUN = mean)
```

The first argument is a formula, similar to what we used for `lm`, saying that we want a summary of `Temp` grouped by `Month`. Similar formulas are used also in other R functions, for instance when building regression models. In the second argument, `data`, we specify in which data frame the variables are found, and in the third, `FUN`, we specify which function should be used to compute the summary.

By default, `mean` returns `NA` if there are missing values. In `airquality`, `Ozone` contains missing values, but when we compute the grouped means, the results are not `NA`:

```
aggregate(Ozone ~ Month, data = airquality, FUN = mean)
```

By default, `aggregate` removes `NA` values before computing the grouped summaries.

It is also possible to compute summaries for multiple variables at the same time. For instance, we can compute the standard deviations (using `sd`) of `Temp` and `Wind`, grouped by `Month`:

```
aggregate(cbind(Temp, Wind) ~ Month, data = airquality, FUN = sd)
```

`aggregate` can also be used to count the number of observations in the groups. For instance, we can count the number of days in each month. In order to do so, we put a variable with no `NA` values on the left-hand side in the formula, and use `length`, which returns the length of a vector:

---

[15]In interval notation, `(50, 70]` means that the interval contains all values between 50 and 70, excluding 50 but including 70; the intervals is *open* on the left but *closed* to the right.

```
aggregate(Temp ~ Month, data = airquality, FUN = length)
```

Another function that can be used to compute grouped summaries is by. The results are the same, but the output is not as nicely formatted. Here's how to use it to compute the mean temperature grouped by month:

```
by(airquality$Temp, airquality$Month, mean)
```

What makes by useful is that unlike aggregate it is easy to use with functions that take more than one variable as input. If we want to compute the correlation between Wind and Temp grouped by month, we can do that as follows:

```
names(airquality)  # Check that Wind and Temp are in columns 3 and 4
by(airquality[, 3:4], airquality$Month, cor)
```

For each month, this outputs a *correlation matrix*, which shows both the correlation between Wind and Temp and the correlation of the variables with themselves (which always is 1).

~

Exercise 2.28. Install the datasauRus package using install.packages("datasauRus") (note the capital R!). It contains the dataset datasaurus_dozen. Check its structure and then do the following:

  1. Compute the mean of x, mean of y, standard deviation of x, standard deviation of y, and correlation between x and y, grouped by dataset. Are there any differences between the 12 datasets?

  2. Make a scatterplot of x against y for each dataset (use facetting!). Are there any differences between the 12 datasets?

2.13   Using |> pipes

Consider the code you used to solve part 1 of Exercise 2.25:

```
bookstore$rev_per_minute <- bookstore$purchase / bookstore$visit_length
```

Wouldn't it be more convenient if you didn't have to write the bookstore$ part each time? To just say once that you are manipulating bookstore, and have R implicitly understand that all the variables involved reside in that data frame? Yes. Yes, it would. Fortunately, R has tools that will let you do just that.

2.13.1   *Ceci n'est pas une pipe*

In most programming languages, data manipulation follows the same pattern as in the previous examples. In addition to that classic way of writing code, R offers an additional option: pipes. Pipes are operators that let you improve your code's readability and restructure your code so that it is read from left to right instead of from the inside out. For this, we'll use the `dplyr` package, which contains specialised functions that make it easier to use pipes for summarising, filtering and manipulating data. Let's install it:

```
install.packages("dplyr")
```

Let's say that we are interested in finding out what the mean wind speed (in m/s rather than mph) on hot days (temperature above 80, say) in the `airquality` data is, aggregated by month. We could do something like this:

```
# Extract hot days:
airquality2 <- airquality[airquality$Temp > 80, ]
# Convert wind speed to m/s:
airquality2$Wind <- airquality2$Wind * 0.44704
# Compute mean wind speed for each month:
hot_wind_means <- aggregate(Wind ~ Month, data = airquality2,
                            FUN = mean)
```

There is nothing wrong with this code per se. We create a copy of `airquality` (because we don't want to change the original data), change the units of the wind speed, and then compute the grouped means. A downside is that we end up with a copy of `airquality` that we maybe won't need again. We could avoid that by putting all the operations inside of `aggregate`:

```
# More compact:
hot_wind_means <-   aggregate(Wind*0.44704 ~ Month,
                              data = airquality[airquality$Temp > 80, ],
                              FUN = mean)
```

The problem with this is that it is a little difficult to follow because we have to read the code from the inside out. When we run the code, R will first extract the hot days, then convert the wind speed to m/s, and then compute the grouped means; so the operations happen in an order that is the opposite of the order in which we wrote them.

R 4.1 introduced a new operator, `|>`, called a *pipe*, which can be used to chain functions together. Calls that you would otherwise write as

```
new_variable <- function_2(function_1(your_data))
```

can be written as

```
your_data |> function_1() |> function_2() -> new_variable
```

so that the operations are written in the order they are performed. Some prefer the former style, which is more like mathematics, but many prefer the latter, which is more like natural language (particularly for those of us who are used to reading from left to right).

Three operations are required to solve the `airquality` wind speed problem:

1. Extract the hot days.
2. Convert the wind speed to m/s.
3. Compute the grouped means.

Where before we used function-less operations like `airquality2$Wind <- airquality2$Wind * 0.44704`, we would now require functions that carried out the same operations if we wanted to solve this problem using pipes.

A function that lets us extract the hot days is `filter` from `dplyr`:

```
library(dplyr)
filter(airquality, Temp > 80)
```

The `dplyr` function `mutate` lets us convert the wind speed:

```
mutate(airquality, Wind = Wind * 0.44704)
```

Note that we don't need to write `airquality$Wind` here; unless we specify otherwise, `dplyr` functions automatically assume that any variable names we reference are in the data frame that we provided as the first argument (much like the `ggplot` function).

And finally, the functions `group_by` and `summarise` can be used to compute the grouped means:

```
airquality <- group_by(airquality, Month)
summarise(airquality, Mean_wind_speed = mean(Wind))
```

We could use these functions step-by-step:

```
# Extract hot days:
airquality2 <- filter(airquality, Temp > 80)
# Convert wind speed to m/s:
airquality2 <- mutate(airquality, Wind = Wind * 0.44704)
# Compute mean wind speed for each month:
airquality <- group_by(airquality, Month)
hot_wind_means <- summarise(airquality, Mean_wind_speed = mean(Wind))

hot_wind_means
```

But, because we have functions to perform the operations, we can instead use |> pipes to chain them together in a *pipeline*. Pipes automatically send the output from the previous function as the first argument to the next, so that the data flows from left to right, which makes the code more concise. The resulting code is:

```
airquality |>
    filter(Temp > 80) |>
    mutate(Wind = Wind * 0.44704) |>
```

```
    group_by(Month) |>
    summarise(Mean_wind_speed = mean(Wind)) ->
    hot_wind_means
```

```
hot_wind_means
```

You can read the |> operator as *then*: take the `airquality` data, *then* filter it, *then* convert the `Wind` variable, *then* specify the groups, *then* compute the grouped means. Once you wrap your head around the idea of reading the operations from left to right, this code is arguably clearer and easier to read. Note that we used the right-assignment operator -> to assign the result to `hot_wind_means`, to keep in line with the idea that the data flows from left to right.

If you look at older examples of R code, you'll frequently find an alternative pipe operator used: `%>%`. This is an older implementation of the pipe, which still works (as long as either `dplyr` or the `magrittr` package is loaded) but is slower than |>.

A convenient way of writing the |> symbol in RStudio is to use the keyboard shortcut Ctrl+Shift+M. If this inserts the old `%>%` pipe, you need to go to the RStudio menu, click *Tools > Global Options*, choose the *Code* tab and tick the *Use native pipe operator* box.

In the remainder of the book, we will use pipes in some situations where they make the code easier to write or read. You'll see plenty of examples of how pipes can be used in Chapters 3-11, and you will learn about other pipe operators in Section 6.2.

~

Exercise 2.29. Using the `bookstore` data:

```
age <- c(28, 48, 47, 71, 22, 80, 48, 30, 31)
purchase <- c(20, 59, 2, 12, 22, 160, 34, 34, 29)
visit_length <- c(5, 2, 20, 22, 12, 31, 9, 10, 11)
bookstore <- data.frame(age, purchase, visit_length)
```

Add a new variable `rev_per_minute` which is the ratio between purchase and the visit length, using a pipe and a function from `dplyr`.

2.13.2 Placeholders and `with`

A pipe sends the output from the left-hand side to the first argument of the right-hand function. This won't work if the first argument of the right-hand function doesn't agree with the output from the left-hand function.

An example of this is what happens when we try to compute the correlation between two variables in a data frame using a pipe:

```
library(ggplot2)
msleep |> cor(sleep_total, sleep_rem, use = "complete.obs")
```

The first argument of `cor` should be one of the variables to compute the correlation for, and not a data frame, which is what the pipe passes on. To use pipes with functions like this, we can wrap `cor` with the `with` function. This takes the data frame from the left-hand side as input and then applies `cor` to variables in that dataset:

```
msleep |> with(cor(sleep_total, sleep_rem, use = "complete.obs"))
```

There are some other functions that have an argument for passing a data frame, but not as their first argument. An example is `aggregate`. Its first argument is a formula specifying what variables we are interested in. We've previously used it as follows:

```
aggregate(Temp ~ Month, data = airquality, FUN = mean)
```

If we wanted to use it in a pipeline, we can do so by writing `data = _` in the function's arguments. In pipelines, `_` is called a *placeholder*, and is a shorthand for "the output from the left-hand side of the pipe operator".

```
airquality |> aggregate(Temp ~ Month,
                        data = _,
                        FUN = mean)
```

2.14   Flavours of R: base and tidyverse

R is a programming *language* and just like any language, it has different dialects. When you read about R online, you'll frequently see people mentioning the words "base" and "tidyverse". These are the two most common dialects of R. Base R is just that, R in its purest form. The tidyverse is a collection of add-on packages for working with different types of data. The two are fully compatible, and you can mix and match as much as you like. Both `ggplot2` and `dplyr` are part of the tidyverse, which is more tailored to using pipes.

In recent years, the tidyverse has been heavily promoted as being "modern" R which "makes data science faster, easier and more fun". You should believe the hype: the tidyverse is marvellous. But if you only learn tidyverse R, you will miss out on much of what R has to offer. Base R is just as marvellous, and can definitely make data science as fast, easy and fun as the tidyverse. Besides, nobody uses just pure base R anyway – there are a ton of non-tidyverse packages that extend and enrich R in exciting new ways. Calling the non-tidyverse dialect "base" is a bit of a misnomer.

Anyone who tells you to just learn one of these dialects is wrong. Both are great, they work extremely well together, and they are similar enough that you shouldn't limit yourself to just mastering one of them. This book will show you both base R and tidyverse solutions to problems, so that you can decide for yourself which is faster, easier, and more fun.

A defining property of the tidyverse is that there are separate functions for everything, which is perfect for code that relies on pipes. In contrast, base R uses fewer functions, but with more parameters, to perform the same tasks. If you use tidyverse solutions there is a good chance that there exists a function which performs exactly the task you're going to do with its default settings. This is great (once again, especially if you want to use pipes), but it means that there are many more functions to master for tidyverse users, whereas you can make do with much fewer in base R. You will spend more time looking up function arguments when working with base R (which fortunately is fairly straightforward using the ? documentation); but, on the other hand, looking up arguments for a function that you know the name of is easier than finding a function that does something very specific that you don't know the name of. There are advantages and disadvantages to both approaches.

2.15 Importing data

So far, we've looked at examples of data that either came shipped with base R or `ggplot2` or `palmerpenguins`, or simple toy examples that we created ourselves, like `bookstore`. While you can do all your data entry work in R, `bookstore` style, it is much more common to load data from other sources. Two important types of files are *comma-separated value files*, `.csv`, and Excel spreadsheets, `.xlsx`. `.csv` files are spreadsheets stored as text files – basically Excel files stripped down to the bare minimum – no formatting, no formulas, no macros. You can open and edit them in spreadsheet software like LibreOffice Calc, Google Sheets, or Microsoft Excel. Many devices and databases can export data in `.csv` format, making it a commonly used file format that you are likely to encounter sooner rather than later.

2.15.1 Importing files through the RStudio menus

You can import data into R either through writing code or through using the RStudio menus. In the latter case, RStudio will generate the code for importing the data for you. You can then copy and save that code, in case you need to use it again later.

Let's start with a `csv` file: `philosophers.csv`. Open it with a spreadsheet software to have a quick look at it. Then open it in a text editor (for instance, Notepad for Windows, TextEdit for Mac, or Gedit for Linux). Note how commas are used to separate the columns of the data:

```
"Name","Description","Born","Deceased","Rating"
"Aristotle","Pretty influential, as philosophers go.",-384,"322 BC",
"4.8"
"Basilides","Denied the existence of incorporeal entities.",-175,
"125 BC",4
"Cercops","An Orphic poet",,,"3.2"
"Dexippus","Neoplatonic!",235,"375 AD","2.7"
"Epictetus","A stoic philosopher",50,"135 AD",5
"Favorinus","Sceptic",80,"160 AD","4.7"
```

To import data from a csv file, choose *File > Import dataset > From Text (base)*. You then get to choose what file to import the data from. Try `philosophers.csv`. You then get to choose some settings for the file, e.g., that the columns are separated by commas. There is a preview window showing you what the imported data will look like using the current settings. When you're done, click on *Import*. This will generate the code needed to import the data (using a function called `read.csv`) and run it in the Console window. You should now see an object in your Environment panel called `philosophers`.

You can import data from Excel files in an analogous manner. Choose *File > Import dataset > From Excel* in the RStudio menu. You can then click *Browse* to choose a file, and choose some settings. Click *Import* when you're done.

If you feel that this was complicated enough, you can now skip the remaining sections on importing data, and move on to Section 2.10. But if you want to learn more about how to write code for importing data, and about what the different settings in the menus really mean, read on.

2.15.2   Importing csv files

In order to manually load data from a file into R, you need its *path*, i.e, you need to tell R where to find the file. Unless you specify otherwise, R will look for files in its current *working directory*. To see what your current working directory is, run the following code in the Console pane:

```
getwd()
```

In RStudio, your working directory will usually be shown in the Files pane. If you have opened RStudio by opening a .R file, the working directory will be the directory in which the file is stored. You can change the working directory by using the function setwd or selecting *Session > Set Working Directory > Choose Directory* in the RStudio menu.

Before we discuss paths further, let's look at how you can import data from a file that is in your working directory. The data files that we'll use in examples in this book can be downloaded from the book's web page[16]. They are stored in a zip file (data.zip); open it and copy/extract the files to the folder that is your current working directory. Open philosophers.csv with a spreadsheet software to have a quick look at it. Then open it in a text editor (for instance, Notepad for Windows, TextEdit for Mac or Gedit for Linux). Note how commas are used to separate the columns of the data:

```
"Name","Description","Born","Deceased","Rating"
"Aristotle","Pretty influential, as philosophers go.",-384,"322 BC",
"4.8"
"Basilides","Denied the existence of incorporeal entities.",-175,
"125 BC",4
"Cercops","An Orphic poet",,,"3.2"
"Dexippus","Neoplatonic!",235,"375 AD","2.7"
"Epictetus","A stoic philosopher",50,"135 AD",5
"Favorinus","Sceptic",80,"160 AD","4.7"
```

Then run the following code to import the data using the read.csv function and store it in a variable named imported_data:

```
imported_data <- read.csv("philosophers.csv")
```

---

[16]http://www.modernstatisticswithr.com/data.zip

If you get an error message that says:

```
Error in file(file, "rt") : cannot open the connection
In addition: Warning message:
In file(file, "rt") :
  cannot open file 'philosophers.csv': No such file or directory
```

...it means that `philosophers.csv` is not in your working directory. Either move the file to the right directory (remember, you can use run `getwd()` to see what your working directory is) or change your working directory, as described above.

Now, let's have a look at `imported_data`:

```
View(imported_data)
str(imported_data)
```

The columns `Name` and `Description` both contain text, and have been imported as `character` vectors[17]. The `Rating` column contains numbers with decimals and has been imported as a `numeric` vector. The column `Born` only contain integer values and has been imported as an `integer` vector. The missing value is represented by an `NA`. The `Deceased` column contains years formatted like 125 `BC` and 135 `AD`. These have been imported into a `character` vector – because numbers and letters are mixed in this column, R treats it as a text string (in Chapter 5 we will see how we can convert it to numbers or proper dates). In this case, the missing value is represented by an empty string, `""`, rather than by `NA`.

So, what can you do in case you need to import data from a file that is not in your working directory? This is a common problem, as many of us store script files and data files in separate folders (or even on separate drives). One option is to use `file.choose`, which opens a pop-up window that lets you choose which file to open using a graphical interface:

```
imported_data2 <- read.csv(file.choose())
```

This is fine if you just want to open a single file once. But if you want to reuse your code or run it multiple times, you probably don't want to have to click and select your file each time. Instead, you can specify the path to your file in the call to `read.csv`.

2.15.3 File paths

File paths look different in different operating systems. If the user `Mans` has a file named `philosophers.csv` stored in a folder called `MyData` on his desktop, its path on an English-language Windows system would be:

```
C:\Users\Mans\Desktop\MyData\philosophers.csv
```

---

[17]If you are running an older version of R (specifically, a version older than the 4.0.0 version released in April 2020), the `character` vectors will have been imported as `factor` vectors instead. You can change that behaviour by adding a `stringsAsFactors = FALSE` argument to `read.csv`.

On a Mac, it would be:

```
/Users/Mans/Desktop/MyData/philosophers.csv
```

And on Linux:

```
/home/Mans/Desktop/MyData/philosophers.csv
```

You can copy the path of the file from your file browser: Explorer[18] (Windows), Finder[19] (Mac) or Nautilus/similar[20] (Linux). Once you have copied the path, you can store it in R as a `character` string.

Here's how to do this on Mac and Linux:

```
file_path <- "/Users/Mans/Desktop/MyData/philosophers.csv" # Mac
file_path <- "/home/Mans/Desktop/MyData/philosophers.csv"   # Linux
```

If you're working on a Windows system, file paths are written using backslashes, \, like so:

```
C:\Users\Mans\Desktop\MyData\file.csv
```

You have to be careful when using backslashes in `character` strings in R, because they are used to create special characters (see Section 5.5). If we place the above path in a string, R won't recognise it as a path. Instead, we have to reformat it into one of the following two formats:

```
# Windows example 1:
file_path <- "C:/Users/Mans/Desktop/MyData/philosophers.csv"
# Windows example 2:
file_path <- "C:\\Users\\Mans\\Desktop\\MyData\\philosophers.csv"
```

If you've copied the path to your clipboard, you can also get the path in the second of the formats above by using

```
file_path <- readClipboard()    # Windows example 3
```

Once the path is stored in `file_path`, you can then make a call to `read.csv` to import the data:

```
imported_data <- read.csv(file_path)
```

---

[18]To copy the path, navigate to the file in Explorer. Hold down the Shift key and right-click the file, selecting *Copy as path*.

[19]To copy the path, navigate to the file in Finder and right-click/Control+click/two-finger click on the file. Hold down the Option key, and then select *Copy "file name" as Pathname*.

[20]To copy the path from Nautilus, navigate to the file and press Ctrl+L to show the path, then copy it. If you are using some other file browser or the terminal, my guess is that you're tech-savvy enough that you don't need me to tell you how to find the path of a file.

Try this with your `philosophers.csv` file, to make sure that you know how it works.

Finally, you can read a file directly from a URL, by giving the URL as the file path. Here is an example with data from the WHO Global Tuberculosis Report[21]:

```
# Download WHO tuberculosis burden data:
tb_data <- read.csv("https://tinyurl.com/whotbdata")
```

`.csv` files can differ slightly in how they are formatted – for instance, different symbols can be used to delimit the columns. You will learn how to handle this in the exercises below.

A downside to `read.csv` is that it is very slow when reading large (50 MB or more) csv files. Faster functions are available in add-on packages; see Section 5.7.1. In addition, it is also possible to import data from other statistical software packages, such as SAS and SPSS, from other file formats like JSON, and from databases. We'll discuss most of these in Section 5.14.

2.15.4 Importing Excel files

One common file format we will discuss right away though – `.xlsx` – Excel spreadsheet files. There are several packages that can be used to import Excel files to R. I like the `openxlsx` package, so let's install that:

```
install.packages("openxlsx")
```

Now, download the `philosophers.xlsx` file from the book's web page[22] and save it in a folder of your choice. Then, set `file_path` to the path of the file, just as you did for the `.csv` file. To import data from the Excel file, you can then use:

```
library(openxlsx)
imported_from_Excel <- read.xlsx(file_path)

View(imported_from_Excel)
str(imported_from_Excel)
```

As with `read.csv`, you can replace the file path with `file.choose()` in order to select the file manually.

~

Exercise 2.30. The abbreviation CSV stands for *comma separated values*, i.e., that commas `,` are used to separate the data columns. Unfortunately, the `.csv` format is not standardised, and `.csv` files can use different characters to delimit the columns. Examples include semicolons (;) and tabs (multiple spaces, denoted `\t` in strings in R). Moreover, decimal points can be given either as points (.) or as commas (,). Download the `vas.csv` file from the book's web page[23]. In this dataset, a number of patients with chronic pain have recorded how much pain they experience each day during a period, using the visual analogue scale (VAS, ranging

---

[21]https://www.who.int/tb/country/data/download/en/
[22]http://www.modernstatisticswithr.com/data.zip
[23]http://www.modernstatisticswithr.com/data.zip

from 0 – no pain – to 10 – worst imaginable pain). Inspect the file in a spreadsheet software and a text editor; check which symbol is used to separate the columns and whether a decimal point or a decimal comma is used. Then set `file_path` to its path and import the data from it using the code below:

```
vas <- read.csv(file_path, sep = ";", dec = ",", skip = 4)

View(vas)
str(vas)
```

1. Why are there two variables named X and X.1 in the data frame?
2. What happens if you remove the `sep = ";"` argument?
3. What happens if you instead remove the `dec = ","` argument?
4. What happens if you instead remove the `skip = 4` argument?
5. What happens if you change `skip = 4` to `skip = 5`?

Exercise 2.31. Load the VAS pain data `vas.csv` from Exercise 2.30. Then do the following:

1. Compute the mean VAS for each patient.
2. Compute the lowest and highest VAS recorded for each patient.
3. Compute the number of high-VAS days, defined as days where the VAS was at least 7, for each patient.

Exercise 2.32. Download the `projects-email.xlsx` file from the book's web page[24] and have a look at it in a spreadsheet software. Note that it has three sheets: *Projects*, *Email*, and *Contact*.

1. Read the documentation for `read.xlsx`. How can you import the data from the second sheet, *Email*?
2. Some email addresses are repeated more than once. Read the documentation for `unique`. How can you use it to obtain a vector containing the email addresses without any duplicates?

Exercise 2.33. Download the `vas-transposed.csv` file from the book's web page[25] and have a look at it in a spreadsheet software. It is a *transposed* version of `vas.csv`, where rows represent variables and columns represent observations (instead of the other way around, as is the case in data frames in R). How can we import this data into R?

1. Import the data using `read.csv`. What does the resulting data frame look like?
2. Read the documentation for `read.csv`. How can you make it read the row names that can be found in the first column of the `.csv` file?

---

[24]http://www.modernstatisticswithr.com/data.zip
[25]http://www.modernstatisticswithr.com/data.zip

3. The function t can be applied to transpose (i.e., rotate) your data frame. Try it out on your imported data. Is the resulting object what you were looking for? What happens if you make a call to as.data.frame with your data after transposing it?

2.16 Saving and exporting your data

In many a case, data manipulation is a huge part of statistical work, and of course you want to be able to save a data frame after manipulating it. There are two options for doing this in R: you can either export the data as, e.g., a .csv or a .xlsx file, or save it in R format as an .RData file.

2.16.1 Exporting data

Just as we used the functions read.csv and read.xlsx to import data, we can use write.csv and write.xlsx to export it. The code below saves the bookstore data frame as a .csv file and an .xlsx file. Both files will be created in the current working directory. If you wish to store them somewhere else, you can replace the "bookstore.csv" bit with a full path, e.g., "/home/mans/my-business/bookstore.csv".

```
# Bookstore example
age <- c(28, 48, 47, 71, 22, 80, 48, 30, 31)
purchase <- c(20, 59, 2, 12, 22, 160, 34, 34, 29)
bookstore <- data.frame(age, purchase)

# Export to .csv:
write.csv(bookstore, "bookstore.csv")

# Export to .xlsx (Excel):
library(openxlsx)
write.xlsx(bookstore, "bookstore.xlsx")
```

2.16.2 Saving and loading R data

Being able to export to different spreadsheet formats is very useful, but sometimes you want to save an object that can't be saved in a spreadsheet format. For instance, you may wish to save a machine learning model that you've created. .RData files can be used to store one or more R objects.

To save the objects bookstore and age in a .Rdata file, we can use the save function:

```
save(bookstore, age, file = "myData.RData")
```

To save all objects in your environment, you can use save.image:

```
save.image(file = "allMyData.RData")
```

When we wish to load the stored objects, we use the load function:

```
load(file = "myData.RData")
```

2.17   RStudio projects

It is good practice to create a new folder for each new data analysis project that you are
working on, where you store code, data, and the output from the analysis. In RStudio you
can associate a folder with a Project, which lets you start RStudio with that folder as your
working directory. Moreover, by opening another Project you can have several RStudio
sessions, each with its separate variables and working directories, running simultaneously.

To create a new Project, click *File > New Project* in the RStudio menu. You then get to
choose whether to create a Project associated with a folder that already exists, or to create
a Project in a new folder. After you've created the Project, it will be saved as an .Rproj file.
You can launch RStudio with the Project folder as the working directory by double-clicking
the .Rproj file. If you already have an active RStudio session, this will open another session
in a separate window.

When working in a Project, I recommend that you store your data in a subfolder of the
Project folder. You can then use *relative paths* to access your data files, i.e., paths that are
relative to your working directory. For instance, if the file bookstore.csv is in a folder in
your working directory called Data, its relative path is:

```
file_path <- "Data/bookstore.csv"
```

Much simpler than having to write the entire path, isn't it?

If instead your working directory is contained inside the folder where bookstore.csv is stored,
its relative path would be

```
file_path <- "../bookstore.csv"
```

The beauty of using relative paths is that they are simpler to write, and if you transfer the
entire project folder to another computer, your code will still run, because the relative paths
will stay the same.

2.18   Troubleshooting

Every now and then R will throw an error message at you. Sometimes, these will be
informative and useful, as in this case:

```
age <- c(28, 48, 47, 71, 22, 80, 48, 30, 31)
means(age)
```

where R prints:

```
> means(age)
Error in means(age) : could not find function "means"
```

This tells us that the function that we are trying to use, means, does not exist. There are two possible reasons for this: either we haven't loaded the package in which the function exists, or we have misspelt the function name. In our example, the latter is true; the function that we really wanted to use was of course mean and not means.

At other times, interpreting the error message seems insurmountable, like in these examples:

```
Error in if (str_count(string = f[[j]], pattern = \"\\\\S+\") == 1) { :
  \n  argument is of length zero
```

and

```
Error in if (requir[y] &gt; supply[x]) { : \nmissing value where
  TRUE/FALSE needed
```

When you encounter an error message, I recommend following these steps:

1. Read the error message carefully and try to decipher it. Have you seen it before? Does it point to a particular variable or function? Check Section 13.2 of this book, which deals with common error messages in R.

2. Check your code. Have you misspelt any variable or function names? Are there missing brackets, strange commas, or invalid characters?

3. Copy the error message and do a web search using the message as your search term. It is more than likely that somebody else has encountered the same problem, and that you can find a solution to it online. This is a great shortcut for finding solutions to your problem. In fact, **this may well be the single most important tip in this entire book**.

4. Read the documentation for the function causing the error message, and look at some examples of how to use it (both in the documentation and online, e.g., in blog posts). Have you used it correctly?

5. Use the debugging tools presented in Chapter 13, or try to simplify the example that you are working with (e.g., removing parts of the analysis or the data) and see if that removes the problem.

6. If you still can't find a solution, post a question at a site like Stack Overflow[26] or the RStudio community forums[27]. Make sure to post your code and describe the context in which the error message appears. If at all possible, post a reproducible example, i.e., a piece of code that others can run, that causes the error message. This will make it a lot easier for others to help you.

---

[26]https://stackoverflow.com/
[27]https://community.rstudio.com/

3

The cornerstones of statistics

This chapter is an attempt to introduce modern classical statistics. "Modern classical" may sound like a contradiction, but it is in fact anything but. Classical statistics covers topics like estimation, quantification of uncertainty, and hypothesis testing – all of which are at the heart of data analysis. Since the advent of modern computers, much has happened in this field that has yet to make it to the standard textbooks of introductory courses in statistics. This chapter attempts to bridge part of that gap by dealing with those classical topics, but with a modern approach that uses more recent advances in statistical theory and computational methods.

Whenever it is feasible, the aim of this books is to:

* Use hypothesis tests that are based on permutations or the bootstrap rather than tests based on strict assumptions about the distribution of the data or asymptotic distributions,
* Complement estimates and hypothesis tests with confidence intervals based on sound methods (including the bootstrap), and
* Offer easy-to-use Bayesian methods as an alternative to frequentist tools.

After reading this chapter, you will be able to use R to:

* Create contingency tables,
* Run hypothesis tests including the t-test and the χ^2-test,
* Compute confidence intervals,
* Handle multiple testing,
* Run Bayesian tests, and
* Report statistical results.

3.1 The three cultures

There are three main schools in statistical modelling: the *frequentist* school, the *Bayesian* school, and the *machine learning* school.

The main difference between the frequentist and Bayesian schools is the way in which they approach probability. Frequentist statistics uses the frequency, or long-run proportion, of an event to describe the probability that it occurs. Bayesian statistics incorporates prior knowledge and personal beliefs about the event to compute subjective probabilities.

Many of the best known tools in the statistical toolbox, such as p-values and confidence intervals, stem from frequentist statistics. It has often been considered a more objective approach, suited for analysing experiments. Bayesian statistics has been considered to be more flexible and adaptable, as it allows for the incorporation of prior knowledge and beliefs and can be updated as new evidence becomes available. This makes it well suited for situations where there is a need to incorporate subjective information or to update predictions as new data becomes available.

Both frequentist and Bayesian statistics have a tradition of theoretical statistics and the development of rigorous, formal mathematical methods for analysing and interpreting data, that can be shown to be optimal in certain scenarios. In contrast, the machine learning school is characterised by a focus on developing algorithms and computational tools for automatically identifying patterns in data and making predictions based on those patterns. This culture is associated with more practical, data-driven approaches to solving problems, and much less rigour. Methods are evaluated by checking whether they give good predictions for test datasets, rather than by their theoretical properties. In recent years, a part of this school has been rebranded as artificial intelligence, or AI.

All three schools are essential parts of modern statistics, and there is a lot of interaction between them (for instance, methods from frequentist and Bayesian statistics are often used in machine learning). In this book, we'll make use of tools from all three schools. For statistical analyses in this chapter and in Chapters 7-10, we'll focus on frequentist methods but offer Bayesian methods as a useful alternative and describe their advantages. Chapter 11 is devoted to the machine learning mindset and its use in predictive modelling.

3.2 Frequencies, proportions, and cross-tables

Some statistical tools are equally important in all three statistical schools. Among them are frequency tables and contingency tables,. We'll look at some examples using the `penguins` dataset from the `palmerpenguins` package. Let's start by reading a little about the data:

```
library(palmerpenguins)
?penguins
```

3.2.1 Frequency tables

Frequency tables are used to summarise the distribution of a single variable. They consist of a list of the different values that a variable can take, along with either the number of times (the *frequency*) each value occurs in the data, or the proportion of times it occurs. For example, a frequency table for the variable `species` might show the number of individuals from different species in a dataset, as in this example with the `penguins` data:

Without pipes:

```
table(penguins$species)
```

With pipes:

```
library(dplyr)
penguins |> select(species) |> table()
```

This results in the following output in the Console:

```
##
##    Adelie Chinstrap    Gentoo
##       152        68       124
```

3.2.2   Publication-ready tables

The table in the previous section is fine for data exploration but requires some formatting if you wish to put it in a report. If instead you want something publication-ready that you can put straight into a report or a presentation, I recommend using either the `gtsummary` package or the `ivo.table` package. Let's install them (along with the `flextable` package, which we'll also need), and see how they can be used to create a frequency table for species using the `penguins` data:

```
# Install the packages:
install.packages(c("gtsummary", "flextable", "ivo.table"))
```

Let's start with an example using `gtsummary` and its `tbl_summary` function:

Without pipes:

```
library(gtsummary)
tbl_summary(penguins[,"species"])
```

With pipes:

```
library(dplyr)
library(gtsummary)
penguins |>
    select(species) |>
    tbl_summary()
```

We now get a better-looking table. It is shown in the Viewer pane in RStudio. You can highlight it with your mouse cursor to copy and paste it into, e.g., Word or PowerPoint, or use functions from the `flextable` package to export it to various file formats. Here is an example of how we can export it as a Word document:

```
library(flextable)
penguins |>
    select(species) |>
    tbl_summary() |>
    as_flex_table() |> # Convert the table to a format that can be exported
    save_as_docx(path = "my_table.docx")
```

The `ivo.table` package and the `ivo_table` function offer a similar table constructed in the same way:

Without pipes:

```
library(ivo.table)
ivo_table(penguins[,"species"])
```

With pipes:

```
library(dplyr)
library(ivo.table)
penguins |>
    select(species) |>
    ivo_table()
```

You can modify the settings to change the colours and fonts used, and to show percentages instead of counts:

```
penguins |>
  select(species) |>
  ivo_table(color = "darkred",
            font = "Garamond",
            percent_by = "row")
```

If you prefer, you can also get the table in a long format that is useful when your variable has many different levels:

```
penguins |>
  select(species) |>
  ivo_table(long_table = TRUE)
```

Finally, the table can be exported to Word as follows:

```
library(flextable)
penguins |>
  select(species) |>
  ivo_table() |>
  save_as_docx(path = "my_table.docx")
```

3.2.3   Contingency tables

Contingency tables, also known as cross-tabulations or cross-tables, are used to summarise the relationship between two or more variables. They consist of a table with rows and columns that represent the different values of the variables, and the entries in the table show the number or proportion of times each combination of values occurs in the data. For example, a contingency table for the variables species and island in the penguins data shows the number of individuals of different species at different islands.

Without pipes:

```
ftable(penguins$species,
       penguins$island)
```

With pipes:

```
penguins |>
  select(species, island) |>
  ftable()
```

The resulting table is:

```
##               Adelie Chinstrap Gentoo
##
## Biscoe            44         0    124
## Dream             56        68      0
## Torgersen         52         0      0
```

Again, we can create a nicely formatted publication-ready table. With ivo_table, we follow the same logic as before:

```
library(ivo.table)
penguins |> select(species, island) |>
        ivo_table()
```

The settings and export options that we used before are still available:

```
# Change colours and fonts and export to a Word file named "penguins.docx":
library(flextable)
penguins |> select(species, island) |>
        ivo_table(color = "darkred",
                    font = "Garamond",
                    percent_by = "tot") |>
        save_as_docx(path = "penguins.docx")
```

We can also highlight cells of particular interest. For instance, the cell in the second column (the column with island names counts as a column here) of the third row:

```
penguins |> select(species, island) |>
        ivo_table(highlight_cols = 2,
                    highlight_rows = 3)
```

To create a contingency table using `tbl_summary`, we use the following syntax:

Without pipes:

```
library(gtsummary)
tbl_summary(penguins[,c("species",
                        "island")],
            by = species)
```

With pipes:

```
library(dplyr)
library(gtsummary)
penguins |>
    select(species, island) |>
    tbl_summary(by = species)
```

For tables like the ones we just created, it is common to talk about $R \times C$ contingency tables, where R denotes the number of rows and C denotes the number of columns (not including any rows or columns displaying the margin sums). The table in the `penguins` example above is thus a 3×3 contingency table.

Contingency tables are great for presenting data, and in many cases we can draw conclusions directly from looking at such a table. For instance, in the example above, it is clear that Adelie penguins are the only species present at all three islands (or at least, the only species sampled at all three islands). In other cases, the results aren't as clear-cut. That's when statistical hypothesis testing becomes useful – a powerful set of tools that lets us determine for instance whether there is statistical evidence for differences between groups. We'll get to that soon, but first we'll look at some examples of tables for three and four variables.

3.2.4 Three-way and four-way tables

Three-way and four-way tables are contingency tables showing the distribution of three and four variables, respectively. They are straightforward to create using `ftable` or `ivo_table`.

We simply select the variables we wish to include in the table, and use `ftable` or `ivo_table` as in previous examples. Here are some examples using `ivo_table`:

```
# A three-way table:
library(ivo.table)
penguins |> select(sex, species, island) |>
        ivo_table()


# Exclude missing values:
penguins |> select(sex, species, island) |>
        ivo_table(exclude_missing = TRUE)


# A four-way table:
penguins |> select(sex, species, island, year) |>
        ivo_table()
```

You can't use `gtsummary` to construct three-way and four-way tables, but you can use it for a similar type of table, presenting several variables at once, stratified by another variable. For instance, we can show the frequencies of islands and sexes stratified by species, as in the example below. This type of table, showing the distributions of different categorical variables, is often referred to as *Table 1* in scientific papers.

Without pipes:

```
library(gtsummary)
tbl_summary(penguins[,c("species",
                        "sex",
                        "island")],
            by = species)
```

With pipes:

```
library(dplyr)
library(gtsummary)
penguins |>
    select(species, sex, island) |>
    tbl_summary(by = species)
```

3.3 Hypothesis testing and p-values

Statistical hypothesis testing is used to determine which of two complementary *hypotheses* that is true. In statistics, a hypothesis is a statement about a parameter in a population, such as the population mean value. The two hypotheses in a hypothesis testing problem are:

- The *null hypothesis* H_0: corresponding to "no effect", "no difference", or "no relationship".
- The *alternative hypothesis* H_1: corresponding to "there is an effect", "there is a difference", or "there is a relationship".

To make this more concrete, let's look at some examples.

The effect of a treatment. To study how a new drug affects systolic blood pressure, we measure the blood pressure of a number of patients twice: before and after they take the drug. The population parameter that we're interested in is Δ, the average change in blood pressure between the two measurements. The hypotheses are as follows:

- H_0: there is no effect; $\Delta = 0$.
- H_1: there is an effect; $\Delta \neq 0$.

Comparing two groups. To find out whether flipper length differs between male and female Chinstrap penguins, we measure the flipper lengths of a number of penguins. If the average length in the female population is μ_1 and the average length in the male population is μ_2, then the population parameter that we are interested in is the difference $\mu_1 - \mu_2$. The hypotheses are:

- H_0: there is no difference; $\mu_1 - \mu_2 = 0$.
- H_1: there is a difference; $\mu_1 - \mu_2 \neq 0$.

The relationship between two variables X and Y. We are interested in finding out whether the sex distribution for Adelie penguins is the same on three different islands. For a randomly selected Adelie penguin, let $P(X = x)$ denote the probability that the sex X takes the value x (possible values being male and female), and $P(Y = y)$ that the island Y takes the value y (with three different possible values). The hypotheses are:

- H_0: the variables are independent; $P(X = x \text{ and } Y = y) = P(X = x)P(Y = y)$ for all x, y.
- H_1: the variables are dependent; there is at least one pair x, y such that $P(X = x \text{ and } Y = y) \neq P(X = x)P(Y = y)$.

The purpose of hypothesis testing is to determine which of the two hypotheses to believe in. Hypothesis testing is often compared to legal trials: the null hypothesis is considered to be "innocent until proven guilty", meaning that we won't reject it unless there is compelling evidence against it. We can therefore think of the null hypothesis as a sort of default – we'll believe in it until we have enough evidence to say with some confidence that it in fact isn't true.

In frequentist statistics, the strength of the evidence against the null hypothesis is measured using a *p-value*. It is computed using a *test statistic*, a summary statistic computed from the data, designed to measure how much the observations deviate from what can be expected under the null hypothesis. Bayesian approaches to hypothesis testing are fundamentally different, and are covered in Section 3.9.

The p-value quantifies the amount of evidence as *the probability under H_0 of obtaining an outcome that points in the direction of H_1 at least much as the observed outcome*. The lower the p-value is, the greater the evidence against H_0. Being a probability, it ranges from 0 to 1.

How this probability is computed depends on the hypotheses themselves, how we wish to measure how much an outcome deviates from H_0, and what assumptions we are prepared to make about our data. Throughout this chapter, we'll see several examples of how p-values are computed in different situations. We'll start with an example that inspired the invention of p-values in the 1920s.

3.3.1 The lady tasting tea

The quintessential example of a statistical hypothesis test is that of the lady tasting tea, described by Sir Ronald Fisher in the second chapter of his classic 1935 text *The Design of Experiments*. A colleague of Fisher's, the phycologist Dr. Muriel Bristol, took her tea with milk and claimed that she could tell whether milk or tea was added to the cup first. Fisher did not believe her and devised an experiment to test her claim. It consisted of preparing eight cups of tea out of sight of the lady, four in which the milk was poured first, and four in which the tea was poured first. She would then be presented with the cups in a randomised

	Truth: milk first	Truth: tea first
Guess: milk first	4	0
Guess: tea first	0	4

order, tasked with dividing the cups into two groups of four according to how they were prepared.

The hypotheses that this experiment was designed to test are:

- H_0: the lady's guesses are no better than chance.
- H_1: the lady's guesses are better than chance.

This is a test of independence – if the lady's guesses are no better than chance, then her guess (X) is independent from the content in the cup (Y).

The results of the experiment can be presented in a 2×2 contingency table. For instance, one possible outcome is that the lady gets both sets of four right, which yields the following table:

The design of this experiment is such that all row sums and column sums are 4. Consequently, if we know the count in the upper left cell (the number of cups where the milk was poured first correctly identified as such by the lady), we also know the counts in the other cells[1]. This means that there are 5 possible tables that can result (the count in the upper left cell can be 4, 3, 2, 1, or 0). We can compute the probability of obtaining each table under the null hypothesis[2], with the following results:

- The count is 4 (4 cups with milk poured first right), with probability 1/70,
- The count is 3 (3 cups with milk poured first right and 1 wrong), with probability 16/70,
- The count is 2 (2 cups with milk poured first right and 2 wrong), with probability 36/70,
- The count is 1 (1 cup with milk poured first right and 3 wrong), with probability 16/70,
- The count is 0 (4 cups with milk poured first wrong), with probability 1/70.

A high number of correct guesses would be expected if H_1 were true, but not if H_0 were true. Consequently, high counts can be considered to be evidence against H_0. With this in mind, now that we know what the possible outcomes of the experiment are, and how likely they are if H_0 is true, we can compute the p-values corresponding to the different outcomes.

Getting all 4 cups where the milk poured first right is the most extreme outcome, in the sense that among all possible outcomes, this points the most in the direction of H_1, i.e., that the lady's guesses are better than change. If H_0 is true, this occurs in 1 experiment out of 70. The p-value is therefore $1/70 \approx 0.014$.

Getting 3 cups where the milk poured first right is the second most extreme outcome. If H_0 is true, this occurs in 16 experiments out of 70. In addition, a more extreme outcome (4 cups right) occurs in 1 experiment out of 70. The probability of getting a result that is at least as extreme as this is therefore $\frac{16}{70} + \frac{1}{70} = \frac{17}{70}$, and the p-value given this outcome is $17/70 \approx 0.24$.

Similarly, we find that if the lady gets 2 cups right, the p-value is $\frac{36}{70} + \frac{16}{70} + \frac{1}{70} \approx 0.75$, if she gets 1 cup right the p-value is 0.99 and if she gets 0 cups right the p-value is 1.

---

[1]For instance, if the lady got 3 cups where the milk was poured first right, she must have got 1 cup wrong, so that the upper right cell should have the value 1. Then, because the column sums are 4, the lower left cell has the value 1 and the lower right cell has the value 3.

[2]Either by using combinatorics and the classical definition of probability, or by using a hypergeometric distribution. Both approaches yield the same results.

At this point, you may be asking why we include outcomes that we in fact haven't observed when computing p-values. Why don't we simply compute how unlikely the observed outcome is if H_0 is true? To see why, recall that a low p-value means that we have evidence against H_0, and note that getting 0 cups right is the outcome that is the furthest from what we should expect under H_1. If we used the null probability of this outcome as the p-value, the p-value would become 1/70, i.e., rather low, even though we have absolutely no evidence at all that H_0 is false in this case. Including more extreme outcomes in the computation allows us to quantify how extreme the observed outcome is.

3.3.2   How low does the p-value have to be?

We've now seen how p-values can be computed in the lady tasting tea example, but we still haven't seen how to use p-values to make a decision about what hypothesis to believe in. The lower the p-value, the greater the evidence against H_0. But how low does the p-value have to be for us to say that we have sufficient evidence to reject H_0?

We need some way of determining a cut-off for p-values. Let's call this cut-off α, the *significance level* of our test. It can be any number between 0 and 1. Once we've decided on a value for α, we are ready to make a decision regarding which hypothesis to believe in. If the p-value is less than α, we reject H_0 in favour of H_1 and say the result is statistically *significant*. If the p-value is greater than α, we conclude that there isn't sufficient evidence against H_0, so we'll believe in it for the time being.

How then shall we choose α? There are two types of errors that we can make in hypothesis testing:

* A type I error: falsely rejecting H_0 even though H_0 is true (a false positive result).
* A type II error: not rejecting H_0 even though H_1 is true (a false negative result).

Both of these will depend on what significance level we choose for our p-value. If we choose a low α, we require more evidence before we reject H_0. This lowers the risk of a type I error but increases the risk of a type II error. Conversely, if we choose a higher cut-off, we get a higher risk of a type I error, but a lower risk of a type II error.

The probability of making a type I error is the easiest to control and often also considered to be the most important of the two. If we reject H_0 whenever the p-value is less than α, then the probability of committing a type I error if H_0 is true is also α. For this reason, α is often called the *type I error rate*. A common choice is to use $\alpha = 0.05$ as the cut-off, meaning that the null hypothesis is falsely rejected 5% of all studies where it in fact was true, or that 1 study in 20 finds statistical evidence for alternative hypotheses that are false.

No, wait, let me stop myself right there, because I just lied to you and we probably shouldn't let that slide. I just claimed that if we reject H_0 when the p-value is less than α, the probability of committing a type I error is also α. Indeed, you'll see similar statements in many different texts. This is an oversimplification that only is true in idealised situations that no one has ever encountered outside of a classroom. In the real world, there are two types of tests:

* *Exact tests*, for which the probability of committing a type I error is less than or equal to α.
* *Approximate tests*, for which the probability of committing a type II error is approximately equal to α (but may be greater than α, perhaps substantially so). How close it is to α depends on a number of factors; we'll discuss these for different tests later in this chapter.

The test that we constructed in the lady tasting tea example is an exact test (which can be seen through deeper mathematical analysis of its properties). It is known as *Fisher's exact test*. In the next section, we'll have a look at how we can use it to analyse data in R.

3.3.3   Fisher's exact test

To use Fisher's exact test, we first need some data. The contingency table describing the outcome of the experiment can be constructed in different ways in R, depending on whether our data is stored in a long data frame or already available in tabulated form. If it is stored in a data frame, where each row shows the outcome of a repetition, we can get a contingency table as follows:

```r
# Compute the contingency table from a data frame:
lady_data <- data.frame(Guess = c("Milk first", "Milk first",
                                  "Tea first", "Tea first",
                                  "Milk first", "Tea first",
                                  "Tea first", "Milk first"),
                        Truth = c("Milk first", "Milk first",
                                  "Tea first", "Tea first",
                                  "Milk first", "Tea first",
                                  "Tea first", "Milk first"))
lady_data |> ftable() -> lady_results1
lady_results1
```

If instead the data already is stored as counts, we can create a contingency table by hand by creating a matrix containing the counts:

```r
# Input the contingency table directly, if we only have the counts:
lady_results2 <- matrix(c(4, 0, 0, 4),
                        ncol = 2, nrow = 2,
                        byrow = TRUE,
                        dimnames = list(c("Guess: milk first", "Guess: tea first"),
                                        c("Truth: milk first", "Truth: tea first")))
lady_results2
```

To perform Fisher's exact test, for the lady tasting tea data, we use `fisher.test` as follows:

```r
fisher.test(lady_results1, alternative = "greater")
fisher.test(lady_results2, alternative = "greater")
```

The output contains the p-value (0.01429), and some additional information:

```
##
##  Fisher's Exact Test for Count Data
##
## data:  lady_results2
## p-value = 0.01429
## alternative hypothesis: true odds ratio is greater than 1
## 95 percent confidence interval:
##  2.003768      Inf
```

```
## sample estimates:
## odds ratio
##       Inf
```

3.3.4   One- and two-sided hypotheses

What is that mysterious last argument in `fisher.test`, `alternative = "greater"`? Well, there are three different sets of hypotheses that we may want to test:

1. Whether the lady guesses better than random. Here H_0: the lady's guesses are no better than chance, and H_1: the lady's guesses are better than chance.
2. Whether the lady guesses worse than random. Here H_0: the lady's guesses are no worse than chance, and H_1: the lady's guesses are worse than chance.
3. Whether the lady's guesses are either better than or worse than random. Here H_0: the lady's guesses are equally good as chance, and H_1: the lady's guesses are either worse than or better than chance.

The first two are said to be *one-sided* or *directed*, as they specify a direction in which the lady's ability deviates from chance. The last set of hypotheses is said to be *two-sided*, because the lady's abilities can deviate from chance in either direction. Because the alternative hypotheses are different, the three sets of hypotheses will yield different p-values (which is unsurprising, as we are asking different questions depending on how we specify our hypotheses).

Which of these we choose depends on what question we wish to answer. If, like Fisher, we want to know whether the lady has an uncanny ability to tell which cups were filled first with milk, we'd choose the first set of hypotheses. If instead we believe that maybe there is a difference in flavour that she can detect, but we're not sure whether she can tell which is which, we may go with the third set instead.

We can change the `alternative` argument in `fisher.test` to use the three different sets of hypotheses as follows:

```
fisher.test(lady_results2, alternative = "greater") # 1
fisher.test(lady_results2, alternative = "less") # 2
fisher.test(lady_results2, alternative = "two.sided") # 3
```

3.3.5   The lady binging tea: power and how the sample size affects the analysis

Fisher's exact test can be used to illustrate an important principle: the greater the sample size, the easier it is to detect if H_1 is true.

Let's say that we wish to test whether the lady's guesses are better than chance, using the significance level $\alpha = 0.05$. If she is served 2 cups where the milk was poured first and 2 where the tea was poured first, and guessed all of these correctly, we'd get the following result:

```
lady_results2 <- matrix(c(2, 0, 0, 2),
                        ncol = 2, nrow = 2,
                        byrow = TRUE,
```

```
                            dimnames = list(c("Guess: milk first", "Guess: tea first"),
                                            c("Truth: milk first", "Truth: tea first")))
fisher.test(lady_results2, alternative = "greater")
```

```
##
##   Fisher's Exact Test for Count Data
##
## data:  lady_results2
## p-value = 0.1667
## alternative hypothesis: true odds ratio is greater than 1
## 95 percent confidence interval:
##  0.357675      Inf
## sample estimates:
## odds ratio
##        Inf
```

The p-value is 0.1667, which is greater than α. Because the sample size is too small, we can't reject H_0 even when we get the most extreme outcome possible.

If we have a larger sample size, things are different. If the lady is served 15 cups where the milk was poured first and 15 where the tea was poured first, and guessed 11 of the cups where milked was poured first correctly, we'd get the following result:

```
lady_results2 <- matrix(c(11, 4, 4, 11),
                        ncol = 2, nrow = 2,
                        byrow = TRUE,
                        dimnames = list(c("Guess: milk first", "Guess: tea first"),
                                        c("Truth: milk first", "Truth: tea first")))
fisher.test(lady_results2, alternative = "greater")
```

```
##
##   Fisher's Exact Test for Count Data
##
## data:  lady_results2
## p-value = 0.01342
## alternative hypothesis: true odds ratio is greater than 1
## 95 percent confidence interval:
##  1.500179      Inf
## sample estimates:
## odds ratio
##   6.983892
```

The p-value is 0.01342, which is smaller than α, and we can reject H_0 even though the result was far from the most extreme outcome (which in this case would be 15 cups right).

What we've seen here is that it is easier to detect deviations from H_0 when the sample size is larger.

The probability of rejecting H_0 if H_1 is true is called the *power* of the test. We want this to be as large as possible. The power will depend on the significance level α (remember that this controls how strong evidence we need to reject H_0), how strong the effect is (it is easier to detect the lady's ability if she gets 95% of all cups right than if she gets 55% of all cups right), and how large the sample size is. Generally, the power of a test increases when the sample size increases.

Clearly, the first experiment described above, with just 4 cups of tea, is a poorly designed one, as it never can lead to us rejecting H_0. An important part of designing experiments is to perform power computations, where we calculate the power of the test under different scenarios and for different sample sizes. This lets us check how large a sample we need in order to obtain the desired power at a specific significance level α. We'll see some examples of how this can be done later in this chapter as well as in Sections 7.2-7.3.

3.3.6 Permutation tests

In the lady tasting tea example, the lady was tasked with placing labels (*milk first* or *tea first*) on the cups. To compute the p-value, we calculated in how many ways she could do this, or how many *permutations* of the labels there were. We could then check how many of these permutations that yielded an outcome at least as extreme as that which we observed. A test based on permutations is called a *permutation test*. It is an important class of tests, which we will return to throughout the book.

Not all tests are permutation tests. Next, we'll consider a different approach to testing hypotheses using contingency tables.

3.4 χ^2-tests

When analysing a contingency table containing two variables X and Y, there are two common types of tests. The first is a *test of independence*, where we test whether the two variables are independent:

- H_0: the variables are independent; $P(X = x \text{ and } Y = y) = P(X = x)P(Y = y)$ for all x, y.
- H_1: the variables are dependent; there is at least one pair x, y such that $P(X = x \text{ and } Y = y) \neq P(X = x)P(Y = y)$.

The second is a *test of homogeneity*, where we test whether the distribution of Y is the same regardless of the value of X, which in this case represents different populations:

- H_0: the variables are independent; $P(Y = y | X = x) = P(Y = y)$ for all x, y.
- H_1: the variables are dependent; there is at least one pair x, y such that $P(Y = y | X = x) \neq P(Y = y)$.

Mathematically, a test of homogeneity is equivalent to a test of independence. The difference between the two lies in how the data is sampled. When we plan to perform a test of independence, we sample a single population and then break the observations down into the categories in the contingency table based on their X and Y values. When we plan to perform a test of homogeneity we instead sample several populations, corresponding to the levels of X, and measure the value of Y.

Finally, in some cases, we have a frequency table showing the distribution of a single variable. We may want to test whether this variable follows some specific distribution, F, say. We then do a *goodness-of-fit test* of the hypotheses:

- H_0: the variable follows the distribution F.

- H_1: the variable does not follow the distribution F.

All three of these can be tested using a χ^2-*test* (chi-squared test). Let's say that X has c_x levels and that Y has c_y levels. Then for each of the $c_x \cdot c_y$ cells in the table, we can compare the observed outcome o_{ij} to the outcome e_{ij} that would be expected if the null hypothesis were true. If the observed outcomes deviate a lot from what we would expect under the null hypothesis, we have evidence against H_0.

Formally, the test is performed by computing the statistic

$$X^2 = \sum_{i=1}^{c_x} \sum_{j=1}^{c_y} \frac{(o_{ij} - e_{ij})^2}{e_{ij}}.$$

Large values of this statistic count as evidence against H_0. It can be shown that under the null hypothesis, this statistic is asymptotically χ^2-distributed, i.e., if the sample size is large enough, then the distribution of the statistic can be accurately approximated by the χ^2-distribution. We can therefore compute the probability of obtaining an X^2-statistic that is larger than what we observed in our data, which gives us the p-value for the test.

This differs from Fisher's exact test in that we rely on a mathematical approximation (the asymptotic χ^2-distribution) rather than permutations when computing p-values. The benefit of this is speed: when the sample size becomes large, it takes a very long time to compute all possible permutations, whereas the X^2- statistic can be computed in milliseconds.

To run the test, we use the `chisq.test` function as follows:

Without pipes:

```
chisq.test(ftable(penguins$species,
                  penguins$sex))
```

With pipes:

```
library(dplyr)
penguins |>
    select(species, sex) |>
    ftable() |>
    chisq.test()
```

If we wish to run a goodness-of-fit test, we specify the distribution F given by H_0 using the argument p. If our null hypothesis is that both sexes are equally common, this would look as follows:

Without pipes:

```
chisq.test(ftable(penguins$sex),
           p = c(0.5, 0.5))
```

With pipes:

```
library(dplyr)
penguins |>
    select(sex) |>
    ftable() |>
    chisq.test(p = c(0.5, 0.5))
```

3.4.1 When can we use χ^2-tests?

χ^2-tests are approximate tests that require sufficiently large sample sizes to attain the right type I error rate. What is "large" is largely determined by how high the smallest expected cell counts are. A common rule-of-thumb is that the approximation is adequate as long as all $e_{ij} > 1$ and at most 20% of the cells have $e_{ij} < 5$ (Cochran, 1954).

To assess this, we can check the expected counts of the cells:

```
# Save the results of the test:
penguins |> select(species, sex) |> ftable() |> chisq.test() -> x2res

# Extract the expected counts:
x2res$expected
```

The `chisq.test` output will also give a warning if any cells have $e_{ij} < 5$.

If your sample size is too small to use a χ^2 tests, there are two options in addition to gathering more data. The first is to merge some cells by merging levels of one or more of your categorical variables; see Section 5.4 for more on how this can be done. The second is to use *simulated p-values*, i.e., to use simulation rather than a mathematical approximation to compute the p-value. This turns the test into an approximate permutation test: the X^2 statistic is computed for a large number of randomly selected permutations of the table, and the p-value is then computed as the proportion of permutations for which X^2 was at least as large as for the original table. To use this approach, we add the argument `simulate.p.value = TRUE` to `chisq.test`. The argument `B` can also be added, to control how many permutations should be simulated; `B = 9999` is a reasonable default choice.

Without pipes:

```
chisq.test(ftable(penguins$species, penguins$sex),
           simulate.p.value = TRUE, B = 9999)
```

With pipes:

```
library(dplyr)
penguins |>
  select(species, sex) |>
  ftable() |>
  chisq.test(simulate.p.value = TRUE, B = 9999)
```

You'll learn more about simulation in Chapter 7.

$\sim$

Exercise 3.1. In cases where our data is a contingency table rather than a data frame with one row for each observation, the simplest way to perform a χ^2-test is to create a `matrix` object for the contingency table and run `chisq.test` with that as the input. Here is an example of such data, from a study on antibiotics resistance. In a lab, 18 strains of *E.coli* bacteria and 25 strains of *K.pneumoniae* bacteria were tested for resistance against an antibiotic. The table shows how many strains were resistant:

```
bacteria <- matrix(c(15, 3, 17, 8),
            2, 2,
            byrow = TRUE,
            dimnames = list(c("E.coli", "K.pneumoniae"),
            c("Not resistant", "Resistant")))
```

Perform a χ^2 test of homogeneity to see if the proportion of resistant strains is the same for both species of bacteria. Are the conditions for running a χ^2-test met? If not, what can you do instead?

3.5 Confidence intervals

We use data to compute mean values, sample proportions, and other quantities. Almost always, our end goal is to draw some conclusion about a population parameter: the mean value in the underlying population, the proportion in the population, and so on.

Mean values and sample proportions give us estimates of the corresponding parameters in the population. But, how far off can the estimates be from the population parameters?

Confidence intervals are used to quantify the uncertainty of estimates and show what values of the population parameter are in agreement with the observed data.

Formally, a confidence interval for a parameter is an interval that will cover the true value of the parameter with probability $1 - \alpha$, where α typically is 0.05, 0.01, or 0.1. $1 - \alpha$ is called the *coverage* or *confidence level* of the interval. The confidence level is a proportion but is frequently presented in percentages, i.e., as $(1 - \alpha) \cdot 100\%$. A confidence level of 0.95 or 95% (i.e., with $\alpha = 0.05$) means that the interval will cover the true value of the parameter in 95% of the studies where it is used.

Confidence intervals are much more informative than point estimates. Saying that the sample mean is 86.1 only tells part of the story. If we instead present a confidence interval, we also tell something about the uncertainty of this estimate. "The mean value is in the interval (85.9, 86.3) with 95% confidence" tells us that the uncertainty in the estimate is low, whereas "The mean value is in the interval (65.0, 107.2) with 95% confidence" tells us that the uncertainty is fairly high. In a way, reporting a confidence interval instead of just a point estimate is more honest, because it tells the whole story.

Like tests, confidence intervals can be either *exact* or *approximate*.

- For *exact confidence intervals*, the confidence level is at least $1 - \alpha$.
- For *approximate confidence intervals*, the confidence level is approximately $1 - \alpha$, but it may be lower than $1 - \alpha$.

The higher the confidence level, the wider the interval, because we need wider intervals if we want to be really sure that we cover the true value of the parameter. There is therefore a trade-off between having a confidence level that is high enough and an interval that isn't so wide that it's useless.

For any given parameter, there are different methods for computing a confidence interval. Some have better properties than others. We want an interval that has a coverage that is as close to $1 - \alpha$ as possible, but we also want it to be as short as possible. You'll see examples of how statistical methods, including confidence intervals, can be evaluated in Section 7.2.

Confidence intervals and hypothesis tests are closely connected. For each test, the set of parameter values that wouldn't be rejected if they were the null hypothesis, at significance level α, is a $1 - \alpha$ confidence interval. We will return to this connection in subsequent chapters.

3.5.1   Confidence intervals for proportions

As a first example, let's consider confidence intervals for a proportion. We'll use a function from the `MKinfer` package. Let's install it:

```
install.packages("MKinfer")
```

The `binomCI` function in this package allows us to compute confidence intervals for proportions from binomial experiments using a number of methods. The input is the number of "successes" x, the sample size n, and the `method` to be used.

Let's say that we want to compute a confidence interval for the proportion of herbivore mammals that sleep for more than 7 hours a day. First, we need to compute x and n.

Without pipes:

```
library(ggplot2)
herbivores <- msleep[msleep$vore == "herbi",]

# Compute the number of animals for which we know the sleep time:
n <- sum(!is.na(herbivores$sleep_total))

# Compute the number of "successes", i.e. the number of animals that sleep
# for more than 7 hours:
x <- sum(herbivores$sleep_total > 7, na.rm = TRUE)
```

With pipes:

```
library(ggplot2)
library(dplyr)

# Compute the number of animals for which we know the sleep time and
# the number of "successes", i.e. the number of animals that sleep for
# more than 7 hours:
msleep |>
  filter(vore == "herbi") |>
  select(sleep_total) |>
  na.omit() |>
  summarise(n = n(),
            x = sum(sleep_total > 7)) -> res
x <- res$x; n <- res$n
```

The estimated proportion is x/n, which in this case is 0.625. We'd like to quantify the uncertainty in this estimate by computing a confidence interval. The standard Wald method, taught in most introductory statistics courses, can be computed in the following way:

```
library(MKinfer)
binomCI(x, n, conf.level = 0.95, method = "wald")
```

The confidence interval is printed on the line starting with `prob`: in this case, it is $(0.457, 0.792)$.

Don't use that method though! The Wald interval is known to be severely flawed (Brown et al., 2001), primarily because its coverage can be very far from $1 - \alpha$. Much better options are available. If the proportion can be expected to be close to 0 or 1, the Clopper-Pearson interval is recommended, and otherwise the Wilson interval is the best choice (Thulin, 2014a). We can use either of these methods with `binomCI`:

```
binomCI(x, n, conf.level = 0.95, method = "clopper-pearson")
binomCI(x, n, conf.level = 0.95, method = "wilson")
```

$\sim$

Exercise 3.2. In a survey, 440 out of 998 randomly sampled respondents said that they plan to vote for a particular candidate in an upcoming election. Based on this, compute a 99% Wilson interval for the proportion of voters that plan to vote for this candidate.

Exercise 3.3. The function `binomDiffCI` from `MKinfer` can be used to compute a confidence interval for the *difference* of two proportions. Using the `msleep` data, use it to compute a confidence interval for the difference between the proportion of herbivores that sleep for more than 7 hours a day and the proportion of carnivores that sleep for more than 7 hours a day.

3.5.2 Sample size calculations

Confidence intervals that are too wide aren't that informative. We'd much rather be able to say "the proportion is somewhere between 0.56 and 0.58" than "the proportion is somewhere between 0.37 and 0.78".

How wide a confidence interval for a proportion is depends both on the number of successes x and the sample size n. We don't know x in advance but can usually control n, at least to some extent. It is therefore often useful to perform some computations prior to collecting our data, to make sure that n is large enough that we'll get a confidence interval with a reasonable width.

The `MKpower` package contains several functions that will prove useful when we wish to calculate what sample size we need for a study. Let's install it:

```
install.packages("MKpower")
```

The `ssize.propCI` function in `MKpower` can be used to compute the sample size needed to obtain a confidence interval with a given width – or rather, a given *expected*, or average, width. The width of the interval is a function of a random variable and is therefore also random. The computations rely on asymptotic formulas that are highly accurate, as you later will verify in Exercise 7.9.

When computing the required sample size, we give a rough approximation of what we expect the proportion to be (prop), and specify what width we want our confidence interval to be (width):

```
library(MKpower)
# Compute the sample size required to obtain an interval with
# width 0.1 if the true proportion is 0.4:
ssize.propCI(prop = 0.4, width = 0.1,
             conf.level = 0.95, method = "wilson")
ssize.propCI(prop = 0.4, width = 0.1,
             conf.level = 0.95, method = "clopper-pearson")
```

In the output from these function calls, n is the sample size required to obtain the desired expected width. Try lowering the desired width and see how that effects the required sample size.

$\sim$

Exercise 3.4. What sample size is required to obtain a 95% Wilson confidence interval with expected width 0.05 if the true proportion is 0.7?

3.6 Comparing mean values

In Section 3.3 we imagined a study where we wanted to find out whether flipper length differs between male and female Chinstrap penguins. The penguins dataset can be used for this purpose. It contains measurements of the flipper lengths of a number of penguins. If the average length in the female population is μ_1 and the average length in the male population is μ_2, then the population parameter that we are interested in is the difference $\mu_1 - \mu_2$. The hypotheses that we wish to test are:

- H_0: there is no difference; $\mu_1 - \mu_2 = 0$.
- H_1: there is a difference; $\mu_1 - \mu_2 \neq 0$.

In this section, we will learn about the *t-test*, which can be used to test this and similar hypotheses about means and differences of means. Throughout the section, we'll assume that all observations are *independent*, unless we explicitly say that they aren't.

3.6.1 The t-test for comparing two groups

It seems reasonable to base a test of these hypotheses on the difference between the sample means in the two samples: $\bar{x}_1 - \bar{x}_2$. There is one problem though: said difference is not scale invariant, and so we would get different results depending on whether we measure the flipper lengths in centimetres, metres, or inches. To remedy that, we base our test on the quantity

$$T = \frac{\bar{x}_1 - \bar{x}_2}{s}$$

where s is the standard error of $\bar{x}_1 - \bar{x}_2$ (i.e., an estimate of the standard deviation of the mean difference). How the latter is computed depends on what assumptions we are willing to make. The test based on T is called the *two-sample t-test*. There are two different versions of this test, which differ in how s is computed:

- *The equal-variances t-test*: where the variance is assumed to be the same in both populations.
- *The Welch t-test*: where the variance is allowed to differ between the two populations.

The equal-variances t-test will not work properly when the two populations have differing variances – a low p-value may be due to differences in variances rather than differences in means. In such cases, it no longer tests the hypotheses we were interested in; instead, it tests whether the distribution is the same for the two populations. If that is the question that you're interested in, there are other, better, tests that you can use instead, such as the Kolmogorov-Smirnov test, implemented in `ks.test`.

The Welch t-test works well both when the population variances differ and when they are equal, and it tests the right hypotheses in both cases. It is the default in R, and I strongly recommend using it instead of the equal-variances test.

Some textbooks recommend first performing a test to see whether the population variances differ, and then using that to choose which version of the t-test to use. This is a widespread practice. Indeed, software packages such as SAS and SPSS include such tests in the default output for the t-test. While it may sound like a good idea, this actually leads to a procedure that generally performs worse than the Welch t-test (Rasch et al., 2011; Delacre et al., 2017).

In summary, forget about different versions of the two-sample t-test, and always use the Welch t-test. As it is the default in R, this is easy to do! Here is an example of how to perform a Welch t-test to test whether flipper length differs between male and female Chinstrap penguins, using the `t.test` function:

Without pipes:

```
library(palmerpenguins)
t.test(flipper_length_mm ~ sex,
        data = subset(penguins,
                  species ==
                    "Chinstrap"))
```

With pipes:

```
library(palmerpenguins)
library(dplyr)
penguins |>
  filter(species == "Chinstrap") |>
  t.test(flipper_length_mm ~ sex,
          data = _)
```

In the solution using pipes, recall that `data = _` means that the data used is the output from the previous step in the pipeline (see Section 2.13.2).

The output looks as follows:

```
##
## Welch Two Sample t-test
##
## data: flipper_length_mm by sex
## t = -5.7467, df = 65.905, p-value = 0.0000002535
```

```
## alternative hypothesis: true difference in means between group female and group male
## is not equal to 0
## 95 percent confidence interval:
##  -11.017272  -5.335669
## sample estimates:
## mean in group female mean in group male
##  191.7353  199.9118
```

We are mostly interested in the p-value, which in this case is very low, leading us to reject H_0 and conclude that there is a difference.

In addition to the p-value, a 95% confidence interval for the difference $\mu_1 - \mu_2$ is also presented (remember that there is a close connection between confidence intervals and hypothesis test), along with the means in the two groups. The confidence interval is $(-11.0, -5.3)$, meaning that the average flipper length in females is somewhere between 5.3 and 11.0 mm shorter than in males.

$\sim$

Exercise 3.5. Consider the penguins data again. Run a t-test to see if the average body mass differs between the sexes for Chinstrap penguins.

3.6.2 One-sided hypotheses

The t-test can also be used to test directed hypotheses, such as:

- H_0: $\mu_1 - \mu_2 \geq 0$.
- H_1: $\mu_1 - \mu_2 < 0$.

To specify that a one-sided test should be performed, we use the `alternative` argument, just as we did for Fisher's exact test in Section 3.3.4:

Without pipes:

```
library(palmerpenguins)
t.test(flipper_length_mm ~ sex,
       data = subset(penguins,
                 species ==
                   "Chinstrap"),
       alternative = "less")
```

With pipes:

```
library(palmerpenguins)
library(dplyr)
penguins |>
  filter(species == "Chinstrap") |>
  t.test(flipper_length_mm ~ sex,
         data = _,
         alternative = "less")
```

3.6.3 The t-test for a single sample

In some cases, we wish to test hypotheses about the mean value of a single population. For example:

- H_0: the mean value equals some specific value μ_0; $\mu = \mu_0$.
- H_1: the mean value does not equal μ_0; $\mu \neq \mu_0$.

This can be done using a one-sample t-test, which uses a test statistic based on the sample mean $\bar{x}$. We use `t.test` and specify the value μ_0 of μ under H_0 using the argument `mu`. Here's an example where we test whether the average length of the flippers of male Chinstrap penguins is 200 mm:

Without pipes:

```
library(palmerpenguins)
t.test(flipper_length_mm ~ 1,
       data = subset(penguins, species == "Chinstrap" &  sex == "male"),
       mu = 200)

# Or, in two steps:
new_data <- penguins[penguins$species == "Chinstrap" &  penguins$sex == "male",]
t.test(new_data$flipper_length_mm, mu = 200)
```

With pipes:

```
library(palmerpenguins)
library(dplyr)
penguins |>
  filter(species == "Chinstrap", sex == "male") |>
  t.test(flipper_length_mm ~ 1, data = _, mu = 200)
```

In this case, the p-value is 0.93 and we cannot reject H_0 – that is, we do not have any statistical evidence that the average flipper length isn't 200 mm.

$\sim$

Exercise 3.6. Consider the `penguins` data again. Run a one-sided test to see if the average flipper length for male Chinstrap penguins is greater than 195.

3.6.4   The t-test for paired samples

In some cases, we have *paired* observations in our data, i.e., observations that have been collected in pairs. A common example is when we make measurements on the same subjects before and after a treatment. In such situations, we are often interested in the effect Δ of the treatment, where Δ measures how much the mean value changes because of the treatment. A common set of hypotheses are:

- H_0: there is no effect; $\Delta = 0$.
- H_1: there is an effect; $\Delta \neq 0$.

These can be tested using a paired samples t-test, where the two samples (before and after) are compared using the information about the pairing. Here is an example with measurements before and after a treatment stored in different columns in a table:

```
exdata <- data.frame(before = c(8.5, 4.4, 9.4, 0.0, 2.2, 9.7, 6.2, 8.2),
                      after = c(8.5,  4.8, 10.1,  1.0,  5.6, 11.4, 7.8, 10.2))
```

Without pipes: With pipes:

```
t.test(exdata$after, exdata$before,        exdata |>
       paired = TRUE)                          t.test(Pair(after, before) ~ 1,
                                                      data = _)
```

We could run a two-sample t-test to compare the two samples, but that would result in a test with lower power, as it ignores some of the information available to use (namely that the observations belong in pairs). A paired samples t-test is always preferable when you have paired data.

3.6.5   When can we use the t-test?

The first requirements for the t-test to work is that the observations are independent. The only exception to this is when we have paired samples, in which case we allow the observations within a pair to be dependent. Dependence between observations occur for instance when we have multiple observations for the same individual. Methods for analysing such datasets are discussed in Section 8.8.

And then there is the matter of the normal distribution.

For decades teachers all over the world have been telling the story of William Sealy Gosset: the head brewer at Guinness who derived the formulas used for the t-test and, following company policy, published the results under the pseudonym "Student". By assuming that the variable followed a normal distribution, he was able to derive the distribution of the t-statistic under the null distribution, which could then be used to compute p-values.

The tests (and the associated confidence intervals) described above work well when the variables we are interested in indeed do follow a normal distribution. For the two-sample test, the variable should be normal in each population. For the paired samples test, the differences need to follow a normal distribution.

Unfortunately, real data is almost never normally distributed.

If the variable doesn't follow a normal distribution (you'll learn more about assessing the normality assumption in Section 7.1.3), the tests and confidence intervals may still be valid if you have large sample sizes, or, for the two-sample test, if you have balanced sample sizes, i.e., equally many observations in each group. How large "large" is depends on the shape of the distribution. The less symmetric it is, the larger the sample size needs to be.

But, why do we need the normality assumption in the first place? Is there some way that we can get rid of it? It turns out that there is. Gosset's work was hugely important, but the passing of time has rendered at least parts of it largely obsolete. His distributional formulas were derived out of necessity: lacking the computer power that we have available to us today, he was forced to impose the assumption of normality on the data, in order to derive the formulas he needed to be able to carry out his analyses. Today we can use simulation to carry out analyses with fewer assumptions. As an added bonus, these simulation techniques often happen to result in statistical methods with better performance than Student's t-test and other similar methods. You'll learn about two simulation-based ways to compute p-values for the t-test next.

3.6.6 Permutation t-tests

Continuing our example with the Chinstrap penguin data, we note that there are 34 males and 34 females in the sample – 68 animals in total. If there are no differences in flipper length between the sexes (i.e., if H_0 is true), the sex labels offer no information about how long the flippers are. Under the null hypothesis, the assignment of sex labels to different animals is therefore for all intents and purposes random. To find the distribution of the test statistic under the null hypothesis, we could look at all possible ways to assign 34 animals the label male and 34 animals the label female. That is, look at all permutations of the labels. The probability of a result at least as extreme as that obtained in our sample (in the direction of the alternative), i.e., the p-value, would then be the proportion of permutations that yield a *t*-statistic at least as extreme as that in our sample.

This is a permutation version of the *t*-test. Note that it doesn't rely on any assumptions about normality.

Permutation tests were known to the statisticians that developed the t-test in the first decades of the 20th century, but because the number of permutations of labels often tend to become quite large (28,453,041,475,240,599,552, in our male–female example), they lacked the means to actually use them, and so had to resort to cruder mathematical approximations using assumptions about the distribution of the data. The 28,453,041,475,240,599,552 permutations may be too many even today, but we can obtain very good approximations of the p-values of permutation tests using simulation.

The idea is that we look at a large number of randomly selected permutations, and check for how many of them we obtain a test statistic that is more extreme than the sample test statistic. The law of large numbers guarantees that this proportion will converge to the permutation test p-value as the number of randomly selected permutations increases.

The function that we'll use to perform a simulation-based permutation t-test, `perm.t.test`, works exactly like `t.test`. In all the examples above, we can replace `t.test` with `perm.t.test` to run a (simulation-based) permutation t-test instead. Here's how to do it:

Without pipes:

```
library(palmerpenguins)
library(MKinfer)
perm.t.test(flipper_length_mm ~ sex,
        data = subset(penguins,
                species ==
                "Chinstrap"))
```

With pipes:

```
library(palmerpenguins)
library(dplyr)
library(MKinfer)
penguins |>
    filter(species == "Chinstrap") |>
    perm.t.test(flipper_length_mm ~ sex,
            data = _)
```

Note that two p-values and confidence intervals are presented: one set from the permutations and one from the traditional approach – so make sure that you look at the right ones!

So, you say, how many randomly selected permutations do we need to get an accurate approximation of the permutation test p-value? My answer is that, by default, `perm.t.test` uses 9,999 permutations (you can change that number using the argument `R`), which is widely considered to be a reasonable number. If you are running a permutation test with a much more complex (and computationally intensive) statistic, you may have to use a lower number, but avoid that if you can.

You may ask why we use 9,999 permutations and not 10,000. The reason is that we avoid p-values that are equal to traditional significance levels like 0.05 and 0.01 this way. If we'd used 10,000 permutations, 500 of which yielded a statistics that had a larger absolute value than the sample statistic, then the p-value would have been exactly 0.05, which would cause some difficulties in trying to determine whether or not the result was significant at the 5% level. This cannot happen when we use 9,999 permutations instead (500 statistics with a large absolute value yields the p-value $0.050005 > 0.05$, and 499 yields the p-value $0.0499 < 0.05$).

3.6.7   Bootstrap t-tests

A popular method for computing p-values and confidence intervals that resembles the permutation approach is the bootstrap. Instead of drawing permuted samples, new observations are drawn with replacement from the original sample, and then labels are randomly allocated to them. That means that each randomly drawn sample will differ not only in the permutation of labels, but also in what observations are included – some may appear more than once and some not at all. We can then check what proportion of these *bootstrap samples* that yield a more extreme test statistic than what we observed in our original sample. In effect, the p-value is computed in the same way as in the traditional t-test, but with the theoretical normal distribution replaced with the empirical distribution of the data. We can therefore view the bootstrap t-test as a compromise between the traditional t-test and the permutation t-test.

We will have a closer look at the bootstrap in Section 7.4, where we will learn how to use it for creating confidence intervals and computing p-values for any test statistic. For now, we'll just note that `MKinfer` offers a bootstrap version of the t-test, `boot.t.test` :

Without pipes:

```
library(palmerpenguins)
library(MKinfer)
boot.t.test(flipper_length_mm ~ sex,
      data = subset(penguins,
                    species ==
                  "Chinstrap"))
```

With pipes:

```
library(palmerpenguins)
library(dplyr)
library(MKinfer)
penguins |>
  filter(species == "Chinstrap") |>
  boot.t.test(flipper_length_mm ~ sex,
              data = _)
```

Both `perm.test` and `boot.test` have a useful argument called `symmetric`, the details of which are discussed in depth in Section 14.3.

3.6.8   Publication-ready tables for means

In Section 3.2.2, we used `tbl_summary` from the `gtsummary` package to create summary tables for categorical variables. It can also be used for numeric variables. In the example below, we create a table showing the average flipper length for male and female Chinstrap penguins.

By default, `tbl_summary` displays the median and inter-quartile range for numeric variables:

```
library(palmerpenguins)
```

```
library(dplyr)
library(gtsummary)
penguins |>
    filter(species == "Chinstrap") |>
    select(sex, flipper_length_mm) |>
    tbl_summary(by = sex)
```

We can however change this, and for instance print the mean value and bootstrap confidence intervals instead. This involves creating custom functions, a topic which you'll learn more about in Section 6.1. For now, we'll just use the first two lines of the code chunk below without fully understanding why they work (the first line creates a function that extracts the lower confidence bound, and the second line creates a function that extracts the upper confidence bound), to obtain a table showing what we want:

```
confint1 <- function(x) { MKinfer::boot.t.test(x)$boot.conf.int[1] }
confint2 <- function(x) { MKinfer::boot.t.test(x)$boot.conf.int[2] }
penguins |>
  filter(species == "Chinstrap") |>
  select(sex, flipper_length_mm) |>
  tbl_summary(by = sex, statistic = list(
                  all_continuous() ~ "{mean} ({confint1}, {confint2})",
                  all_categorical() ~ "{n} ({p}%)"))
```

To export the output from t.test to a nice-looking table, we can use tbl_regression:

```
penguins |>
  filter(species == "Chinstrap") |>
  t.test(flipper_length_mm ~ sex,
        data = _) |>
  tbl_regression(label =
                  list(sex.flipper_length_mm = "Flipper length by sex")) |>
  modify_header(estimate = "**Difference**")
```

3.6.9 Sample size computations for the t-test

In any study, it is important to collect enough data for the inference that we wish to make. If we want to use a t-test for a test about a mean or the difference of two means, what constitutes "enough data" is usually measured by the power of the test. The sample is large enough when the test achieves high enough power. If we are comfortable assuming normality (and we may well be, especially as the main goal with sample size computations is to get a ballpark figure), we can use power.t.test to compute what power our test would achieve under different settings. For a two-sample test with unequal variances, we can use power.welch.t.test from MKpower instead. Both functions can be used to either find the sample size required for a certain power, or to find out what power will be obtained from a given sample size.

power.t.test and power.welch.t.test both use delta to denote the mean difference under the alternative hypothesis. In addition, we must supply the standard deviation sd of the distribution. Here are some examples:

```
library(MKpower)

# A one-sided one-sample test with 80 % power:
power.t.test(power = 0.8, delta = 1, sd = 1, sig.level = 0.05,
             type = "one.sample", alternative = "one.sided")

# A two-sided two-sample test with sample size n = 25 and equal
# variances:
power.t.test(n = 25, delta = 1, sd = 1, sig.level = 0.05,
             type = "two.sample", alternative = "two.sided")

# A one-sided two-sample test with 90 % power and equal variances:
power.t.test(power = 0.9, delta = 1, sd = 0.5, sig.level = 0.01,
             type = "two.sample", alternative = "one.sided")

# A one-sided two-sample test with 90 % power and unequal variances:
power.welch.t.test(power = 0.9, delta = 1, sd1 = 0.5, sd2 = 1,
                   sig.level = 0.01,
                   alternative = "one.sided")
```

You may wonder how to choose `delta` and `sd`. If possible, it is good to base these numbers on a pilot study or related previous work. If no such data is available, your guess is as good as mine. For `delta`, some useful terminology comes from medical statistics, where the concept of *clinical significance* is used increasingly often. Make sure that `delta` is large enough to be clinically significant, that is, large enough to actually matter in practice.

If we have reason to believe that the data follows a non-normal distribution, another option is to use simulation to compute the sample size that will be required. We'll do just that in Section 7.3.

~

Exercise 3.7. Return to the one-sided t-test that you performed in Exercise 3.6. Assume that `delta` is 5 (i.e., that the true mean is 200) and that the standard deviation is 6. How large does the sample size n have to be for the power of the test to be 95% at a 5% significance level? What is the power of the test when the sample size is $n = 34$ (which is the case for the `penguins` data)?

3.7 Multiple testing

If we run a single hypothesis test, the risk for a type I error (also called a false positive result) is α. If we run two independent tests, the risk for committing at least one type I error is greater, as there now are two tests for which this can occur. When we run even more tests, the risk for a type I error increases, to the point where we're virtually guaranteed to get at least one significant result, even if the null hypotheses are true for all tests.

The figure below shows the risk of getting at least one type I error when we perform multiple independent tests for which the null hypothesis is true. As you can see, the risk quickly becomes quite high.

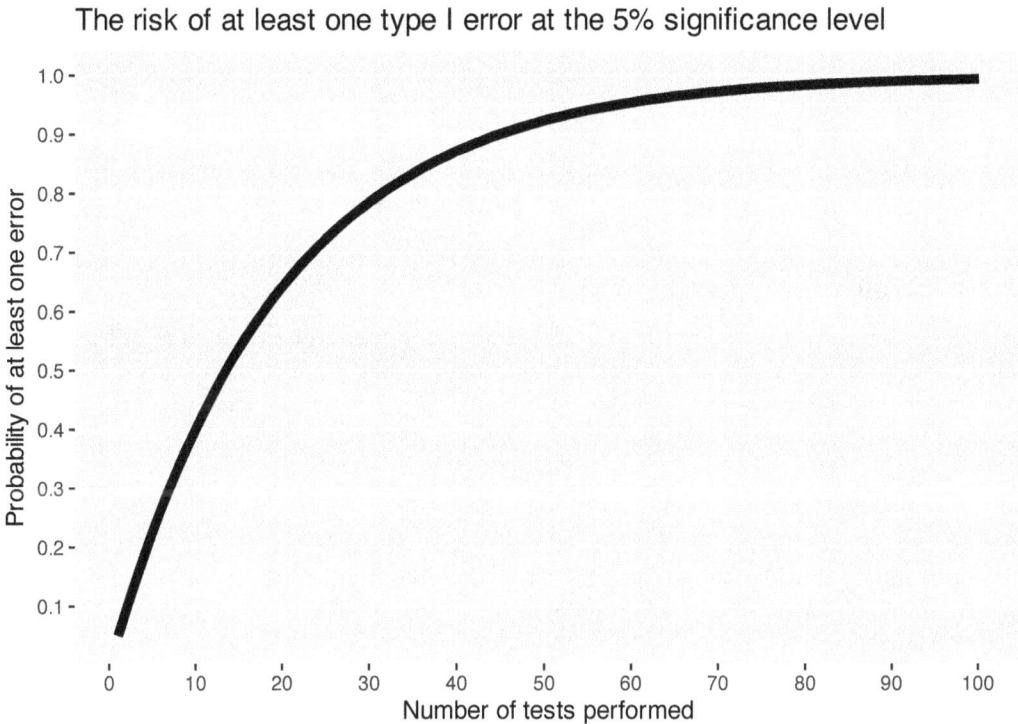

The risk of at least one type I error at the 5% significance level

FIGURE 3.1 The risk of at least one type I error when performing multiple tests.

3.7.1 Adjusting for multiplicity

We can reduce the risk of false positive results by *adjusting the p-values of the tests for multiplicity*. When this is done, the p-values are increased somewhat, to decrease the risk of type I errors.

There are severals methods for adjusting for multiplicity, including for instance Bonferroni correction (a method that often is too conservative and adjusts the p-values too much), Holm's method (an improved version of the standard Bonferroni approach), and the Benjamini-Hochberg approach (which controls the *false discovery rate* and is useful if you for instance are screening a lot of variables for differences).

We can apply these methods by using `p.adjust`:

```
# Some example p-values:
p_values <- c(0.00023, 0.003, 0.021, 0.042, 0.060, 0.2, 0.81)

# Adjusted p-values:
p.adjust(p_values, method = "bonferroni")
p.adjust(p_values, method = "holm")
p.adjust(p_values, method = "BH")
```

The adjusted p-values are then compared to α. If the Bonferroni or Holm methods have been used, the probability of at least one type I error is bounded by α under certain assumptions.

$$\sim$$

Exercise 3.8. Rerun the test from 3.8, once for each penguin species in the `penguins` data. Are the null hypotheses rejected? If so, are the differences still significant after adjusting for multiplicity using Holm's method?

Hint: you can extract the p-values from the `t.test` function using `t.test(...)$p.val` (replace `...` with the usual arguments).

3.7.2   Multivariate testing with Hotelling's T^2

If you are interested in comparing the means of several variables for two groups, using a multivariate test is sometimes a better option than running multiple univariate t-tests. The multivariate generalisation of the t-test, Hotelling's T^2, is available through the `Hotelling` package:

```
install.packages("Hotelling")
```

As an example, consider the `airquality` data. Let's say that we want to test whether the mean ozone, solar radiation, wind speed, and temperature differ between June and July. We could use four separate t-tests to test this, but we could also use Hotelling's T^2 to test the null hypothesis that the mean vector, i.e., the vector containing the four means, is the same for both months. The function used for this is `hotelling.test`:

```
# Subset the data:
airquality_t2 <- subset(airquality, Month == 6 | Month == 7)

# Run the test under the assumption of normality:
library(Hotelling)
t2 <- hotelling.test(Ozone + Solar.R + Wind + Temp ~ Month,
            data = airquality_t2)
t2

# Run a permutation test instead:
t2 <- hotelling.test(Ozone + Solar.R + Wind + Temp ~ Month,
            data = airquality_t2, perm = TRUE)
t2
```

3.8   Correlations

We are often interested in measuring how strong the dependence between two variables is. A common group of such measures are called *correlation measures*. A correlation measure ranges from -1 (perfect negative dependence) to 1 (perfect positive dependence), with 0 meaning that there is no dependence (in the sense measured by the correlation measure).

Values close to 1 indicate strong positive dependence, and values close to -1 indicate strong negative dependence.

3.8.1 Estimation

The most commonly used correlation measure is the *Pearson correlation*. It measures how strong the *linear dependence* is between two variables. This means that the correlation can be close to 0 even if there is a dependence, as long as that dependence is non-linear:

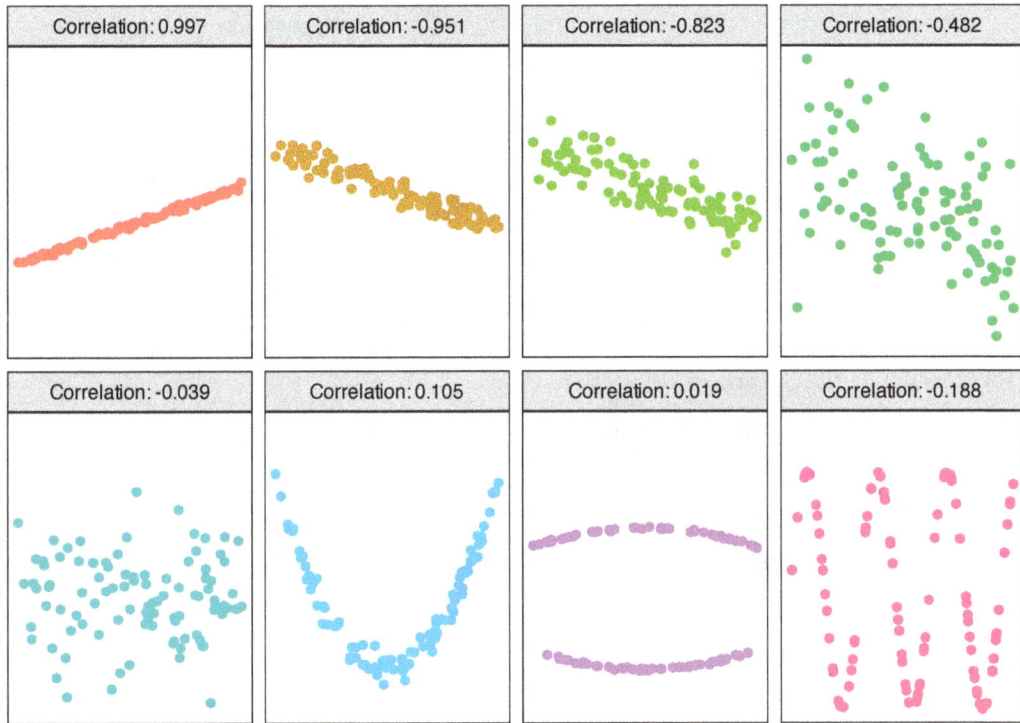

FIGURE 3.2 The Pearson correlation for eight datasets.

We can compute the Pearson correlation between two variables using cor:

Without pipes:

```
library(palmerpenguins)
cor(penguins$flipper_length_mm,
    penguins$body_mass_g,
    use = "pairwise.complete")
```

With pipes:

```
library(palmerpenguins)
penguins |>
  with(cor(flipper_length_mm,
           body_mass_g,
           use = "pairwise.complete"))
```

The setting use = "pairwise.complete" means that NA values are ignored.

The result, a correlation of 0.87, indicates a fairly strong linear dependence between the two variables.

Another common option is the Spearman correlation, which measures the strength of *monotone dependence* between two variables. This means that it's better at finding non-linear dependencies.

Without pipes:

```
library(palmerpenguins)
cor(penguins$flipper_length_mm,
    penguins$body_mass_g,
    method = "spearman",
    use = "pairwise.complete")
```

With pipes:

```
library(palmerpenguins)
penguins |>
   with(cor(flipper_length_mm,
            body_mass_g,
            method = "spearman",
            use = "pairwise.complete"))
```

In this case, the two methods yield similar results.

3.8.2   Hypothesis testing

To test the null hypothesis that two numerical variables are uncorrelated, i.e., that their correlation is 0, we can use `cor.test`. The test can be based on either the Pearson correlation (the default) or the Spearman correlation.

Let's try it with sleep times and brain weight, using the `msleep` data again:

```
library(ggplot2)

# Pearson correlation
cor.test(msleep$sleep_total, msleep$brainwt,
         use = "pairwise.complete")

# Spearman correlation
cor.test(msleep$sleep_total, msleep$brainwt,
         method = "spearman",
         use = "pairwise.complete")
```

These tests are all based on asymptotic approximations, which among other things cause the Pearson correlation test to perform poorly for non-normal data. In Section 7.4 we will create a bootstrap version of the correlation test, which has better performance.

3.9   Bayesian approaches

The Bayesian paradigm differs in many ways from the frequentist approach that we use in the rest of this chapter. In Bayesian statistics, we first define a *prior distribution* for the parameters that we are interested in, representing our beliefs about them (for instance based on previous studies). Bayes' theorem is then used to derive the *posterior distribution*, i.e., the distribution of the coefficients given the prior distribution and the data. Philosophically, this is very different from frequentist estimation, in which we don't incorporate prior beliefs into our models (except for through what variables we include).

In many situations, we don't have access to data that can be used to create an *informative* prior distribution. In such cases, we can use a so-called weakly informative prior instead. These act as a sort of "default priors", representing large uncertainty about the values of the coefficients.

The `rstanarm` package contains methods for using Bayesian estimation to fit some common statistical models. It takes a while to install but is well worth the wait:

```
install.packages("rstanarm")
```

3.9.1 Inference for a proportion

Let's return to the example that we used in Section 3.5.1:

```
library(ggplot2)
herbivores <- msleep[msleep$vore == "herbi",]

# Compute the number of animals for which we know the sleep time:
n <- sum(!is.na(herbivores$sleep_total))

# Compute the number of "successes", i.e. the number of animals
# that sleep for more than 7 hours:
x <- sum(herbivores$sleep_total > 7, na.rm = TRUE)
```

The Bayesian analogue to a confidence interval is a *credible interval*: an interval that covers the population parameter with a posterior probability that is $1 - \alpha$.

An excellent Bayesian credible interval for a proportion is the Jeffreys interval, which uses the weakly informative Jeffreys prior:

```
library(MKinfer)
binomCI(x, n, conf.level = 0.95, method = "jeffreys")
```

The Jeffreys interval is interesting because it also has good frequentist properties (Brown et al., 2001).

3.9.2 Inference for means

To use a Bayesian model with a weakly informative prior to analyse the difference in sleep time between herbivores and carnivores, we load `rstanarm` and use `stan_glm` in complete analogue with how we use `t.test`:

```
library(rstanarm)
library(ggplot2)
m <- stan_glm(sleep_total ~ vore, data =
        subset(msleep, vore == "carni" | vore == "herbi"))

# Print the estimates:
m
```

There are two estimates here: an "intercept" (the average sleep time for carnivores) and `vore-herbi` (the difference between carnivores and herbivores). To plot the posterior distribution of the difference, we can use `plot`:

```
plot(m, "dens", pars = c("voreherbi"))
```

To get a 95% credible interval for the difference, we can use `posterior_interval` as follows:

```
posterior_interval(m,
        pars = c("voreherbi"),
        prob = 0.95)
```

p-values are not a part of Bayesian statistics, so don't expect any. It is however possible to perform a kind of Bayesian test of whether there is a difference by checking whether the credible interval for the difference contains 0. If not, we have evidence that there is a difference (Thulin, 2014c). In this case, 0 is contained in the interval, and there is no evidence of a difference.

In most cases, Bayesian estimation is done using Monte Carlo integration (specifically, a class of methods known as Markov Chain Monte Carlo, MCMC). To check that the model fitting has converged, we can use a measure called $\hat{R}$. It should be less than 1.1 if the fitting has converged:

```
plot(m, "rhat")
```

If the model fitting hasn't converged, you may need to increase the number of iterations of the MCMC algorithm. You can increase the number of iterations by adding the argument `iter` to `stan_glm` (the default is 2,000).

If you want to use a custom prior for your analysis, that is of course possible too. See `?priors` and `?stan_glm` for details about this, and about the default weakly informative prior.

3.10   Reporting statistical results

Carrying out a statistical analysis is only the first step. After that, you probably need to communicate your results to others: your boss, your colleagues, your clients, the public, and so on. This section contains some tips for how best to do that.

3.10.1   What should you include?

When reporting your results, it should always be clear:

- How the data was collected,
- If, how, and why any observations were removed from the data prior to the analysis,
- What method was used for the analysis (including a reference unless it is a routine method),
- If any other analyses were performed/attempted on the data, and if you don't report their results, why.

Let's say that you've estimated some parameter, for instance the mean sleep time of mammals, and want to report the results. The first thing to think about is that you shouldn't include too many decimals: don't give the mean with 5 decimals if sleep times only were measured with one decimal.

BAD: The mean sleep time of mammals was found to be 10.43373.

GOOD: The mean sleep time of mammals was found to be 10.4.

It is common to see estimates reported with standard errors or standard deviations:

BAD: The mean sleep time of mammals was found to be 10.3 ($\sigma = 4.5$).

---

or

BAD: The mean sleep time of mammals was found to be 10.3 (standard error 0.49).

---

or

BAD: The mean sleep time of mammals was found to be 10.3 ± 0.49.

Although common, this isn't a very good practice. Standard errors/deviations are included to give some indication of the uncertainty of the estimate but are very difficult to interpret. In most cases, they will probably cause the reader to either overestimate or underestimate the uncertainty in your estimate. A much better option is to present the estimate with a confidence interval, which quantifies the uncertainty in the estimate in an interpretable manner:

GOOD: The mean sleep time of mammals was found to be 10.3 (95% percentile bootstrap confidence interval: 9.5-11.4).

Similarly, it is common to include error bars representing standard deviations and standard errors, e.g., in bar charts. This questionable practice becomes even more troublesome because a lot of people fail to indicate what the error bars represent. If you wish to include error bars in your figures, they should always represent confidence intervals, unless you have a very strong reason for them to represent something else. In the latter case, make sure that you clearly explain what the error bars represent.

If the purpose of your study is to describe differences between groups, you should present a confidence interval for the difference between the groups, rather than one confidence interval (or error bar) for each group. It is possible for the individual confidence intervals to overlap even if there is a significant difference between the two groups, so reporting group-wise confidence intervals will only lead to confusion. If you are interested in the difference, then of course *the difference* is what you should report a confidence interval for.

BAD: There was no significant difference between the sleep times of carnivores (mean 10.4, 95% percentile bootstrap confidence interval: 8.4-12.5) and herbivores (mean 9.5, 95% percentile bootstrap confidence interval: 8.1-12.6).

GOOD: There was no significant difference between the sleep times of carnivores (mean 10.4) and herbivores (mean 9.5), with the 95% percentile bootstrap confidence interval for the difference being (-1.8, 3.5).

3.10.2   Citing R packages

In statistical reports, it is often a good idea to specify what version of a software or a package you used, for the sake of reproducibility (indeed, this is a requirement in some scientific journals). To get the citation information for the version of R that you are running, simply type `citation()`. To get the version number, you can use `R.Version` as follows:

```
citation()
R.Version()$version.string
```

To get the citation and version information for a package, use `citation` and `packageVersion` as follows:

```
citation("ggplot2")
packageVersion("ggplot2")
```

3.11 Ethics and good statistical practice

Throughout this book, there will be sections devoted to ethics. Good statistical practice is intertwined with good ethical practice. Both require transparent assumptions, reproducible results, and valid interpretations.

3.11.1 Ethical guidelines

One of the most commonly cited ethical guidelines for statistical work is The American Statistical Association's (ASA) *Ethical Guidelines for Statistical Practice* (Committee on Professional Ethics of the American Statistical Association, 2018), a shortened version of which is presented below[3]. The full ethical guidelines are available at: https://www.amstat .org/ASA/Your-Career/Ethical-Guidelines-for-Statistical-Practice.aspx

- **Professional Integrity and Accountability**. The ethical statistician uses methodology and data that are relevant and appropriate; without favoritism or prejudice; and in a manner intended to produce valid, interpretable, and reproducible results. The ethical statistician does not knowingly accept work for which he/she is not sufficiently qualified, is honest with the client about any limitation of expertise, and consults other statisticians when necessary or in doubt. It is essential that statisticians treat others with respect.
- **Integrity of Data and Methods**. The ethical statistician is candid about any known or suspected limitations, defects, or biases in the data that may affect the integrity or reliability of the statistical analysis. Objective and valid interpretation of the results requires that the underlying analysis recognises and acknowledges the degree of reliability and integrity of the data.
- **Responsibilities to Science/Public/Funder/Client**. The ethical statistician supports valid inferences, transparency, and good science in general, keeping the interests of the public, funder, client, or customer in mind (as well as professional colleagues, patients, the public, and the scientific community).
- **Responsibilities to Research Subjects**. The ethical statistician protects and respects the rights and interests of human and animal subjects at all stages of their involvement in a project. This includes respondents to the census or to surveys, those whose data are contained in administrative records, and subjects of physically or psychologically invasive research.
- **Responsibilities to Research Team Colleagues**. Science and statistical practice are often conducted in teams made up of professionals with different professional standards. The statistician must know how to work ethically in this environment.
- **Responsibilities to Other Statisticians or Statistics Practitioners**. The practice of statistics requires consideration of the entire range of possible explanations for observed

---

[3]The excerpt is from the version of the guidelines dated April 2018 and presented here with permission from the ASA.

phenomena, and distinct observers drawing on their own unique sets of experiences can arrive at different and potentially diverging judgments about the plausibility of different explanations. Even in adversarial settings, discourse tends to be most successful when statisticians treat one another with mutual respect and focus on scientific principles, methodology, and the substance of data interpretations.

- **Responsibilities Regarding Allegations of Misconduct**. The ethical statistician understands the differences between questionable statistical, scientific, or professional practices and practices that constitute misconduct. The ethical statistician avoids all of the above and knows how each should be handled.
- **Responsibilities of Employers, Including Organizations, Individuals, Attorneys, or Other Clients Employing Statistical Practitioners**. Those employing any person to analyse data are implicitly relying on the profession's reputation for objectivity. However, this creates an obligation on the part of the employer to understand and respect statisticians' obligation of objectivity.

Similar ethical guidelines for statisticians have been put forward by the International Statistical Institute (https://www.isi-web.org/about-isi/policies/professional-ethics), the United Nations Statistics Division (https://unstats.un.org/unsd/dnss/gp/fundprincipl es.aspx), and the Data Science Association (http://www.datascienceassn.org/code-of-conduct.html). For further reading on ethics in statistics, see Franks (2020) and Fleming & Bruce (2021).

$$\sim$$

Exercise 3.9. *Discuss the following.* In the introduction to American Statistical Association's *Ethical Guidelines for Statistical Practice*, it is stated that "using statistics in pursuit of unethical ends is inherently unethical". What is considered unethical depends on social, moral, political, and religious values, and ultimately you must decide for yourself what you consider to be unethical ends. Which (if any) of the following do you consider to be unethical?

1. Using statistical analysis to help a company that harm the environment through their production processes. Does it matter to you what the purpose of the analysis is?
2. Using statistical analysis to help a tobacco or liquor manufacturing company. Does it matter to you what the purpose of the analysis is?
3. Using statistical analysis to help a bank identify which loan applicants that are likely to default on their loans.
4. Using statistical analysis of social media profiles to identify terrorists.
5. Using statistical analysis of social media profiles to identify people likely to protest against the government.
6. Using statistical analysis of social media profiles to identify people to target with political adverts.
7. Using statistical analysis of social media profiles to target ads at people likely to buy a bicycle.
8. Using statistical analysis of social media profiles to target ads at people likely to gamble at a new online casino. Does it matter to you if it's an ad for the casino or for help for gambling addiction?

The use and misuse of statistical inference offer many ethical dilemmas. Some common issues related to ethics and good statistical practice are discussed below. As you read them and work with the associated exercises, consider consulting the ASA's ethical guidelines, presented above.

3.11.2 p-hacking and the file-drawer problem

Hypothesis tests are easy to misuse. If you run enough tests on your data, you are almost guaranteed to end up with significant results – either due to chance or because some of the null hypotheses you test are false. The process of trying lots of different tests (different methods, different hypotheses, different subgroups) in search of significant results is known as *p-hacking*, *p-fishing*, or *data dredging*. This greatly increases the risk of false findings and can often produce misleading results.

Many practitioners inadvertently resort to p-hacking, by mixing exploratory data analysis and hypothesis testing, or by coming up with new hypotheses to test as they work with their data. This can be avoided by planning your analyses in advance, a practice that in fact is required in medical trials.

On the other end of the spectrum, there is the *file-drawer problem*, in which studies with negative (i.e., not statistically significant) results aren't published or reported but instead are stored in the researcher's file drawers. There are many reasons for this, one being that negative results usually are seen as less important and less worthy of spending time on. Simply put, negative results just aren't news. If your study shows that eating kale every day significantly reduces the risk of cancer, then that is news, something that people are interested in learning, and something that can be published in a prestigious journal. However, if your study shows that a daily serving of kale has no impact on the risk of cancer, that's not news, people aren't really interested in hearing it, and it may prove difficult to publish your findings.

But what if 100 different researchers carried out the same study? If eating kale doesn't affect the risk of cancer, then we can still expect five out of these researchers to get significant results (using a 5% significance level). If only those researchers publish their results, that may give the impressions that there is strong evidence of the cancer-preventing effect of kale backed up by several papers, even though the majority of studies actually indicated that there was no such effect.

~

Exercise 3.10. *Discuss the following.* You are helping a research team with statistical analysis of data that they have collected. You agree on five hypotheses to test. None of the tests turns out to be significant. Fearing that all their hard work won't lead anywhere, your collaborators then ask you to carry out five new tests. Neither turns out to be significant. Your collaborators closely inspect the data and then ask you to carry out 10 more tests, two of which are significant. The team wants to publish these significant results in a scientific journal. Should you agree to publish them? If so, what results should be published? Should you have put your foot down and told them not to run more tests? Does your answer depend on how long it took the research team to collect the data? What if the team won't get funding for new projects unless they publish a paper soon? What if other research teams competing for the same grants do their analyses like this?

Exercise 3.11. *Discuss the following.* You are working for a company that is launching a new product, a hair loss treatment. In a small study, the product worked for 19 out of 22 participants (86%). You compute a 95% Clopper-Pearson confidence interval (Section

3.5.1) for the proportion of successes and find that it is (0.65, 0.97). Based on this, the company wants to market the product as being 97% effective. Is that acceptable to you? If not, how should it be marketed? Would your answer change if the product was something else (new running shoes that make you faster, a plastic film that protects smartphone screens from scratches, or contraceptives)? What if the company wanted to market it as being 86% effective instead?

Exercise 3.12. *Discuss the following.* You have worked long and hard on a project. In the end, to see if the project was a success, you run a hypothesis test to check if two variables are correlated. You find that they are not (p = 0.15). However, if you remove three outliers, the two variables are significantly correlated (p = 0.03). What should you do? Does your answer change if you only have to remove one outlier to get a significant result? If you have to remove 10 outliers? 100 outliers? What if the p-value is 0.051 before removing the outliers and 0.049 after removing the outliers?

Exercise 3.13. *Discuss the following.* You are analysing data from an experiment to see if there is a difference between two treatments. You estimate[4] that given the sample size and the expected difference in treatment effects, the power of the test that you'll be using, i.e., the probability of rejecting the null hypothesis if it is false, is about 15%. Should you carry out such an analysis? If not, how high does the power need to be for the analysis to be meaningful?

3.11.3   Reproducibility

An analysis is *reproducible* if it can be reproduced by someone else. By producing reproducible analyses, we make it easier for others to scrutinise our work. We also make all the steps in the data analysis transparent. This can act as a safeguard against data fabrication and data dredging.

In order to make an analysis reproducible, we need to provide at least two things. First, the data – all unedited data files in their original format. This also includes *metadata* with information required to understand the data (e.g., codebooks explaining variable names and codes used for categorical variables). Second, the computer code used to prepare and analyse the data. This includes any wrangling and preliminary testing performed on the data.

As long as we save our data files and code, data wrangling and analyses in R are inherently reproducible, in contrast to the same tasks carried out in menu-based software such as Excel. However, if reports are created using a word processor, there is always a risk that something will be lost along the way. Perhaps numbers are copied by hand (which may introduce errors), or maybe the wrong version of a figure is pasted into the document. R Markdown (Section 4.1) is a great tool for creating completely reproducible reports, as it allows you to integrate R code for data wrangling, analyses, and graphics in your report-writing. This reduces the risk of manually inserting errors and allows you to share your work with others easily.

~

Exercise 3.14. *Discuss the following.* You are working on a study at a small-town hospital. The data involves biomarker measurements for a number of patients, and you show that patients with a sexually transmitted disease have elevated levels of some of the biomarkers. The data also includes information about the patients: their names, ages, zip codes, heights,

---

[4]We'll discuss methods for producing such estimates in Section 7.2.3.

and weights. The research team wants to publish your results and make the analysis reproducible. Is it ethically acceptable to share all your data? Can you make the analysis reproducible without violating patient confidentiality?

4

Exploratory data analysis and unsupervised learning

Exploratory data analysis (EDA) is a process in which we summarise and visually explore a dataset. An important part of EDA is unsupervised learning, which is a collection of methods for finding interesting subgroups and patterns in our data. Unlike statistical hypothesis testing, which is used to reject hypotheses, EDA can be used to *generate* hypotheses (which can then be confirmed or rejected by new studies). Another purpose of EDA is to find outliers and incorrect observations, which can lead to a cleaner and more useful dataset. In EDA we ask questions about our data and then try to answer them using summary statistics and graphics. Some questions will prove to be important, and some will not. The key to finding important questions is to ask a lot of questions. This chapter will provide you with a wide range of tools for question-asking.

After working with the material in this chapter, you will be able to use R to:

- Create reports using R Markdown,
- Customise the look of your plots,
- Visualise the distribution of a variable,
- Create interactive plots,
- Detect and label outliers,
- Investigate patterns in missing data,
- Visualise trends,
- Plot time series data,
- Visualise multiple variables at once using scatterplot matrices, correlograms, and bubble plots,
- Visualise multivariate data using principal component analysis, and
- Use unsupervised learning techniques for clustering.

4.1 Reports with R Markdown

R Markdown files can be used to create nicely formatted documents using R that are easy to export to other formats, like HTML, Word or PDF. They allow you to mix R code with results and text. They can be used to create reproducible reports that are easy to update with new data, because they include the code for making tables and figures. Additionally, they can be used as notebooks for keeping track of your work and your thoughts as you carry out an analysis. You can even use them for writing books; in fact, this entire book was written using R Markdown.

It is often a good idea to use R Markdown for exploratory analyses, as it allows you to write down your thoughts and comments as the analysis progresses, as well as to save the results of the exploratory journey. For that reason, we'll start this chapter by looking at some examples of what you can do using R Markdown. According to your preference, you can use either R Markdown or ordinary R scripts for the analyses in the remainder of the chapter. The R code used is the same and the results are identical; but, if you use R Markdown, you can also save the output of the analysis in a nicely formatted document.

4.1.1 A first example

When you create a new R Markdown document in RStudio by clicking *File > New File > R Markdown* in the menu, a document similar to the one below is created:

```
---
title: "Untitled"
author: "Måns Thulin"
date: "10/20/2020"
output: html_document
---

```{r setup, include=FALSE}
knitr::opts_chunk$set(echo = TRUE)
```

R Markdown

This is an R Markdown document. Markdown is a simple formatting syntax
for authoring HTML, PDF, and MS Word documents. For more details on
using R Markdown see
<http://rmarkdown.rstudio.com>.

When you click the **Knit** button a document will be generated that
includes both content as well as the output of any embedded R code
chunks within the document. You can embed an R code chunk like this:

```{r cars}
summary(cars)
```

Including Plots

You can also embed plots, for example:

```{r pressure, echo=FALSE}
plot(pressure)
```

Note that the `echo = FALSE` parameter was added to the code chunk to
prevent printing of the R code that generated the plot.
```

Press the *Knit* button at the top of the Script panel to create an HTML document using this Markdown file. It will be saved in the same folder as your Markdown file. Once the HTML document has been created, it will open so that you can see the results. You may have to install additional packages for this to work, in which case RStudio will prompt you.

Now, let's have a look at what the different parts of the Markdown document do. The first part is called the *document header* or *YAML header*. It contains information about the document, including its title, the name of its author, and the date on which it was first created:

```
---
title: "Untitled"
author: "Måns Thulin"
date: "10/20/2020"
output: html_document
---
```

The part that says `output: html_document` specifies what type of document should be created when you press *Knit*. In this case, it's set to `html_document`, meaning that an HTML document will be created. By changing this to `output: word_document` you can create a `.docx` Word document instead. By changing it to `output: pdf_document`, you can create a `.pdf` document using LaTeX (you'll have to install LaTeX if you haven't already – RStudio will notify you if that is the case).

The second part sets the default behaviour of *code chunks* included in the document, specifying that the output from running the chunks should be printed unless otherwise specified:

```
```{r setup, include=FALSE}
knitr::opts_chunk$set(echo = TRUE)
```
```

The third part contains the first proper section of the document. First, a header is created using ##. Then there is some text with formatting: < > is used to create a link and ** is used to get **bold text**. Finally, there is a code chunk, delimited by ```:

```
## R Markdown

This is an R Markdown document. Markdown is a simple formatting syntax
for authoring HTML, PDF, and MS Word documents. For more details on
using R Markdown see
<http://rmarkdown.rstudio.com>.

When you click the **Knit** button a document will be generated that
includes both content as well as the output of any embedded R code
chunks within the document. You can embed an R code chunk like this:

```{r cars}
summary(cars)
```
```

The fourth and final part contains another section, this time with a figure created using R. A setting is added to the code chunk used to create the figure, which prevents the underlying code from being printed in the document:

```
## Including Plots
```

```
You can also embed plots, for example:
```

```
```{r pressure, echo=FALSE}
plot(pressure)
```
```

```
Note that the `echo = FALSE` parameter was added to the code chunk to
prevent printing of the R code that generated the plot.
```

In the next few sections, we will look at how formatting and code chunks work in R Markdown.

4.1.2   Formatting text

To create plain text in a Markdown file, you simply have to write plain text. If you wish to add some formatting to your text, you can use the following:

* _italics_ or *italics*: to create text in *italics*.
* __bold__ or **bold**: to create **bold** text.
* [linked text](http://www.modernstatisticswithr.com): to create linked text[1].
* `code`: to include inline code in your document.
* $a^2 + b^2 = c^2$: to create inline equations like $a^2 + b^2 = c^2$ using LaTeX syntax.
* $$a^2 + b^2 = c^2$$: to create a centred equation on a new line.

To add headers and sub-headers, and to divide your document into section, start a new line with #'s as follows:

```
# Header text
## Sub-header text
### Sub-sub-header text
...and so on.
```

4.1.3   Lists, tables, and images

To create a bullet list, you can use * as follows. Note that you need a blank line between your list and the previous paragraph to begin a list.

```
* Item 1
* Item 2
    + Sub-item 1
    + Sub-item 2
* Item 3
```

yielding:

* Item 1
* Item 2
    - Sub-item 1
    - Sub-item 2
* Item 3

---

[1] http://www.modernstatisticswithr.com

To create an ordered list, use:

```
1. First item
2. Second item
   i) Sub-item 1
   ii) Sub-item 2
3. Item 3
```

yielding:

1. First item
2. Second item

    i) Sub-item 1
    ii) Sub-item 2

3. Item 3

To create a table, use | and --------- as follows:

```
Column 1  | Column 2
--------- | ---------
Content   | More content
Even more | And some here
Even more? | Yes!
```

which yields the following output:

Column 1	Column 2
Content	More content
Even more	And some here
Even more?	Yes!

To include an image, use the same syntax as when creating linked text with a link to the image path (either local or on the web), but with a ! in front:

```
![](https://www.r-project.org/Rlogo.png)
```

```
![R logo.](https://www.r-project.org/Rlogo.png)
```

which yields:

FIGURE 4.1 R logo.

4.1.4   Code chunks

The simplest way to define a code chunk is to write:

```
```{r}
plot(pressure)
```
```

In RStudio, Ctrl+Alt+I is a keyboard shortcut for inserting this kind of code chunk.

We can add a name and a caption to the chunk, which lets us reference objects created by the chunk:

```
```{r pressureplot, fig.cap = "Plot of the pressure data."}
plot(pressure)
```
```

```
As we can see in Figure \\ref{fig:pressureplot}, the relationship between
temperature and pressure resembles a banana.
```

This yields the following output:

```
plot(pressure)
```

As we can see in Figure 4.2, the relationship between temperature and pressure resembles a banana.

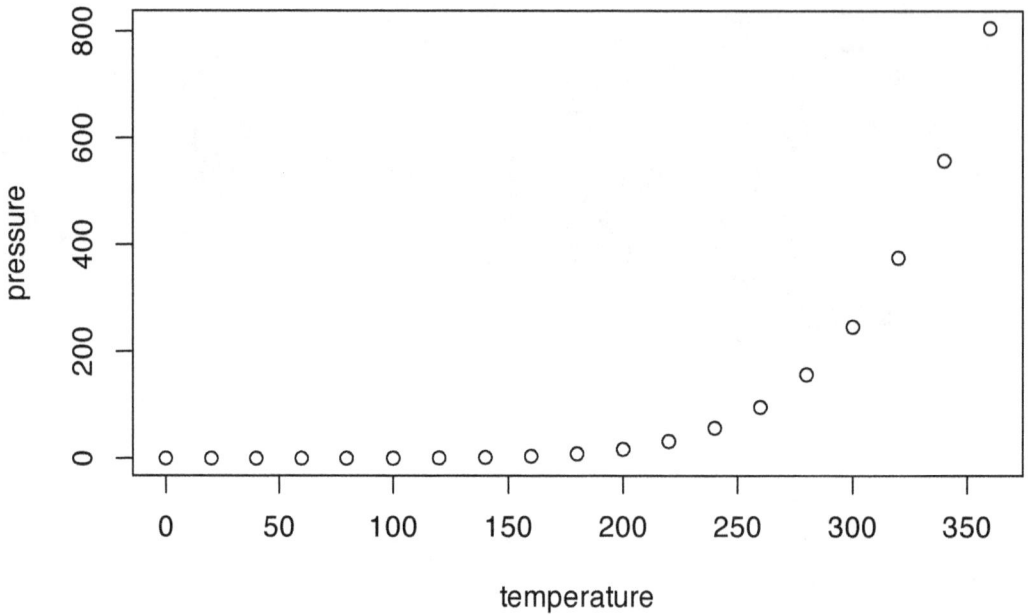

FIGURE 4.2 Plot of the pressure data.

In addition, you can add settings to the chunk header to control its behaviour. For instance, you can include a code chunk without running it by adding `echo = FALSE`:

```{r, eval = FALSE}
plot(pressure)
```

You can add the following settings to your chunks:

- `echo = FALSE` to run the code without printing it,
- `eval = FALSE` to print the code without running it,
- `results = "hide"` to hide printed output,
- `fig.show = "hide"` to hide plots,
- `warning = FALSE` to suppress warning messages from being printed in your document,
- `message = FALSE` to suppress other messages from being printed in your document,
- `include = FALSE` to run a chunk without showing the code or results in the document,
- `error = TRUE` to continue running your R Markdown document even if there is an error in the chunk (by default, the documentation creation stops if there is an error).

Data frames can be printed either as in the Console or as a nicely formatted table. For example,

```{r, echo = FALSE}
head(airquality)
```

yields:

```
##   Ozone Solar.R Wind Temp Month Day
## 1    41     190  7.4   67     5   1
## 2    36     118  8.0   72     5   2
## 3    12     149 12.6   74     5   3
```

```
## 4      18      313 11.5    62      5   4
## 5      NA       NA 14.3    56      5   5
## 6      28       NA 14.9    66      5   6
```

whereas

```
```{r, echo = FALSE}
knitr::kable(
  head(airquality),
  caption = "Some data I found."
)
```
```

yields a nicely formatted table.

Further help and documentation for R Markdown can be found through the RStudio menus, by clicking *Help > Cheatsheets > R Markdown Cheat Sheet* or *Help > Cheatsheets > R Markdown Reference Guide*.

4.2 Customising `ggplot2` plots

We'll be using `ggplot2` a lot in this chapter; so, before we get started with exploratory analyses, we'll take some time to learn how we can customise the look of `ggplot2`-plots.

Consider the following facetted plot from Section 2.7.4:

```
library(ggplot2)

ggplot(msleep, aes(brainwt, sleep_total)) +
    geom_point() +
    labs(x = "Brain weight (logarithmic scale)",
        y = "Total sleep time") +
    scale_x_log10() +
    facet_wrap(~ vore)
```

It looks nice, sure, but there may be things that you'd like to change. Maybe you'd like the plot's background to be white instead of grey, or perhaps you'd like to use a different font. These, and many other things, can be modified using *themes* and *palettes*. Before we look at that, we'll take a quick look at how to modify the labels and axes of the plot.

4.2.1 Modifying labels

As we've already seen, the `labs` function allows us to change the labels for the aesthethics, like the header of the legend showing the different colours in the figure:

```
ggplot(msleep, aes(brainwt, sleep_total, colour = vore)) +
    geom_point() +
    labs(x = "Brain weight (logarithmic scale)",
        y = "Total sleep time",
```

```
        colour = "Feeding behaviour") +
    scale_x_log10()
```

4.2.2   Modifying axis scales

We've seen how functions with names beginning with `scale_x`, such as `scale_x_log10` can be used to modify the scale of the x-axis (and `scale_y`-functions used to modify the y-axis). In addition to log transforms, we can for instance control where the tick marks are located by using the `breaks` argument. If we only want to modify the tick marks (without doing a log transform), we can use `scale_x_continuous`:

```
# Default:
ggplot(msleep, aes(brainwt, sleep_total, colour = vore)) +
    geom_point() +
    labs(x = "Brain weight",
         y = "Total sleep time",
         colour = "Feeding behaviour")

# Tick marks at all integers from 0 to 6:
ggplot(msleep, aes(brainwt, sleep_total, colour = vore)) +
    geom_point() +
    labs(x = "Brain weight",
         y = "Total sleep time",
         colour = "Feeding behaviour") +
    scale_x_continuous(breaks = 0:6)

# Tick mark at specific values:
ggplot(msleep, aes(brainwt, sleep_total, colour = vore)) +
    geom_point() +
    labs(x = "Brain weight",
         y = "Total sleep time",
         colour = "Feeding behaviour") +
    scale_x_continuous(breaks = c(0, 2.5, 4.25))
```

The y-axis can be modified analogously using `scale_y_continuous`.

4.2.3   Using themes

`ggplot2` comes with a number of basic themes. All are fairly similar but differ in things like background colour, grids, and borders. You can add them to your plot using `theme_themeName`, where `themeName` is the name of the theme[2]. Here are some examples:

```
p <- ggplot(msleep, aes(brainwt, sleep_total, colour = vore)) +
    geom_point() +
    labs(x = "Brain weight (logarithmic scale)",
         y = "Total sleep time",
```

---

[2]See `?theme_grey` for a list of available themes.

```
                colour = "Feeding behaviour") +
    scale_x_log10() +
    facet_wrap(~ vore)

# Create plot with different themes:
p + theme_grey() # The default theme
p + theme_bw()
p + theme_linedraw()
p + theme_light()
p + theme_dark()
p + theme_minimal()
p + theme_classic()
```

There are several packages available that contain additional themes. Let's try a few:

```
install.packages("ggthemes")
library(ggthemes)

# Create plot with different themes from ggthemes:
p + theme_tufte() # Minimalist Tufte theme
p + theme_wsj() # Wall Street Journal
p + theme_solarized() + scale_colour_solarized() # Solarized colours


#############################

install.packages("hrbrthemes")
library(hrbrthemes)

# Create plot with different themes from hrbrthemes:
p + theme_ipsum() # Ipsum theme
p + theme_ft_rc() # Suitable for use with dark RStudio themes
p + theme_modern_rc() # Suitable for use with dark RStudio themes
```

4.2.4   Colour palettes

Unlike, e.g., background colours, the *colour palette*, i.e., the list of colours used for plotting, is not part of the theme that you're using. Next, we'll have a look at how to change the colour palette used for your plot.

Let's start by creating a ggplot object:

```
p <- ggplot(msleep, aes(brainwt, sleep_total, colour = vore)) +
    geom_point() +
    labs(x = "Brain weight (logarithmic scale)",
         y = "Total sleep time",
         colour = "Feeding behaviour") +
    scale_x_log10()
```

You can change the colour palette using `scale_colour_brewer`. Three types of colour palettes are available:

- **Sequential palettes**: these range from a colour to white. These are useful for representing ordinal (i.e., ordered) categorical variables and numerical variables.
- **Diverging palettes**: these range from one colour to another, with white in between. Diverging palettes are useful when there is a meaningful middle or 0 value (e.g., when your variables represent temperatures or profit/loss), which can be mapped to white.
- **Qualitative palettes**: these contain multiple distinct colours. They are useful for nominal (i.e., with no natural ordering) categorical variables.

See `?scale_colour_brewer` or http://www.colorbrewer2.org for a list of the available palettes. Here are some examples:

```
# Sequential palette:
p + scale_colour_brewer(palette = "OrRd")
# Diverging palette:
p + scale_colour_brewer(palette = "RdBu")
# Qualitative palette:
p + scale_colour_brewer(palette = "Set1")
```

In this case, because `vore` is a nominal categorical variable, a qualitative palette is arguably the best choice.

In addition to these ready-made palettes, you can create your own custom palettes. Here's an example with colours that have been recommended as being easy to distinguish for people who are colour-blind:

```
# Create a vector with colours:
colour_blind_palette <- c("#000000", "#E69F00", "#56B4E9", "#009E73",
                          "#0072B2", "#D55E00", "#CC79A7")

# Use the palette with the plot:
p + scale_colour_manual(values = colour_blind_palette)
```

Use `scale_colour_manual` to change the palette used for `colour` and `scale_fill_manual` to change the palette used for `fill` in your plots.

Finally, if you want to use your custom palette in many different plots, and you don't want to have to add `scale_colour_manual` to every single plot, you can change the default colours for all `ggplot2` plots in your R session as follows:

```
options(ggplot2.discrete.colour = colour_blind_palette,
        ggplot2.discrete.fill = colour_blind_palette)
```

Some examples of themes are shown in Figure 4.3.

4.2.5   Theme settings

The point of using a theme is that you get a combination of colours, fonts, and other choices that are supposed to go well together, meaning that you don't have to spend too much time

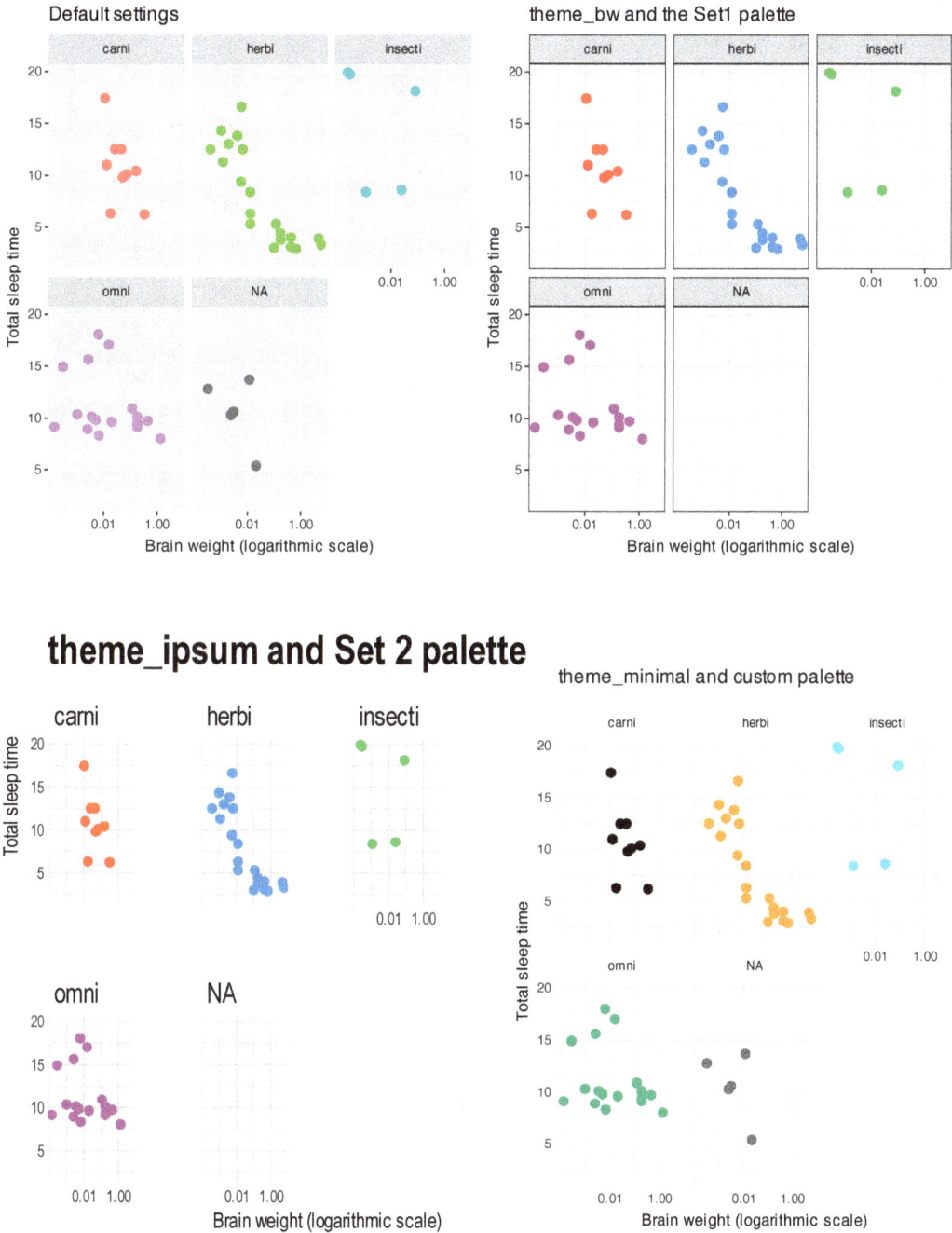

FIGURE 4.3 Examples of ggplot themes and colour palettes.

picking combinations. But if you like, you can override the default options and customise any and all parts of a theme.

The theme controls all visual aspects of the plot not related to the aesthetics. You can change the theme settings using the `theme` function. For instance, you can use `theme` to remove the legend or change its position:

```
# No legend:
p + theme(legend.position = "none")

# Legend below figure:
p + theme(legend.position = "bottom")

# Legend inside plot:
p + theme(legend.position = c(0.9, 0.7))
```

In the last example, the vector `c(0.9, 0.7)` gives the relative coordinates of the legend, with `c(0 0)` representing the bottom left corner of the plot and `c(1, 1)` the upper right corner. Try to change the coordinates to different values between `0` and `1` and see what happens.

The base size of a theme controls the scale of the entire figure. This makes it useful when you want to rescale all elements of your plot at the same time:

```
p + theme_grey(base_size = 8)
p + theme_grey(base_size = 20)
```

`theme` has a lot of other settings, including for the colours of the background, the grid, and the text in the plot. Here are a few examples that you can use as a starting point for experimenting with the settings:

```
p + theme(panel.grid.major = element_line(colour = "black"),
          panel.grid.minor = element_line(colour = "purple",
                                          linetype = "dotted"),
          panel.background = element_rect(colour = "red", size = 2),
          plot.background = element_rect(fill = "yellow"),
          axis.text = element_text(family = "mono", colour = "blue"),
          axis.title = element_text(family = "serif", size = 14))
```

To find a complete list of settings, see `?theme`, `?element_line` (lines), `?element_rect` (borders and backgrounds), `?element_text` (text), and `element_blank` (for suppressing plotting of elements). As before, you can use `colors()` to get a list of built-in colours, or use colour hex codes.

∼

Exercise 4.1. Use the documentation for `theme` and the `element_...` functions to change the plot object `p` created above as follows:

1. Change the background colour of the entire plot to `lightblue`.

2. Change the font of the legend to `serif`.

3. Remove the grid.

4. Change the colour of the axis ticks to `orange` and make them thicker.

4.3 Exploring distributions

It is often useful to visualise the distribution of a numerical variable. Comparing the distributions of different groups can lead to important insights. Visualising distributions is also essential when checking assumptions used for various statistical tests (sometimes called *initial data analysis*). In this section, we will illustrate how this can be done using the `diamonds` data from the `ggplot2` package. You can read more about it by running `?diamonds`.

4.3.1 Density plots and frequency polygons

We already know how to visualise the distribution of the data by dividing it into bins and plotting a histogram:

```
library(ggplot2)
ggplot(diamonds, aes(carat)) +
     geom_histogram(colour = "black")
```

A similar plot is created using frequency polygons, which uses lines instead of bars to display the counts in the bins:

```
ggplot(diamonds, aes(carat)) +
     geom_freqpoly()
```

An advantage with frequency polygons is that they can be used to compare groups, e.g., diamonds with different cuts, without facetting:

```
ggplot(diamonds, aes(carat, colour = cut)) +
     geom_freqpoly()
```

It is clear from this figure that there are more diamonds with ideal cuts than diamonds with fair cuts in the data. The polygons have roughly the same shape, except perhaps for the polygon for diamonds with fair cuts.

In some cases, we are more interested in the shape of the distribution than in the actual counts in the different bins. Density plots are similar to frequency polygons but show an estimate of the density function of the underlying random variable. These estimates are smooth curves that are scaled so that the area below them is 1 (i.e., scaled to be proper density functions):

```
ggplot(diamonds, aes(carat, colour = cut)) +
     geom_density()
```

From this figure, shown in Figure 4.4, it becomes clear that low-carat diamonds tend to have better cuts, which wasn't obvious from the frequency polygons. However, the plot does not provide any information about *how* common different cuts are. Use density plots if you're more interested in the shape of a variable's distribution, and frequency polygons if you're more interested in counts.

There are several settings that can help improve the look of your density plot, which you'll explore in the next exercise.

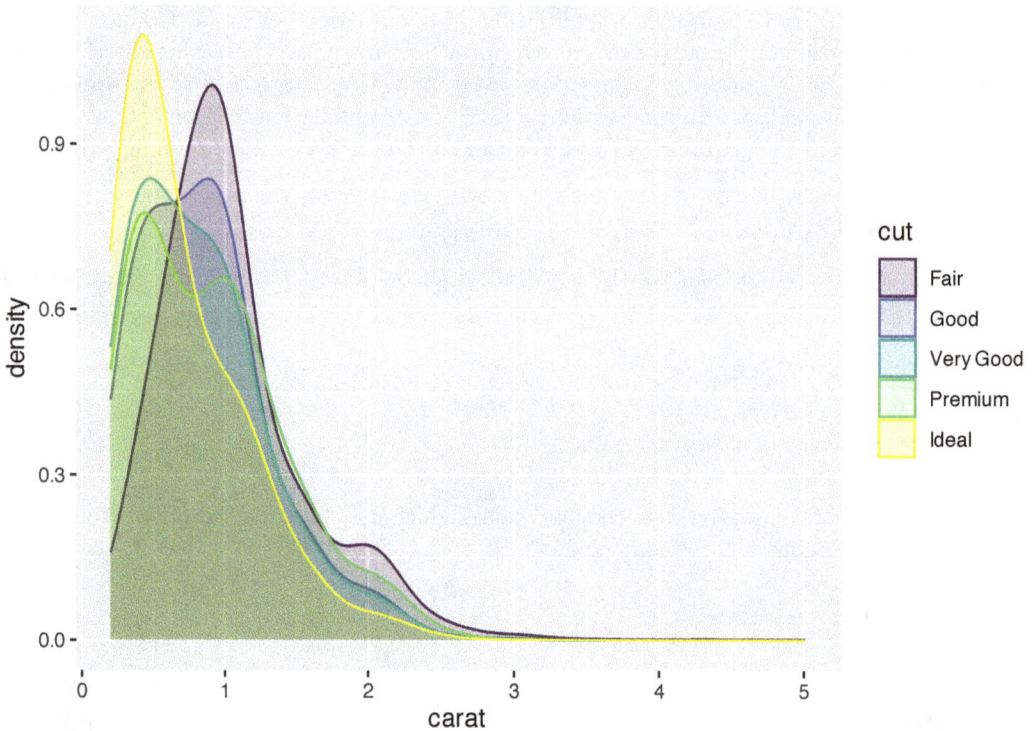

FIGURE 4.4 Density plot for diamond carats.

~

Exercise 4.2. Using the density plot we just created above (Figure 4.4) and the documentation for `geom_density`, do the following:

1. Increase the smoothness of the density curves.

2. Fill the area under the density curves with the same colour as the curves themselves.

3. Make the colours that fill the areas under the curves transparent.

4. The figure still isn't that easy to interpret. Install and load the `ggridges` package, an extension of `ggplot2` that allows you to make so-called ridge plots (density plots that are separated along the y-axis, similar to facetting). Read the documentation for `geom_density_ridges` and use it to make a ridge plot of diamond prices for different cuts.

Exercise 4.3. Return to the histogram created by `ggplot(diamonds, aes(carat)) + geom_histogram()` above. As there are very few diamonds with carat greater than 3, cut the x-axis at 3. Then decrease the bin width to 0.01. Do any interesting patterns emerge?

4.3.2 Asking questions

Exercise 4.3 causes us to ask why diamonds with carat values that are multiples of 0.25 are more common than others. Perhaps the price is involved? Unfortunately, a plot of `carat` versus `price` is not that informative:

```
ggplot(diamonds, aes(carat, price)) +
    geom_point(size = 0.05)
```

Maybe we could compute the average price in each bin of the histogram? In that case, we need to extract the bin breaks from the histogram somehow. We could then create a new categorical variable using the breaks with `cut` (as we did in Exercise 2.27). It turns out that extracting the bins is much easier using base graphics than `ggplot2`, so let's do that:

```
# Extract information from a histogram with bin width 0.01,
# which corresponds to 481 breaks:
carat_br <- hist(diamonds$carat, breaks = 481, right = FALSE,
                 plot = FALSE)
# Of interest to us are:
# carat_br$breaks, which contains the breaks for the bins
# carat_br$mid, which contains the midpoints of the bins
#                (useful for plotting!)

# Create categories for each bin:
diamonds$carat_cat <- cut(diamonds$carat, 481, right = FALSE)
```

We now have a variable, `carat_cat`, that shows to which bin each observation belongs. Next, we'd like to compute the mean for each bin. This is a grouped summary – mean by category. After we've computed the bin means, we could then plot them against the bin midpoints. Let's try it:

```
means <- aggregate(price ~ carat_cat, data = diamonds, FUN = mean)

plot(carat_br$mid, means$price)
```

That didn't work as intended. We get an error message when attempting to plot the results:

```
Error in xy.coords(x, y, xlabel, ylabel, log) :
  'x' and 'y' lengths differ
```

The error message implies that the number of bins and the number of mean values that have been computed differ. But we've just computed the mean for each bin, haven't we? So what's going on?

By default, `aggregate` ignores groups for which there are no values when computing grouped summaries. In this case, there are a lot of empty bins – there is for instance no observation in the [4.99,5) bin. In fact, only 272 out of the 481 bins are non-empty.

We can solve this in different ways. One way is to remove the empty bins. We can do this using the `match` function, which returns the positions of *matching values* in two vectors. If we use it with the bins from the grouped summary and the vector containing all bins, we can find the indices of the non-empty bins. This requires the use of the `levels` function, which you'll learn more about in Section 5.4:

```
means <- aggregate(price ~ carat_cat, data = diamonds, FUN = mean)

id <- match(means$carat_cat, levels(diamonds$carat_cat))
```

Finally, we'll also add some vertical lines to our plot, to call attention to multiples of 0.25.

Using base graphics is faster here:

```
plot(carat_br$mid[id], means$price,
     cex = 0.5)

# Add vertical lines at multiples
# of 0.25:
abline(v = c(0.5, 0.75, 1,
             1.25, 1.5))
```

But we can of course stick to `ggplot2` if we like:

```
library(ggplot2)

d2 <- data.frame(
        bin = carat_br$mid[id],
        mean = means$price)

ggplot(d2, aes(bin, mean)) +
    geom_point() +
    geom_vline(xintercept =
            c(0.5, 0.75, 1,
              1.25, 1.5))
# geom_vline add vertical lines at
# multiples of 0.25
```

It appears that there are small jumps in the prices at some of the 0.25-marks. This explains why there are more diamonds just above these marks than just below.

The above example illustrates three crucial things regarding exploratory data analysis:

- Plots (in our case, the histogram) often lead to new questions.

- Often, we must transform, summarise, or otherwise manipulate our data to answer a question. Sometimes this is straightforward, and sometimes it means diving deep into R code.

- Sometimes the thing that we're trying to do doesn't work straight away. There is almost always a solution though (and oftentimes more than one!). The more you work with R, the more problem-solving tricks you will learn.

4.3.3   Violin plots

Density curves can also be used as alternatives to boxplots. In Exercise 2.17, you created boxplots to visualise price differences between diamonds of different cuts:

```
ggplot(diamonds, aes(cut, price)) +
    geom_boxplot()
```

Instead of using a boxplot, we can use a violin plot. Each group is represented by a "violin", given by a rotated and duplicated density plot:

```
ggplot(diamonds, aes(cut, price)) +
    geom_violin()
```

Compared to boxplots, violin plots capture the entire distribution of the data rather than just a few numerical summaries. If you like numerical summaries (and you should) you can add the median and the quartiles (corresponding to the borders of the box in the boxplot) using the `draw_quantiles` argument:

```
ggplot(diamonds, aes(cut, price)) +
    geom_violin(draw_quantiles =  c(0.25, 0.5, 0.75))
```

~

Exercise 4.4. Using the first boxplot created above, i.e., `ggplot(diamonds, aes(cut, price)) + geom_violin()`, do the following:

1. Add some colour to the plot by giving different colours to each violin.

2. Because the categories are shown along the x-axis, we don't really need the legend. Remove it.

3. Both boxplots and violin plots are useful. Maybe we can have the best of both worlds? Add the corresponding boxplot inside each violin. Hint: the `width` and `alpha` arguments in `geom_boxplot` are useful for creating a nice-looking figure here.

4. Flip the coordinate system to create horizontal violins and boxes instead.

4.4   Combining multiple plots into a single graphic

When exploring data with many variables, you'll often want to make the same kind of plot (e.g., a violin plot) for several variables. It will frequently make sense to place these side-by-side in the same plot window. The `patchwork` package extends `ggplot2` by letting you do just that. Let's install it:

```
install.packages("patchwork")
```

To use it, save each plot as a plot object and then add them together:

```
plot1 <- ggplot(diamonds, aes(cut, carat, fill = cut)) +
        geom_violin() +
        theme(legend.position = "none")
plot2 <- ggplot(diamonds, aes(cut, price, fill = cut)) +
        geom_violin() +
        theme(legend.position = "none")

library(patchwork)
plot1 + plot2
```

You can also arrange the plots on multiple lines, with different numbers of plots on each line. This is particularly useful if you are combining different types of plots in a single plot window. In this case, you separate plots that are one the same line by | and mark the beginning of a new line with /:

```
# Create two more plot objects:
plot3 <- ggplot(diamonds, aes(cut, depth, fill = cut)) +
        geom_violin() +
        theme(legend.position = "none")
plot4 <- ggplot(diamonds, aes(carat, fill = cut)) +
        geom_density(alpha = 0.5) +
        theme(legend.position = c(0.9, 0.6))

# One row with three plots and one row with a single plot:
(plot1 | plot2 | plot3) / plot4

# One column with three plots and one column with a single plot:
(plot1 / plot2 / plot3) | plot4
```

(You may need to enlarge your plot window for this to look good!)

4.5   Outliers and missing data

4.5.1   Detecting outliers

Both boxplots and scatterplots are helpful in detecting deviating observations – often called outliers. Outliers can be caused by measurement errors or errors in the data input but can also be interesting rare cases that can provide valuable insights about the process that generated the data. Either way, it is often of interest to detect outliers, for instance because that may influence the choice of what statistical tests to use.

Let's draw a scatterplot of diamond carats versus prices:

```
ggplot(diamonds, aes(carat, price)) +
    geom_point()
```

There are some outliers which we may want to study further. For instance, there is a surprisingly cheap 5-carat diamond, and some cheap 3-carat diamonds. Note that it is not just the prices nor just the carats of these diamonds that make them outliers, but the unusual combinations of prices and carats. But how can we identify those points?

One option is to use the `plotly` package to make an interactive version of the plot, where we can hover interesting points to see more information about them. Start by installing it:

```r
install.packages("plotly")
```

To use `plotly` with a ggplot graphic, we store the graphic in a variable and then use it as input to the `ggplotly` function. The resulting (interactive!) plot takes a little longer than usual to load. Try hovering the points:

```r
myPlot <- ggplot(diamonds, aes(carat, price)) +
    geom_point()

library(plotly)
ggplotly(myPlot)
```

By default, `plotly` only shows the carat and price of each diamond. But we can add more information to the box by adding a `text` aesthetic:

```r
myPlot <- ggplot(diamonds, aes(carat, price, text = paste("Row:",
                                  rownames(diamonds)))) +
                                  geom_point()

ggplotly(myPlot)
```

We can now find the row numbers of the outliers visually, which is very useful when exploring data.

~

Exercise 4.5. The variables x and y in the `diamonds` data describe the length and width of the diamonds (in millimetres). Use an interactive scatterplot to identify outliers in these variables. Check prices, carat, and other information and think about if any of the outliers can be due to data errors.

4.5.2   Labelling outliers

Interactive plots are great when exploring a dataset but are not always possible to use in other contexts, e.g., for printed reports and some presentations. In these other cases, we can instead annotate the plot with notes about outliers. One way to do this is to use a geom called `geom_text`.

For instance, we may want to add the row numbers of outliers to a plot. To do so, we use `geom_text` along with a condition that specifies for which points we should add annotations. As in Section 2.11.3, if we, e.g., wish to add row numbers for diamonds with carats greater than four, our condition would be `carat > 4`. The `ifelse` function, which we'll look closer at in

Section 6.3, is perfect to use here. The syntax will be ifelse(condition, what text to write if the condition is satisfied, what text to write else). To add row names for observations that fulfill the condition but not for other observations, we use ifelse(condition, rownames(diamonds), ""). If, instead, we wanted to print the price of the diamonds, we'd use ifelse(condition, price, "").

Here are some different examples of conditions used to plot text:

```
# Using the row number (the 5 carat diamond is on row 27,416)
ggplot(diamonds, aes(carat, price)) +
      geom_point() +
      geom_text(aes(label = ifelse(rownames(diamonds) == 27416,
                                     rownames(diamonds), "")),
                hjust = 1.1)
# (hjust=1.1 shifts the text to the left of the point)


# Plot text next to all diamonds with carat>4
ggplot(diamonds, aes(carat, price)) +
      geom_point() +
      geom_text(aes(label = ifelse(carat > 4, rownames(diamonds), "")),
                hjust = 1.1)


# Plot text next to 3 carat diamonds with a price below 7500
ggplot(diamonds, aes(carat, price)) +
      geom_point() +
      geom_text(aes(label = ifelse(carat == 3 & price < 7500,
                                     rownames(diamonds), "")),
                hjust = 1.1)
```

~

Exercise 4.6. Create a static (i.e., non-interactive) scatterplot of x versus y from the diamonds data. Label the diamonds with suspiciously high y-values.

4.5.3   Missing data

Like many datasets, the mammal sleep data msleep contains a lot of missing values, represented by NA (Not Available) in R. This becomes evident when we have a look at the data:

```
library(ggplot2)
View(msleep)
```

We can check if a particular observation is missing using the is.na function:

```
is.na(msleep$sleep_rem[4])
is.na(msleep$sleep_rem)
```

We can count the number of missing values for each variable using:

```
colSums(is.na(msleep))
```

Here, `colSums` computes the sum of `is.na(msleep)` for each column of `msleep` (remember that in summation, `TRUE` counts as 1 and `FALSE` as 0), yielding the number of missing values for each variable. In total, there are 136 missing values in the dataset:

```
sum(is.na(msleep))
```

You'll notice that `ggplot2` prints a warning in the Console when you create a plot with missing data:

```
ggplot(msleep, aes(brainwt, sleep_total)) +
     geom_point() +
     scale_x_log10()
```

Sometimes, data are missing simply because the information is not yet available (for instance, the brain weight of the mountain beaver could be missing because no one has ever weighed the brain of a mountain beaver). In other cases, data can be missing because something about them is different (for instance, values for a male patient in a medical trial can be missing because the patient died, or because some values only were collected for female patients). Therefore, it is of interest to see if there are any differences in non-missing variables between subjects that have missing data and subjects that don't.

In `msleep`, all animals have recorded values for `sleep_total` and `bodywt`. To check if the animals that have missing `brainwt` values differ from the others, we can plot them in a different colour in a scatterplot:

```
ggplot(msleep, aes(bodywt, sleep_total, colour = is.na(brainwt))) +
     geom_point() +
     scale_x_log10()
```

(If `is.na(brainwt)` is `TRUE`, then the brain weight is missing in the dataset.) In this case, there are no apparent differences between the animals with missing data and those without.

~

Exercise 4.7. Create a version of the diamonds dataset where the x value is missing for all diamonds with $x > 9$. Make a scatterplot of carat versus price, in which points where the x value is missing are plotted in a different colour. How would you interpret this plot?

4.5.4 Exploring data

The `nycflights13` package contains data about flights to and from three airports in New York, USA, in 2013. As a summary exercise, we will study a subset of these, namely all flights departing from New York on 1 January of that year:

```
install.packages("nycflights13")
library(nycflights13)
flights2 <- flights[flights$month == 1 & flights$day == 1,]
```

~

Exercise 4.8. Explore the `flights2` dataset, focusing on delays and the amount of time spent in the air. Are there any differences between the different carriers? Are there missing data? Are there any outliers?

4.6 Trends in scatterplots

Let's return to a familiar example: the relationship between animal brain size and sleep times:

```
ggplot(msleep, aes(brainwt, sleep_total)) +
    geom_point() +
    labs(x = "Brain weight (logarithmic scale)",
         y = "Total sleep time") +
    scale_x_log10()
```

There appears to be a decreasing trend in the plot. To aid the eye, we can add a smoothed line by adding a new geom, `geom_smooth`, to the figure:

```
ggplot(msleep, aes(brainwt, sleep_total)) +
    geom_point() +
    geom_smooth() +
    labs(x = "Brain weight (logarithmic scale)",
         y = "Total sleep time") +
    scale_x_log10()
```

This technique is useful for bivariate data as well as for time series, which we'll delve into next.

By default, `geom_smooth` adds a line fitted using either LOESS[3] or GAM[4], as well as the corresponding 95% confidence interval describing the uncertainty in the estimate. There are several useful arguments that can be used with `geom_smooth`. You will explore some of these in the exercise below.

~

---

[3]LOESS, LOcally Estimated Scatterplot Smoothing, is a nonparametric regression method that fits a polynomial to local areas of the data.

[4]GAM, Generalised Additive Model, is a generalised linear model where the response variable is a linear function of smooth functions of the predictors.

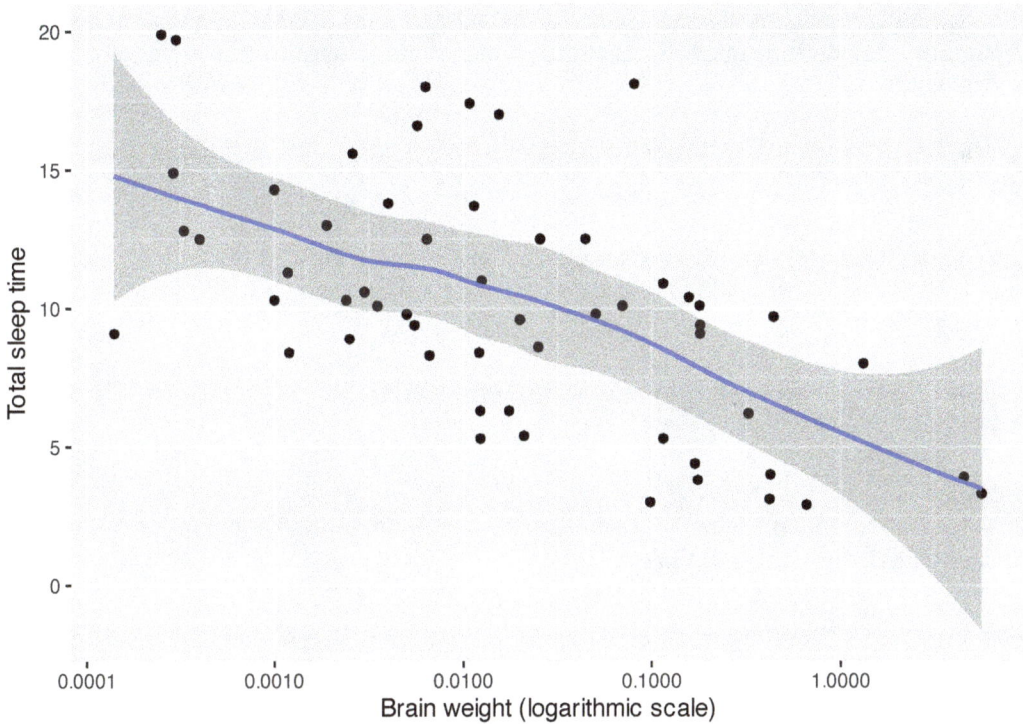

FIGURE 4.5 Scatterplot with a smooth trend curve.

Exercise 4.9. Check the documentation for `geom_smooth`. Starting with the plot of brain weight vs. sleep time created above, do the following:

1. Decrease the degree of smoothing for the LOESS line that was fitted to the data. What is better in this case, more or less smoothing?

2. Fit a straight line to the data instead of a non-linear LOESS line.

3. Remove the confidence interval from the plot.

4. Change the colour of the fitted line to red.

4.7 Exploring time series

Before we have a look at time series, you should install four useful packages: `forecast`, `nlme`, `fma` and `fpp2`. The first contains useful functions for plotting time series data, and the latter three contain datasets that we'll use.

```
install.packages(c("nlme", "forecast", "fma", "fpp2"),
                 dependencies = TRUE)
```

The `a10` dataset contains information about the monthly anti-diabetic drug sales in Australia from July 1991 to June 2008. By checking its structure, we see that it is saved as a time series object[5]:

```
library(fpp2)
str(a10)
```

ggplot2 requires that data is saved as a data frame in order for it to be plotted. In order to plot the time series, we could first convert it to a data frame and then plot each point using geom_points:

```
a10_df <- data.frame(time = time(a10), sales = a10)
ggplot(a10_df, aes(time, sales)) +
    geom_point()
```

It is however usually preferable to plot time series using lines instead of points. This is done using a different geom: `geom_line`:

```
ggplot(a10_df, aes(time, sales)) +
    geom_line()
```

Having to convert the time series object to a data frame is a little awkward. Luckily, there is a way around this. ggplot2 offers a function called `autoplot` that automatically draws an appropriate plot for certain types of data. `forecast` extends this function to time series objects:

```
library(forecast)
autoplot(a10)
```

We can still add other geoms, axis labels, and other things just as before. `autoplot` has simply replaced the `ggplot(data, aes()) + geom` part that would be the first two rows of the ggplot2 figure and has implicitly converted the data to a data frame.

~

Exercise 4.10. Using the `autoplot(a10)` figure, do the following:

1. Add a smoothed line describing the trend in the data. Make sure that it is smooth enough *not* to capture the seasonal variation in the data.

2. Change the label of the x-axis to "Year" and the label of the y-axis to "Sales ($ million)".

3. Check the documentation for the `ggtitle` function. What does it do? Use it with the figure.

4. Change the colour of the time series line to red.

---

[5]Time series objects are a special class of vectors, with data sampled at equispaced points in time. Each observation can have a year/date/time associated with it.

4.7.1 Annotations and reference lines

We sometimes wish to add text or symbols to plots, for instance to highlight interesting observations. Consider the following time series plot of daily morning gold prices, based on the `gold` data from the `forecast` package:

```
library(forecast)
autoplot(gold)
```

There is a sharp spike a few weeks before day 800, which is due to an incorrect value in the data series. We'd like to add a note about that to the plot. First, we wish to find out on which day the spike appears. This can be done by checking the data manually or using some code:

```
spike_date <- which.max(gold)
```

To add a circle around that point, we add a call to `annotate` to the plot:

```
autoplot(gold) +
      annotate(geom = "point", x = spike_date, y = gold[spike_date],
               size = 5, shape = 21,
               colour = "red",
               fill = "transparent")
```

`annotate` can be used to annotate the plot with both geometrical objects and text (and can therefore be used as an alternative to `geom_text`).

~

Exercise 4.11. Using the figure created above and the documentation for `annotate`, do the following:

1. Add the text "Incorrect value" next to the circle.

2. Create a second plot where the incorrect value has been removed.

3. Read the documentation for the geom `geom_hline`. Use it to add a red reference line to the plot, at $y = 400$.

4.7.2 Longitudinal data

Multiple time series with identical time points, known as longitudinal data or panel data, are common in many fields. One example of this is given by the `elecdaily` time series from the `fpp2` package, which contains information about electricity demand in Victoria, Australia during 2014. As with a single time series, we can plot these data using `autoplot`:

```
library(fpp2)
autoplot(elecdaily)
```

In this case, it is probably a good idea to facet the data, i.e., to plot each series in a different figure:

```
autoplot(elecdaily, facets = TRUE)
```

~

Exercise 4.12. Make the following changes to the `autoplot(elecdaily, facets = TRUE)`:

1. Remove the `WorkDay` variable from the plot (it describes whether or not a given date is a workday, and while it is useful for modelling purposes, we do not wish to include it in our figure).

2. Add smoothed trend lines to the time series plots.

4.7.3   Path plots

Another option for plotting multiple time series is path plots. A path plot is a scatterplot where the points are connected with lines in the order they appear in the data (which, for time series data, should correspond to time). The lines and points can be coloured to represent time.

To make a path plot of Temperature versus Demand for the `elecdaily` data, we first convert the time series object to a data frame and create a scatterplot:

```
library(fpp2)
ggplot(as.data.frame(elecdaily), aes(Temperature, Demand)) +
    geom_point()
```

Next, we connect the points by lines using the `geom_path` geom:

```
ggplot(as.data.frame(elecdaily), aes(Temperature, Demand)) +
    geom_point() +
    geom_path()
```

The resulting figure is quite messy. Using colour to indicate the passing of time helps a little. For this, we need to add the day of the year to the data frame. To get the values right, we use `nrow`, which gives us the number of rows in the data frame.

```
elecdaily2 <- as.data.frame(elecdaily)
elecdaily2$day <- 1:nrow(elecdaily2)

ggplot(elecdaily2, aes(Temperature, Demand, colour = day)) +
    geom_point() +
    geom_path()
```

It becomes clear from the plot that temperatures were the highest at the beginning of the year and lower in the winter months (July-August).

~

Exercise 4.13. Make the following changes to the plot you created above:

1. Decrease the size of the points.

2. Add text annotations showing the dates of the highest and lowest temperatures, next to the corresponding points in the figure.

4.7.4 Spaghetti plots

In cases where we've observed multiple subjects over time, we often wish to visualise their individual time series together using so-called spaghetti plots. With `ggplot2` this is done using the `geom_line` geom. To illustrate this, we use the `Oxboys` data from the `nlme` package, showing the heights of 26 boys over time.

```
library(nlme)
ggplot(Oxboys, aes(age, height, group = Subject)) +
      geom_point() +
      geom_line()
```

The first two `aes` arguments specify the x- and y-axes, and the third specifies that there should be one line per subject (i.e., per boy) rather than a single line interpolating all points. The latter would be a rather useless figure that looks like this:

```
ggplot(Oxboys, aes(age, height)) +
      geom_point() +
      geom_line() +
      ggtitle("A terrible plot")
```

Returning to the original plot, if we wish to be able to identify which time series corresponds to which boy, we can add a colour aesthetic:

```
ggplot(Oxboys, aes(age, height, group = Subject, colour = Subject)) +
      geom_point() +
      geom_line()
```

Note that the boys are ordered by height, rather than subject number, in the legend.

Now, imagine that we wish to add a trend line describing the general growth trend for all boys. The growth appears approximately linear, so it seems sensible to use `geom_smooth(method = "lm")` to add the trend:

```
ggplot(Oxboys, aes(age, height, group = Subject, colour = Subject)) +
      geom_point() +
      geom_line() +
      geom_smooth(method = "lm", colour = "red", se = FALSE)
```

Unfortunately, because we have specified in the aesthetics that the data should be grouped by `Subject`, `geom_smooth` produces one trend line for each boy. The "problem" is that when we specify an aesthetic in the `ggplot` call, it is used for all geoms.

~

Exercise 4.14. Figure out how to produce a spaghetti plot of the Oxboys data with a single red trend line based on the data from all 26 boys.

4.7.5   Seasonal plots and decompositions

The forecast package includes a number of useful functions when working with time series. One of them is ggseasonplot, which allows us to easily create a spaghetti plot of different periods of a time series with seasonality, i.e., with patterns that repeat seasonally over time. It works similar to the autoplot function, in that it replaces the ggplot(data, aes) + geom part of the code.

```
library(forecast)
library(fpp2)
ggseasonplot(a10)
```

This function is very useful when visually inspecting seasonal patterns.

The year.labels and year.labels.left arguments remove the legend in favour of putting the years at the end and beginning of the lines:

```
ggseasonplot(a10, year.labels = TRUE, year.labels.left = TRUE)
```

As always, we can add more things to our plot if we like:

```
ggseasonplot(a10, year.labels = TRUE, year.labels.left = TRUE) +
    labs(y = "Sales ($ million)") +
    ggtitle("Seasonal plot of anti-diabetic drug sales")
```

When working with seasonal time series, it is common to decompose the series into a seasonal component, a trend component, and a remainder. In R, this is typically done using the stl function, which uses repeated LOESS smoothing to decompose the series. There is an autoplot function for stl objects:

```
autoplot(stl(a10, s.window = 365))
```

This plot can too be manipulated in the same way as other ggplot objects. You can access the different parts of the decomposition as follows:

```
stl(a10, s.window = 365)$time.series[,"seasonal"]
stl(a10, s.window = 365)$time.series[,"trend"]
stl(a10, s.window = 365)$time.series[,"remainder"]
```

~

Exercise 4.15. Investigate the writing dataset from the fma package graphically. Make a time series plot with a smoothed trend line, a seasonal plot and an stl-decomposition plot. Add appropriate plot titles and labels to the axes. Can you see any interesting patterns?

4.7.6 Detecting changepoints

The `changepoint` package contains a number of methods for detecting *changepoints* in time series, i.e., time points at which either the mean or the variance of the series changes. Finding changepoints can be important for detecting changes in the process underlying the time series. The `ggfortify` package extends `ggplot2` by adding `autoplot` functions for a variety of tools, including those in `changepoint`. Let's install the packages:

```
install.packages(c("changepoint", "ggfortify"))
```

We can now look at some examples with the anti-diabetic drug sales data:

```
library(forecast)
library(fpp2)
library(changepoint)
library(ggfortify)

# Plot the time series:
autoplot(a10)

# Remove the seasonal part and plot the series again:
a10_ns <- a10 - stl(a10, s.window = 365)$time.series[,"seasonal"]
autoplot(a10_ns)

# Plot points where there are changes in the mean:
autoplot(cpt.mean(a10_ns))

# Choosing a different method for finding changepoints
# changes the result:
autoplot(cpt.mean(a10_ns, method = "BinSeg"))

# Plot points where there are changes in the variance:
autoplot(cpt.var(a10_ns))

# Plot points where there are changes in either the mean or
# the variance:
autoplot(cpt.meanvar(a10_ns))
```

As you can see, the different methods from `changepoint` all yield different results. The results for changes in the mean are a bit dubious – which isn't all that strange as we are using a method that looks for jumps in the mean on a time series where the increase actually is more or less continuous. The changepoint for the variance looks more reliable – there is a clear change toward the end of the series where the sales become more volatile. We won't go into details about the different methods here but mention that the documentation at `?cpt.mean`, `?cpt.var`, and `?cpt.meanvar` contains descriptions of and references for the available methods.

~

Exercise 4.16. Are there any changepoints for variance in the `Demand` time series in `elecdaily`? Can you explain why the behaviour of the series changes?

4.7.7   Interactive time series plots

The `plotly` packages can be used to create interactive time series plots. As before, you create a `ggplot2` object as usual, assigning it to a variable and then call the `ggplotly` function. Here is an example with the `elecdaily` data:

```
library(plotly)
library(fpp2)
myPlot <- autoplot(elecdaily[,"Demand"])

ggplotly(myPlot)
```

When you hover the mouse pointer over a point, a box appears, displaying information about that data point. Unfortunately, the date formatting isn't great in this example – dates are shown as weeks with decimal points. We'll see how to fix this in Section 5.6.

4.8   Using polar coordinates

Most plots are made using *Cartesian coordinate systems*, in which the axes are orthogonal to each other and values are placed in an even spacing along each axis. In some cases, non-linear axes (e.g., log-transformed) are used instead, as we have already seen.

Another option is to use a *polar* coordinate system, in which positions are specified using an angle and a (radial) distance from the origin. Here is an example of a polar scatterplot, where `sleep_rem` is represented by the angle and `sleep_total` by the radial distance:

```
ggplot(msleep, aes(sleep_rem, sleep_total, colour = vore)) +
    geom_point() +
    labs(x = "REM sleep (circular axis)",
        y = "Total sleep time (radial axis)") +
    coord_polar()
```

4.8.1   Visualising periodic data

Polar coordinates are particularly useful when the data is periodic. Consider for instance the following dataset, describing monthly weather averages for Cape Town, South Africa:

```
Cape_Town_weather <- data.frame(
Month = 1:12,
Temp_C = c(22, 23, 21, 18, 16, 13, 13, 13, 14, 16, 18, 20),
Rain_mm = c(20, 20, 30, 50, 70, 90, 100, 70, 50, 40, 20, 20),
Sun_h = c(11, 10, 9, 7, 6, 6, 5, 6, 7, 9, 10, 11))
```

We can visualise the monthly average temperature using lines in a Cartesian coordinate system:

```
ggplot(Cape_Town_weather, aes(Month, Temp_C)) +
    geom_line()
```

What this plot doesn't show is that the 12th month and the 1st month actually are consecutive months. If we instead use polar coordinates, this becomes clearer:

```
ggplot(Cape_Town_weather, aes(Month, Temp_C)) +
    geom_line() +
    coord_polar()
```

To improve the presentation, we can change the scale of the x-axis (i.e., the circular axis) so that January and December aren't plotted at the same angle:

```
ggplot(Cape_Town_weather, aes(Month, Temp_C)) +
    geom_line() +
    coord_polar() +
    xlim(0, 12)
```

$\sim$

Exercise 4.17. In the plot that we just created, the last and first month of the year aren't connected. You can fix manually this by adding a cleverly designed faux data point to `Cape_Town_weather`. How?

4.8.2 Pie charts

Consider the stacked bar chart that we plotted in Section 2.8:

```
ggplot(msleep, aes(factor(1), fill = vore)) +
    geom_bar()
```

What would happen if we plotted this figure in a polar coordinate system instead? If we map the height of the bars (the y-axis of the Cartesian coordinate system) to both the angle and the radial distance, we end up with a pie chart:

```
ggplot(msleep, aes(factor(1), fill = vore)) +
    geom_bar() +
    coord_polar(theta = "y")
```

There are many arguments against using pie charts for visualisations. Most boil down to the fact that the same information is easier to interpret when conveyed as a bar chart. This is at least partially due to the fact that most people are more used to reading plots in Cartesian coordinates than in polar coordinates.

If we make a similar transformation of a grouped bar chart, we get a different type of pie chart, in which the height of the bars is mapped to both the angle and the radial distance[6]:

---

[6]Florence Nightingale, who famously pioneered the use of the pie chart, drew her pie charts this way.

```
# Cartesian bar chart:
ggplot(msleep, aes(vore, fill = vore)) +
    geom_bar() +
    theme(legend.position = "none")

# Polar bar chart:
ggplot(msleep, aes(vore, fill = vore)) +
    geom_bar() +
    coord_polar() +
    theme(legend.position = "none")
```

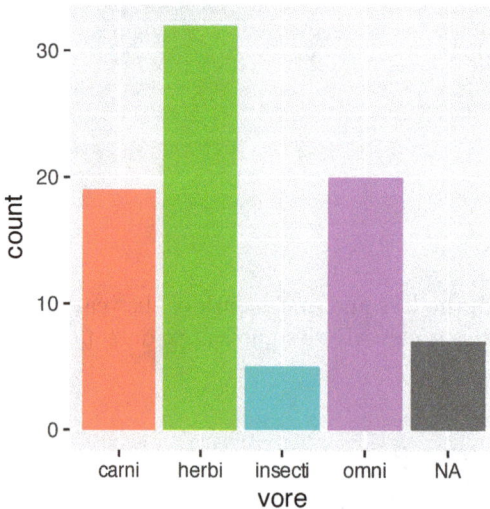

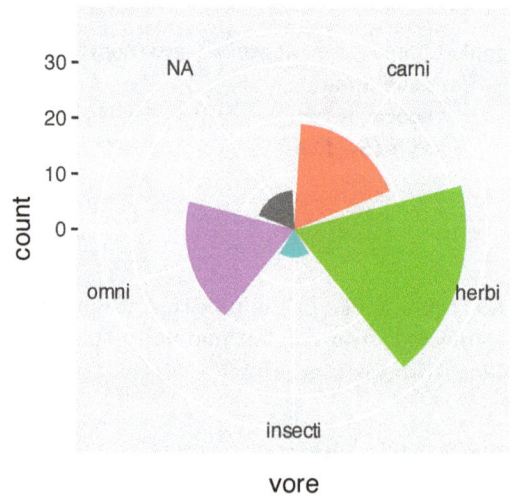

FIGURE 4.6 Bar charts vs. pie charts.

4.9 Visualising multiple variables

4.9.1 Scatterplot matrices

When we have a large enough number of numeric variables in our data, plotting scatterplots of all pairs of variables becomes tedious. Luckily there are some R functions that speed up this process.

The GGally package is an extension to ggplot2 which contains several functions for plotting multivariate data. They work similarly to the autoplot functions that we have used in previous sections. One of these is ggpairs, which creates a scatterplot matrix, a grid with scatterplots of all pairs of variables in data. In addition, it also plots density estimates (along the diagonal) and shows the (Pearson) correlation for each pair. Let's start by installing GGally:

```
install.packages("GGally")
```

To create a scatterplot matrix for the `airquality` dataset, simply write:

```
library(GGally)
ggpairs(airquality)
```

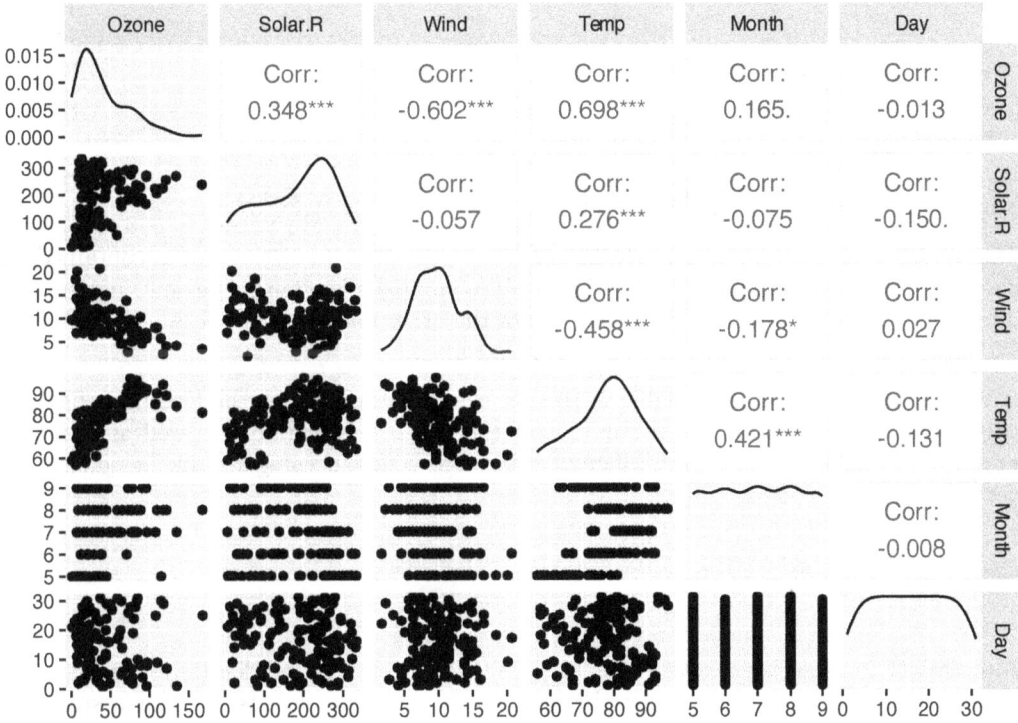

FIGURE 4.7 Scatterplot matrix for the airquality data.

(Enlarging your Plot window can make the figure look better.)

If we want to create a scatterplot matrix but only want to include some of the variables in a dataset, we can do so by providing a vector with variable names. Here is an example for the animal sleep data `msleep`.

Without pipes:

```
ggpairs(msleep[, c("sleep_total", "sleep_rem", "sleep_cycle", "awake",
                   "brainwt", "bodywt")])
```

With pipes:

```
library(dplyr)
msleep |>
  select(sleep_total, sleep_rem, sleep_cycle, awake, brainwt, bodywt) |>
  ggpairs()
```

Optionally, if we wish to create a scatterplot involving all `numeric` variables, we can replace the vector with variable names with some R code that extracts the columns containing `numeric` variables:

Without pipes:

```
ggpairs(msleep[, which(sapply(msleep, class) == "numeric")])
```

With pipes:

```
msleep |>
  select(where(is.numeric)) |>
  ggpairs()
```

You'll learn more about the `sapply` function in Section 6.5.

The resulting plot is identical to the previous one, because the list of names contained all `numeric` variables. The grab-all-numeric-variables approach is often convenient, because we don't have to write all the variable names. On the other hand, it's not very helpful in case we only want to include some of the `numeric` variables.

If we include a categorical variable in the list of variables (such as the feeding behaviour `vore`), the matrix will include a bar plot of the categorical variable as well as boxplots and facetted histograms to show differences between different categories in the continuous variables:

```
ggpairs(msleep[, c("vore", "sleep_total", "sleep_rem", "sleep_cycle",
                  '"awake", "brainwt", "bodywt")])
```

Alternatively, we can use a categorical variable to colour points and density estimates using `aes(colour = ...)`. The syntax for this follows the same pattern as that for a standard ggplot call - `ggpairs(data, aes)`. The only exception is that if the categorical variable is not included in the `data` argument, we must specify which data frame it belongs to:

```
ggpairs(msleep[, c("sleep_total", "sleep_rem", "sleep_cycle", "awake",
                  "brainwt", "bodywt")],
      aes(colour = msleep$vore, alpha = 0.5))
```

As a side note, if all variables in your data frame are `numeric`, and if you only are looking for a quick-and-dirty scatterplot matrix without density estimates and correlations, you can also use the base R `plot`:

```
plot(airquality)
```

~

Exercise 4.18. Create a scatterplot matrix for all `numeric` variables in `diamonds`. Differentiate different cuts by colour. Add a suitable title to the plot. (`diamonds` is a fairly large dataset, and it may take a minute or so for R to create the plot.)

4.9.2   3D scatterplots

The `plotly` package lets us make three-dimensional scatterplots with the `plot_ly` function, which can be a useful alternative to scatterplot matrices in some cases. Here is an example using the `airquality` data:

```
library(plotly)
plot_ly(airquality, x = ~Ozone, y = ~Wind, z = ~Temp,
        color = ~factor(Month))
```

Note that you can drag and rotate the plot, to see it from different angles.

4.9.3   Correlograms

Scatterplot matrices are not a good choice when we have too many variables, partially because the plot window needs to be very large to fit all variables and partially because it becomes difficult to get a good overview of the data. In such cases, a correlogram, where the strength of the correlation between each pair of variables is plotted instead of scatterplots, can be used instead. It is effectively a visualisation of the correlation matrix of the data, where the strengths and signs of the correlations are represented by different colours.

The `GGally` package contains the function `ggcorr`, which can be used to create a correlogram:

```
ggcorr(msleep[, c("sleep_total", "sleep_rem", "sleep_cycle", "awake",
                  "brainwt", "bodywt")])
```

~

Exercise 4.19. Using the `diamonds` dataset and the documentation for `ggcorr`, do the following:

1. Create a correlogram for all `numeric` variables in the dataset.

2. The Pearson correlation that `ggcorr` uses by default isn't always the best choice. A commonly used alternative is the Spearman correlation. Change the type of correlation used to create the plot to the Spearman correlation.

3. Change the colour scale from a categorical scale with five categories.

4. Change the colours on the scale to go from yellow (low correlation) to black (high correlation).

4.9.4   Adding more variables to scatterplots

We have already seen how scatterplots can be used to visualise two continuous and one categorical variable by plotting the two continuous variables against each other and using the categorical variable to set the colours of the points. There are, however, more ways we can incorporate information about additional variables into a scatterplot.

So far, we have set three aesthetics in our scatterplots: x, y, and `colour`. Two other important aesthetics are `shape` and `size`, which, as you'd expect, allow us to control the shape and size

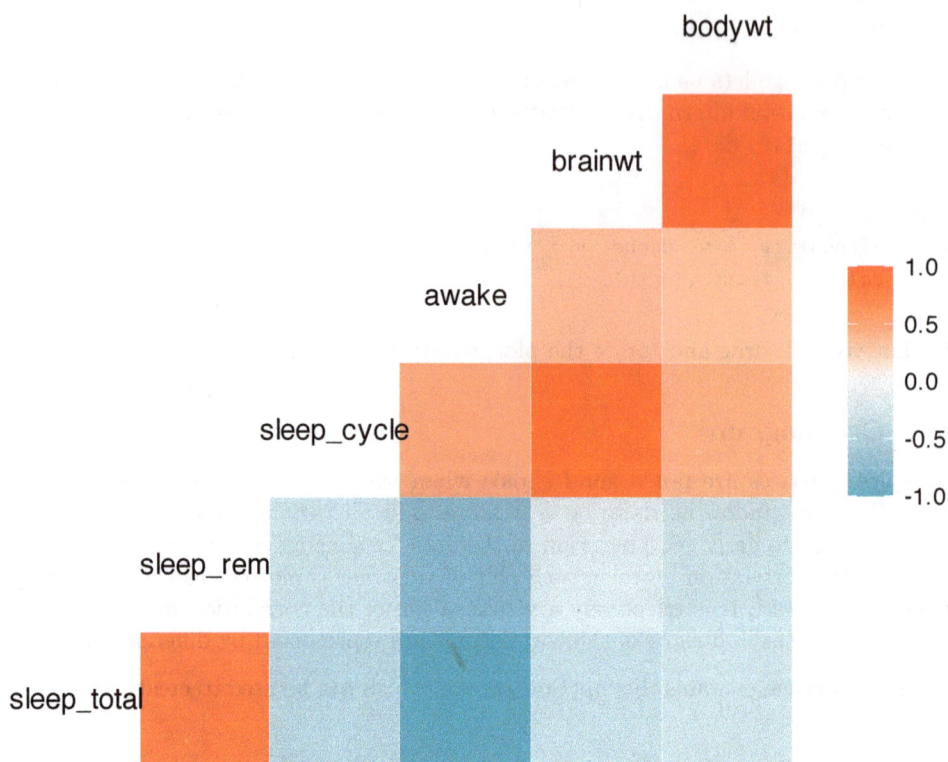

FIGURE 4.8 Correlogram for variables in the mammal sleep data.

of the points. As a first example using the `msleep` data, we use feeding behaviour (`vore`) to set the shapes used for the points:

```
ggplot(msleep, aes(brainwt, sleep_total, shape = vore)) +
    geom_point() +
    scale_x_log10()
```

The plot looks a little nicer if we increase the `size` of the points:

```
ggplot(msleep, aes(brainwt, sleep_total, shape = vore, size = 2)) +
    geom_point() +
    scale_x_log10()
```

Another option is to let `size` represent a continuous variable, in what is known as a bubble plot:

```
ggplot(msleep, aes(brainwt, sleep_total, colour = vore,
                   size = bodywt)) +
    geom_point() +
    scale_x_log10()
```

The size of each "bubble" now represents the weight of the animal. Because some animals are much heavier (i.e., have higher `bodywt` values) than most others, almost all points are quite small. There are a couple of things we can do to remedy this. First, we can transform `bodywt`, e.g., using the square root transformation `sqrt(bodywt)`, to decrease the differences between large and small animals. This can be done by adding `scale_size(trans = "sqrt")` to the plot. Second, we can also use `scale_size` to control the range of point sizes (e.g., from size 1 to size 20). This will cause some points to overlap, so we add `alpha = 0.5` to the geom, to make the points transparent:

```
ggplot(msleep, aes(brainwt, sleep_total, colour = vore,
                   size = bodywt)) +
      geom_point(alpha = 0.5) +
      scale_x_log10() +
      scale_size(range = c(1, 20), trans = "sqrt")
```

This produces a fairly nice-looking plot, but it'd look even better if we changed the axes labels and legend texts. We can change the legend text for the size scale by adding the argument `name` to `labs`. Including a `\n` in the text lets us create a line break – you'll learn more tricks like that in Section 5.5.

```
ggplot(msleep, aes(brainwt, sleep_total, colour = vore,
                   size = bodywt)) +
      geom_point(alpha = 0.5) +
      labs(x = "Brain weight (logarithmic scale)",
           y = "Total sleep time",
           size = "Square root of\nbody weight",
           colour = "Feeding behaviour") +
      scale_x_log10() +
      scale_size(range = c(1, 20), trans = "sqrt")
```

∼

Exercise 4.20. Using the bubble plot created above, do the following:

    1. Replace `colour = vore` in the `aes` by `fill = vore` and add `colour = "black"`, `shape = 21` to `geom_point`. What happens?

    2. Use `ggplotly` to create an interactive version of the bubble plot above, where variable information and the animal name are displayed when you hover a point.

4.9.5 Overplotting

Let's make a scatterplot of `table` versus `depth` based on the `diamonds` dataset:

```
ggplot(diamonds, aes(table, depth)) +
      geom_point()
```

This plot is cluttered. There are too many points, which makes it difficult to see if, for instance, high `table` values are more common than low `table` values. In this section, we'll look at some ways to deal with this problem, known as overplotting.

The first thing we can try is to decrease the point size:

```
ggplot(diamonds, aes(table, depth)) +
    geom_point(size = 0.1)
```

This helps a little, but now the outliers become a bit difficult to spot. We can try changing the opacity using `alpha` instead:

```
ggplot(diamonds, aes(table, depth)) +
    geom_point(alpha = 0.2)
```

This is also better than the original plot, but neither plot is great. Instead of plotting each individual point, maybe we can try plotting the counts or densities in different regions of the plot instead? Effectively, this would be a two-dimensional version of a histogram. There are several ways of doing this in `ggplot2`.

First, we bin the points and count the numbers in each bin, using `geom_bin2d`:

```
ggplot(diamonds, aes(table, depth)) +
    geom_bin2d()
```

By default, `geom_bin2d` uses 30 bins. Increasing that number can sometimes give us a better idea about the distribution of the data:

```
ggplot(diamonds, aes(table, depth)) +
    geom_bin2d(bins = 50)
```

If you prefer, you can get a similar plot with hexagonal bins by using `geom_hex` instead:

```
ggplot(diamonds, aes(table, depth)) +
    geom_hex(bins = 50)
```

As an alternative to bin counts, we could create a two-dimensional density estimate and create a contour plot showing the levels of the density:

```
ggplot(diamonds, aes(table, depth)) +
    stat_density_2d(aes(fill = ..level..), geom = "polygon",
                    colour = "white")
```

The `fill = ..level..` bit above probably looks a little strange to you. It means that an internal function (the level of the contours) is used to choose the fill colours. It also means that we've reached a point where we're reaching deep into the depths of `ggplot2`!

We can use a similar approach to show a summary statistic for a third variable in a plot. For instance, we may want to plot the average price as a function of `table` and `depth`. This is called a tile plot:

```
ggplot(diamonds, aes(table, depth, z = price)) +
    geom_tile(binwidth = 1, stat = "summary_2d", fun = mean) +
    ggtitle("Mean prices for diamonds with different depths and
            tables")
```

~

Exercise 4.21. The following tasks involve the `diamonds` dataset:

1. Create a tile plot of `table` versus `depth`, showing the highest price for a diamond in each bin.

2. Create a bin plot of `carat` versus `price`. What type of diamonds have the highest bin counts?

4.9.6 Categorical data

When visualising a pair of categorical variables, plots similar to those in the previous section prove to be useful. One way of doing this is to use the `geom_count` geom. We illustrate this with an example using `diamonds`, showing how common different combinations of colours and cuts are:

```
ggplot(diamonds, aes(color, cut)) +
    geom_count()
```

However, it is often better to use colour rather than point size to visualise counts, which we can do using a tile plot. First, we have to compute the counts though, using `aggregate`. We now wish to have two grouping variables, `color` and `cut`, which we can put on the right-hand side of the formula as follows:

```
diamonds2 <- aggregate(carat ~ cut + color, data = diamonds,
                       FUN = length)
diamonds2
```

`diamonds2` is now a data frame containing the different combinations of `color` and `cut` along with counts of how many diamonds belong to each combination (labelled `carat`, because we put `carat` in our formula). Let's change the name of the last column from `carat` to `Count`:

```
names(diamonds2)[3] <- "Count"
```

Next, we can plot the counts using `geom_tile`:

```
ggplot(diamonds2, aes(color, cut, fill = Count)) +
    geom_tile()
```

It is also possible to combine point size and colours:

```
ggplot(diamonds2, aes(color, cut, colour = Count, size = Count)) +
    geom_count()
```

~

Exercise 4.22. Using the `diamonds` dataset, do the following:

1. Use a plot to find out what the most common combination of cut and clarity is.

2. Use a plot to find out which combination of cut and clarity has the highest average price.

4.9.7 Putting it all together

In the next two exercises, you will repeat what you have learned so far by investigating the `gapminder` and `planes` datasets. First, load the corresponding libraries and have a look at the documentation for each dataset:

```
install.packages("gapminder")
library(gapminder)
?gapminder

library(nycflights13)
?planes
```

~

Exercise 4.23. Do the following using the `gapminder` dataset:

1. Create a scatterplot matrix showing life expectancy, population, and GDP per capita for all countries, using the data from the year 2007. Use colours to differentiate countries from different continents. Note: you'll probably need to add the argument `upper = list(continuous = "na")` when creating the scatterplot matrix. By default, correlations are shown above the diagonal, but the fact that there only are two countries from Oceania will cause a problem there – at least three points are needed for a correlation test.

2. Create an interactive bubble plot, showing information about each country when you hover the points. Use data from the year 2007. Put log(GDP per capita) on the x-axis and life expectancy on the y-axis. Let population determine point size. Plot each country in a different colour and facet by continent. Tip: the `gapminder` package provides a pretty colour scheme for different countries, called `country_colors`. You can use that scheme by adding `scale_colour_manual(values = country_colors)` to your plot.

Exercise 4.24. Use graphics to answer the following questions regarding the `planes` dataset:

1. What is the most common combination of manufacturer and plane type in the dataset?

2. Which combination of manufacturer and plane type has the highest average number of seats?

3. Do the numbers of seats on planes change over time? Which plane had the highest number of seats?

4. Does the type of engine used change over time?

4.10 Sankey diagrams

A Sankey diagram is a type of flow diagram used to show flows from one state to another. We'll consider an example with data from a medical trial. Patients were recruited from four hospitals and assigned to one of two treatments: either surgery or physiotherapy. At the end of the study, they had either recovered or not. The data is in the `surgphys.csv` file, which can be downloaded from the book's web page[7].

Set `file_path` to the path of `surgphys.csv` to load the data:

```
surgphys <- read.csv(file_path)
View(surgphys)
```

We'll use the `ggsankey` package for creating the diagram, so let's install that:

```
install.packages("remotes")
remotes::install_github("davidsjoberg/ggsankey")
```

First, we need to reformat the data by pointing out the order of the different "states". In this case, patients are first recruited at a hospital, then assigned a treatment, and then an outcome is observed. The `make_longer` function from `ggsankey` formats the data to reflect this:

```
library(ggsankey)
surgphys |> make_long(Hospital, Treatment, Outcome) -> surgphys_sankey
```

Next, we create a ggplot using the variables we just created, and add the `geom_sankey` geom:

```
ggplot(surgphys_sankey, aes(x = x,
                        next_x = next_x,
                        node = node,
                        next_node = next_node,
                        fill = factor(node),
```

---

[7]http://www.modernstatisticswithr.com/data.zip

```
                              label = node)) +
    geom_sankey()
```

To make this a little prettier, we can change some settings for `geom_sankey`, add labels with `geom_sankey_label`, and change the theme settings and colour palette:

```
ggplot(surgphys_sankey, aes(x = x,
                            next_x = next_x,
                            node = node,
                            next_node = next_node,
                            fill = factor(node),
                            label = node)) +
    geom_sankey(flow.alpha = 0.5,
                node.color = "black",
                show.legend = FALSE) +
    geom_sankey_label(size = 3,
                      colour = "black",
                      fill = "white",
                      hjust = -0.3) +
    theme_minimal() +
    theme(axis.title = element_blank(),
          axis.text.y = element_blank(),
          axis.ticks = element_blank(),
          panel.grid = element_blank()) +
    scale_fill_manual(values = c("darkorange", "skyblue",
                                 "forestgreen", "#00AFBB",
                                 "#E7B800", "#FC4E07",
                                 "deeppink", "purple"))
```

If we want to show the number of patients in each state, we can add counts to the data, and then add them to the node labels as follows (see Section 5.12 for an explanation of what `right_join` does):

```
library(dplyr)
surgphys_sankey |> group_by(node) |>
                   count() |>
                   right_join(surgphys_sankey, by = "node") -> surgphys_sankey

ggplot(surgphys_sankey, aes(x = x,
                            next_x = next_x,
                            node = node,
                            next_node = next_node,
                            fill = factor(node),
                            label = paste0(node," (n=", n, ")"))) +
    geom_sankey(flow.alpha = 0.5,
                node.color = "black",
                show.legend = FALSE) +
    geom_sankey_label(size = 3,
                      colour = "black",
```

```
                              fill = "white",
                              hjust = -0.15) +
theme_minimal() +
theme(axis.title = element_blank(),
      axis.text.y = element_blank(),
      axis.ticks = element_blank(),
      panel.grid = element_blank()) +
scale_fill_manual(values = c("darkorange", "skyblue",
                             "forestgreen", "#00AFBB",
                             "#E7B800", "#FC4E07",
                             "deeppink", "purple"))
```

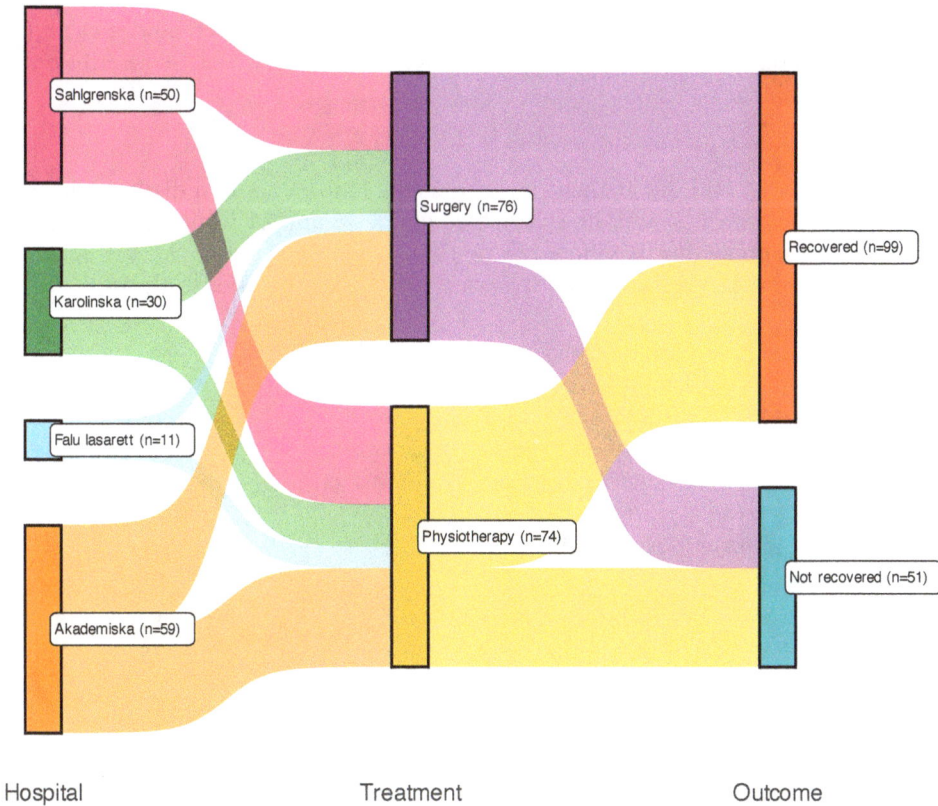

FIGURE 4.9 Sankey diagram comparing surgery and physiotherapy.

4.11 Principal component analysis

If there are many variables in your data, it can often be difficult to detect differences between groups or create a perspicuous visualisation. A useful tool in this context is *principal component analysis* (PCA), which can reduce high-dimensional data to a lower number of variables that can be visualised in one or two scatterplots. The idea is to compute new variables, called *principal components*, that are linear combinations of the original variables[8]. These are constructed with two goals in mind: the principal components should capture as much of the variance in the data as possible, and each principal component should be uncorrelated to the other components. You can then plot the principal components to get a low-dimensional representation of your data, which hopefully captures most of its variation.

By design, the number of principal components computed are as many as the original number of variables, with the first having the largest variance, the second having the second largest variance, and so on. We hope that it will suffice to use just the first few of these to represent most of the variation in the data, but this is not guaranteed. Principal component analysis is more likely to yield a useful result if several variables are correlated.

4.11.1   Running a principal component analysis

To illustrate the principles of PCA, we will use a dataset from Charytanowicz et al. (2010), containing measurements on wheat kernels for three varieties of wheat. A description of the variables is available at:

http://archive.ics.uci.edu/ml/datasets/seeds

We are interested to find out if these measurements can be used to distinguish between the varieties. The data is stored in a `.txt` file, which we import using `read.table` (which works just like `read.csv`, but is tailored to text files) and convert the `Variety` column to a categorical `factor` variable (which you'll learn more about in Section 5.4):

```
# The data is downloaded from the UCI Machine Learning Repository:
# http://archive.ics.uci.edu/ml/datasets/seeds
seeds <- read.table("https://tinyurl.com/seedsdata",
        col.names = c("Area", "Perimeter", "Compactness",
        "Kernel_length", "Kernel_width", "Asymmetry",
        "Groove_length", "Variety"))
seeds$Variety <- factor(seeds$Variety)
```

If we make a scatterplot matrix of all variables, it becomes evident that there are differences between the varieties, but that no single pair of variables is enough to separate them:

```
library(ggplot2)
library(GGally)
ggpairs(seeds[, -8], aes(colour = seeds$Variety, alpha = 0.2))
```

Moreover, for presentation purposes, the amount of information in the scatterplot matrix is a bit overwhelming. It would be nice to be able to present the data in a single scatterplot, without losing too much information. We'll therefore compute the principal components using the `prcomp` function. It is usually recommended that PCA is performed using standardised data, i.e., using data that has been scaled to have mean 0 and standard deviation 1. The reason for this is that it puts all variables on the same scale. If we don't standardise our data, then variables with a high variance will completely dominate the principal components. This isn't desirable, as variance is affected by the scale of the measurements, meaning that the choice of measurement scale would influence the results (as an example, the variance of

---

[8]A linear combination is a weighted sum of the form $a_1x_1 + a_2x_2 + \cdots + a_px_p$. If you like, you can think of principal components as weighted averages of variables, computed for each row in your data.

kernel length will be a million times greater if lengths are measured in millimetres instead of in metres).

We don't have to standardise the data ourselves, but can let `prcomp` do that for us using the arguments `center = TRUE` (to get mean 0) and `scale. = TRUE` (to get standard deviation 1):

```
# Compute principal components:
pca <- prcomp(seeds[,-8], center = TRUE, scale. = TRUE)
```

To see the *loadings* of the components, i.e., how much each variable contributes to the components, simply type the name of the object `prcomp` created:

```
pca
```

The first principal component is more or less a weighted average of all variables but has stronger weights on `Area`, `Perimeter`, `Kernel_length`, `Kernel_width`, and `Groove_length`, all of which are measures of size. We can therefore interpret it as a size variable. The second component has higher loadings for `Compactness` and `Asymmetry`, meaning that it mainly measures those shape features. In Exercise 4.26 you'll see how the loadings can be visualised in a *biplot*.

4.11.2 Choosing the number of components

To see how much of the variance each component represents, use `summary`:

```
summary(pca)
```

The first principal component accounts for 71.87% of the variance, and the first three combined account for 98.67%.

To visualise this, we can draw a *scree plot*, which shows the variance of each principal component – the total variance of the data is the sum of the variances of the principal components:

```
screeplot(pca, type = "lines")
```

We can use this to choose how many principal components to use when visualising or summarising our data. In that case, we look for an "elbow", i.e., a bend in the curve after which increasing the number of components doesn't increase the amount of variance explained much. In this case, we see an "elbow" somewhere between two and four components.

4.11.3 Plotting the results

We can access the values of the principal components using `pca$x`. Let's check that the first two components really are uncorrelated:

```
cor(pca$x[,1], pca$x[,2])
```

In this case, almost all of the variance is summarised by the first two or three principal components. It appears that we have successfully reduced the data from seven variables

pca

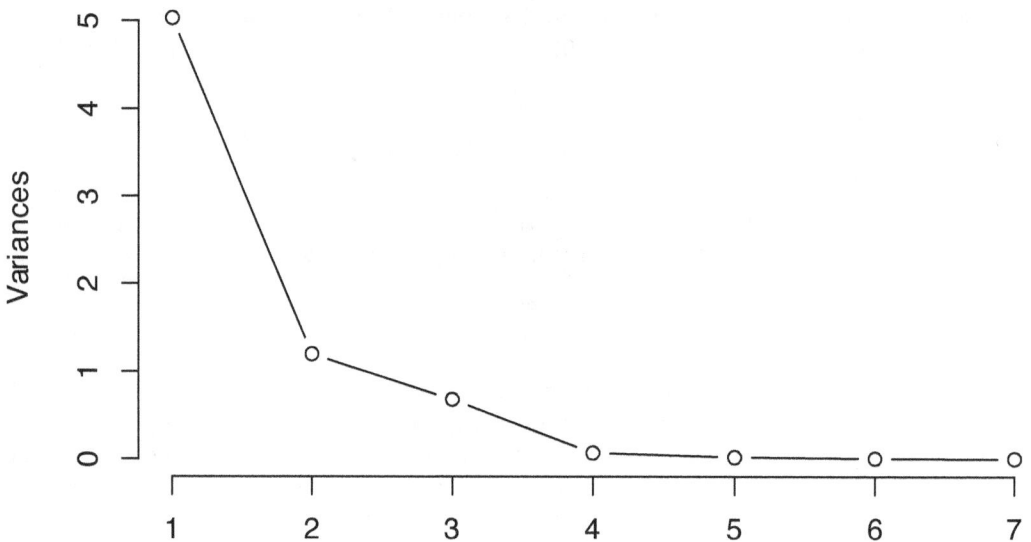

FIGURE 4.10 Screeplot for principal component analysis.

to between two and three, which should make visualisation much easier. The `ggfortify` package contains an `autoplot` function for PCA objects that creates a scatterplot of the first two principal components:

```
library(ggfortify)
autoplot(pca, data = seeds, colour = "Variety")
```

That is much better! The groups are almost completely separated, which shows that the variables can be used to discriminate between the three varieties. The first principal component accounts for 71.87% of the total variance in the data, and the second for 17.11%.

If you like, you can plot other pairs of principal components than just components 1 and 2. In this case, component 3 may be of interest, as its variance is almost as high as that of component 2. You can specify which components to plot with the x and y arguments:

```
# Plot 2nd and 3rd PC:
autoplot(pca, data = seeds, colour = "Variety",
         x = 2, y = 3)
```

Here, the separation is nowhere near as clear as in the previous figure. In this particular example, plotting the first two principal components is the better choice.

Judging from these plots, it appears that the kernel measurements can be used to discriminate between the three varieties of wheat. In Chapters 7 and 11 you'll learn how to use R to build models that can be used to do just that, e.g., by predicting which variety of wheat a kernel comes from given its measurements. If we wanted to build a statistical model that could be used for this purpose, we could use the original measurements. But we could also try using the first two principal components as the only input to the model. Principal component

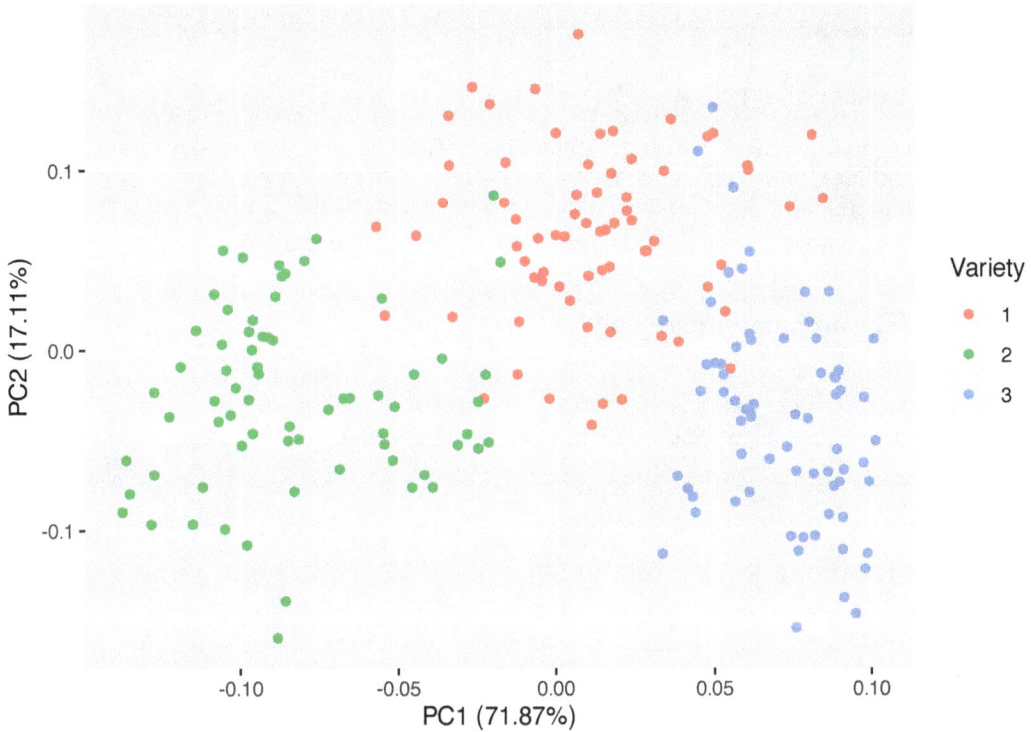

FIGURE 4.11 Plot of the first two principal components for the seeds data.

analysis is very useful as a pre-processing tool used to create simpler models based on fewer variables (or ostensibly simpler, because the new variables are typically more difficult to interpret than the original ones).

~

Exercise 4.25. Use principal components on the `carat`, `x`, `y`, `z`, `depth`, and `table` variables in the `diamonds` data, and answer the following questions:

    1. How much of the total variance does the first principal component account for? How many components are needed to account for at least 90% of the total variance?

    2. Judging by the loadings, what do the first two principal components measure?

    3. What is the correlation between the first principal component and `price`?

    4. Can the first two principal components be used to distinguish between diamonds with different cuts?

Exercise 4.26. Return to the scatterplot of the first two principal components for the `seeds` data created above. Adding the arguments `loadings = TRUE` and `loadings.label = TRUE` to the `autoplot` call creates a *biplot*, which shows the loadings for the principal components on top of the scatterplot. Create a biplot and compare the result to those obtained by looking at the loadings numerically. Do the conclusions from the two approaches agree?

4.12   Cluster analysis

Cluster analysis is concerned with grouping observations into groups, *clusters*, that in some sense are similar. Numerous methods are available for this task, approaching the problem from different angles. Many of these are available in the `cluster` package, which ships with R. In this section, we'll look at a smorgasbord of clustering techniques.

4.12.1   Hierarchical clustering

As a first example where clustering can be of interest, we'll consider the `votes.repub` data from `cluster`. It describes the proportion of votes for the Republican candidate in US presidential elections from 1856 to 1976 in 50 different states:

```
library(cluster)
?votes.repub
View(votes.repub)
```

We are interested in finding subgroups – clusters – of states with similar voting patterns.

To find clusters of similar observations (states, in this case), we could start by assigning each observation to its own cluster. We'd then start with 50 clusters, one for each observation. Next, we could merge the two clusters that are the most similar, yielding 49 clusters, one of which consisted of two observations and 48 consisting of a single observation. We could repeat this process, merging the two most similar clusters in each iteration until only a single cluster was left. This would give us a *hierarchy* of clusters, which could be plotted in a tree-like structure, where observations from the same cluster would be shown one the same branch. Like this:

```
clusters_agnes <- agnes(votes.repub)
plot(clusters_agnes, which = 2)
```

This type of plot is known as a *dendrogram*.

We've just used `agnes`, a function from `cluster` that can be used to carry out *hierarchical clustering* in the manner described above. There are a couple of things that need to be clarified, though.

First, how do we measure how similar two p-dimensional observations x and y are? `agnes` provides two measures of distance between points:

- `metric = "euclidean"` (the default), uses the Euclidean L_2 distance $||x - y|| = \sqrt{\sum_{i=1}^{p}(x_i - y_i)^2}$,
- `metric = "manhattan"`, uses the Manhattan L_1 distance $||x - y|| = \sum_{i=1}^{p}|x_i - y_i|$.

Note that neither of these work if you have categorical variables in your data. If all your variables are binary, i.e., categorical with two values, you can use `mona` instead of `agnes` for hierarchical clustering.

Second, how do we measure how similar two clusters of observations are? `agnes` offers a number of options here. Among them are:

Dendrogram of agnes(x = votes.repub)

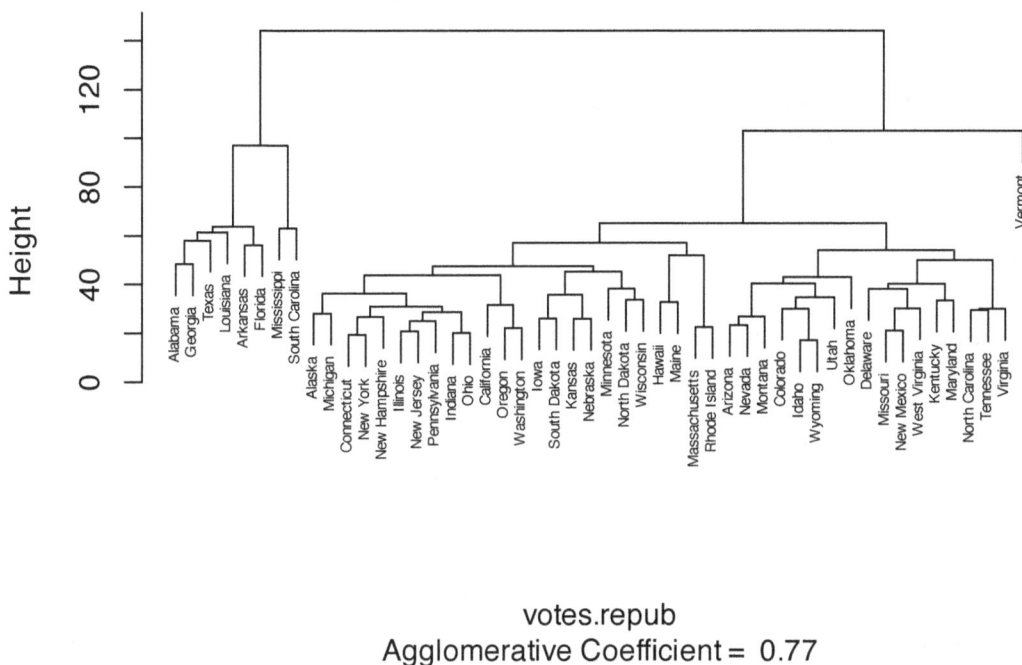

votes.repub
Agglomerative Coefficient = 0.77

FIGURE 4.12 Dendrogram for hierarchical clustering.

- `method = "average"` (the default), unweighted average linkage, uses the average distance between points from the two clusters,
- `method = "single"`, single linkage, uses the smallest distance between points from the two clusters,
- `method = "complete"`, complete linkage, uses the largest distance between points from the two clusters,
- `method = "ward"`, Ward's method, uses the within-cluster variance to compare different possible clusterings, with the clustering with the lowest within-cluster variance chosen.

Regardless of which of these you use, it is often a good idea to standardise the numeric variables in your dataset so that they all have the same variance. If you don't, your distance measure is likely to be dominated by variables with larger variance, while variables with low variances will have little or no impact on the clustering. To standardise your data, you can use `scale`:

```
# Perform clustering on standardised data:
clusters_agnes <- agnes(scale(votes.repub))
# Plot dendrogram:
plot(clusters_agnes, which = 2)
```

At this point, we're starting to use several functions one after another, and so this looks like a perfect job for a pipeline. To carry out the same analysis using the |> pipe, we write:

```
votes.repub |> scale() |>
            agnes() |>
            plot(which = 2)
```

We can now try changing the metric and clustering method used as described above. Let's use the Manhattan distance and complete linkage:

```
votes.repub |> scale() |>
            agnes(metric = "manhattan", method = "complete") |>
            plot(which = 2)
```

We can change the look of the dendrogram by adding hang = -1, which causes all observations to be placed at the same level:

```
votes.repub |> scale() |>
            agnes(metric = "manhattan", method = "complete") |>
            plot(which = 2, hang = -1)
```

As an alternative to agnes, we can consider diana. agnes is an *agglomerative* method, which starts with a lot of clusters and then merge them step-by-step. diana, in contrast, is a *divisive* method, which starts with one large cluster and then step-by-step splits it into several smaller clusters.

```
votes.repub |> scale() |>
            diana() |>
            plot(which = 2)
```

You can change the distance measure used by setting metric in the diana call. Euclidean distance is the default.

To wrap this section up, we'll look at two packages that are useful for plotting the results of hierarchical clustering: dendextend and factoextra. We installed factoextra in the previous sectio, but still need to install dendextend:

```
install.packages("dendextend")
```

To compare the dendrograms produced by different methods (or the same method with different settings), in a *tanglegram*, where the dendrograms are plotted against each other, we can use tanglegram from dendextend:

```
library(dendextend)
# Create clusters using agnes:
votes.repub |> scale() |>
            agnes() -> clusters_agnes
# Create clusters using diana:
votes.repub |> scale() |>
            diana() -> clusters_diana
```

```
# Compare the results:
tanglegram(as.dendrogram(clusters_agnes),
           as.dendrogram(clusters_diana))
```

Some clusters are quite similar here, whereas others are very different.

Often, we are interested in finding a comparatively small number of clusters, k. In hierarchical clustering, we can reduce the number of clusters by "cutting" the dendrogram tree. To do so using the `factoextra` package, we first use `hcut` to cut the tree into k parts, and then `fviz_dend` to plot the dendrogram, with each cluster plotted in a different colour. If, for instance, we want $k = 5$ clusters[9] and want to use `agnes` with average linkage and Euclidean distance for the clustering, we'd do the following:

```
library(factoextra)
votes.repub |> scale() |>
            hcut(k = 5, hc_func = "agnes",
                hc_method = "average",
                hc_metric = "euclidean") |>
            fviz_dend()
```

There is no inherent meaning to the colours – they are simply a way to visually distinguish between clusters.

Hierarchical clustering is especially suitable for data with named observations. For other types of data, other methods may be better. We will consider some alternatives next.

~

Exercise 4.27. Continue the last example above by changing the clustering method to complete linkage with the Manhattan distance.

  1. Do any of the five coloured clusters remain the same?

  2. How can you produce a tanglegram with five coloured clusters, to better compare the results from the two clusterings?

Exercise 4.28. The `USArrests` data contains statistics on violent crime rates in 50 US states. Perform a hierarchical cluster analysis of the data. With which states is Maryland clustered?

4.12.2 Heatmaps and clustering variables

When looking at a dendrogram, you may ask why and how different observations are similar. Similarities between observations can be visualised using a *heatmap*, which displays the levels of different variables using colour hues or intensities. The `heatmap` function creates a heatmap from a `matrix` object. Let's try it with the `votes.repub` voting data. Because `votes.repub` is a `data.frame` object, we have to convert it to a matrix with `as.matrix` first (see Section 2.10.2):

---

[9]Just to be clear, 5 is just an arbitrary number here. We could of course want 4, 14, or any other number of clusters.

```
library(cluster)
votes.repub |> as.matrix() |> heatmap()
```

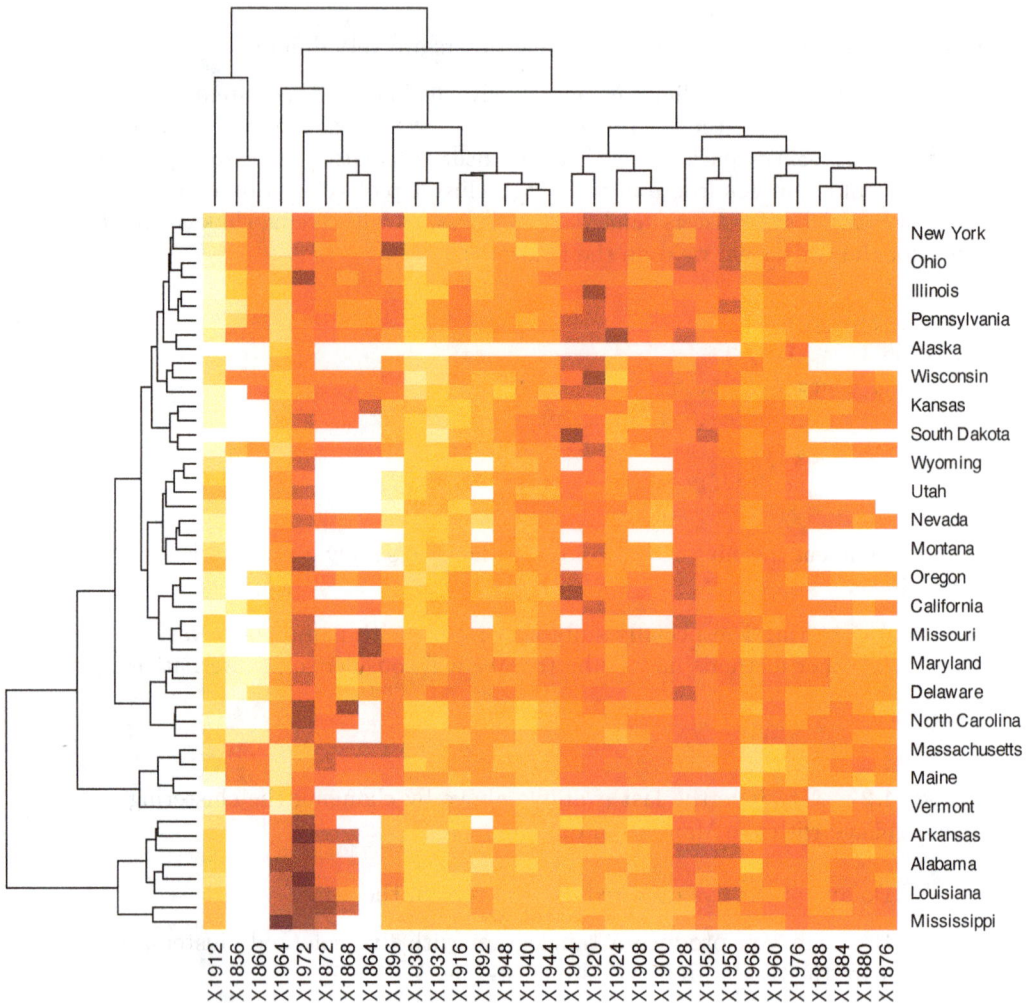

FIGURE 4.13 Heatmap showing similarities between observations.

You may want to increase the height of your Plot window so that the names of all states are displayed properly. Using the default colours, low values are represented by a light yellow and high values by a dark red. White represents missing values.

You'll notice that dendrograms are plotted along the margins. heatmap performs hierarchical clustering (by default, agglomerative with complete linkage) of the observations as well as of the variables. In the latter case, variables are grouped together based on *similarities between observations*, creating *clusters of variables*. In essence, this is just a hierarchical clustering of the transposed data matrix, but it does offer a different view of the data, which at times can be very revealing. The rows and columns are sorted according to the two hierarchical clusterings.

As per usual, it is a good idea to standardise the data before clustering, which can be done using the `scale` argument in `heatmap`. There are two options for scaling, either in the row direction (preferable if you wish to cluster variables) or the column direction (preferable if you wish to cluster observations):

```
# Standardisation suitable for clustering variables:
votes.repub |> as.matrix() |> heatmap(scale = "row")

# Standardisation suitable for clustering observations:
votes.repub |> as.matrix() |> heatmap(scale = "col")
```

Looking at the first of these plots, we can see which elections (i.e., which variables) had similar outcomes in terms of Republican votes. For instance, we can see that the elections in 1960, 1976, 1888, 1884, 1880, and 1876 all had similar outcomes, with the large number of orange rows indicating that the Republicans neither did great nor did poorly.

If you like, you can change the colour palette used. As in Section 4.2.4, you can choose between palettes from http://www.colorbrewer2.org. `heatmap` is not a `ggplot2` function, so this is done in a slightly different way than what you're used to from other examples. Here are two examples, with the white-blue-purple sequential palette `"BuPu"` and the red-white-blue diverging palette `"RdBu"`:

```
library(RColorBrewer)
col_palette <- colorRampPalette(brewer.pal(8, "BuPu"))(25)
votes.repub |> as.matrix() |>
    heatmap(scale = "row", col = col_palette)

col_palette <- colorRampPalette(brewer.pal(8, "RdBu"))(25)
votes.repub |> as.matrix() |>
    heatmap(scale = "row", col = col_palette)
```

~

Exercise 4.29. Draw a heatmap for the `USArrests` data. Have a look at Maryland and the states with which it is clustered. Do they have high or low crime rates?

4.12.3   Centroid-based clustering

Let's return to the `seeds` data that we explored in Section 4.11:

```
# Download the data:
seeds <- read.table("https://tinyurl.com/seedsdata",
        col.names = c("Area", "Perimeter", "Compactness",
          "Kernel_length", "Kernel_width", "Asymmetry",
          "Groove_length", "Variety"))
seeds$Variety <- factor(seeds$Variety)
```

We know that there are three varieties of seeds in this dataset, but what if we didn't? Or what if we'd lost the labels and didn't know what seeds are of what type? There are no row names for this data, and plotting a dendrogram may therefore not be that useful. Instead, we can use k-means clustering, where the points are clustered into k clusters based on their distances to the cluster means, or *centroids*.

When performing k-means clustering (using the algorithm of Hartigan & Wong (1979) that is the default in the function that we'll use), the data is split into k clusters based on their distance to the mean of all points. Points are then moved between clusters, one at a time, based on how close they are (as measured by Euclidean distance) to the mean of each cluster. The algorithm finishes when no point can be moved between clusters without increasing the average distance between points and the means of their clusters.

To run a k-means clustering in R, we can use kmeans. Let's start by using $k = 3$ clusters:

```
# First, we standardise the data, and then we do a k-means
# clustering.
# We ignore variable 8, Variety, which is the group label.
seeds[, -8] |> scale() |>
            kmeans(centers = 3) -> seeds_cluster

seeds_cluster
```

To visualise the results, we'll plot the first two principal components. We'll use colour to show the clusters. Moreover, we'll plot the different varieties in different shapes to see if the clusters found correspond to different varieties:

```
# Compute principal components:
pca <- prcomp(seeds[,-8])
library(ggfortify)
autoplot(pca, data = seeds, colour = seeds_cluster$cluster,
        shape = "Variety", size = 2, alpha = 0.75)
```

In this case, the clusters more or less overlap with the varieties! Of course, in a lot of cases, we don't know the number of clusters beforehand. What happens if we change k?

First, we try $k = 2$:

```
seeds[, -8] |> scale() |>
            kmeans(centers = 2) -> seeds_cluster
autoplot(pca, data = seeds, colour = seeds_cluster$cluster,
        shape = "Variety", size = 2, alpha = 0.75)
```

Next, $k = 4$:

```
seeds[, -8] |> scale() |>
            kmeans(centers = 4) -> seeds_cluster
autoplot(pca, data = seeds, colour = seeds_cluster$cluster,
        shape = "Variety", size = 2, alpha = 0.75)
```

And finally, a larger number of clusters, say $k = 12$:

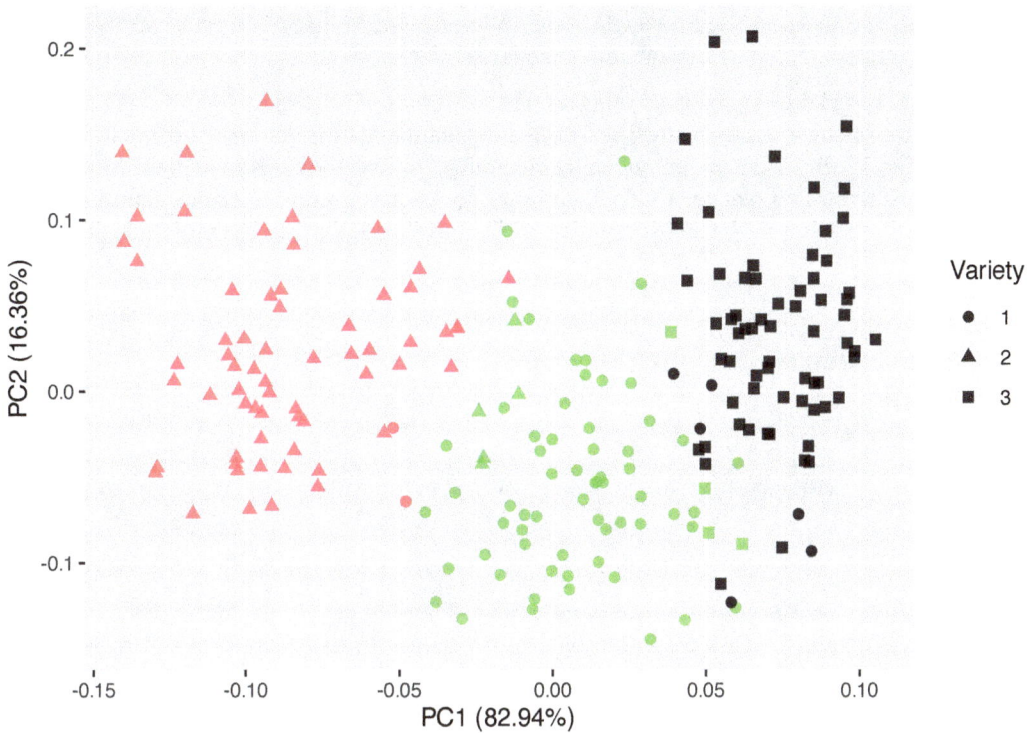

FIGURE 4.14 Clustering using k-means.

```
seeds[, -8] |> scale() |>
                kmeans(centers = 12) -> seeds_cluster
autoplot(pca, data = seeds, colour = seeds_cluster$cluster,
        shape = "Variety", size = 2, alpha = 0.75)
```

If it weren't for the fact that the different varieties were shown as different shapes, we'd have no way to say, based on this plot alone, which choice of k is preferable here. Before we go into methods for choosing k though, we'll mention pam. pam is an alternative to k-means that works in the same way but uses median-like points, *medoids* instead of cluster means. This makes it more robust to outliers. Let's try it with $k = 3$ clusters:

```
seeds[, -8] |> scale() |>
                pam(k = 3) -> seeds_cluster
autoplot(pca, data = seeds, colour = seeds_cluster$clustering,
        shape = "Variety", size = 2, alpha = 0.75)
```

For both kmeans and pam, there are visual tools that can help us choose the value of k in the factoextra package. Let's install it:

```
install.packages("factoextra")
```

The fviz_nbclust function in factoextra can be used to obtain plots that can guide the choice of k. It takes three arguments as input: the data, the clustering function (e.g., kmeans or `pam`) and the method used for evaluating different choices of k. There are three options for the latter: `"wss"`, `"silhouette"` and `"gap_stat"`.

`method = "wss"` yields a plot that relies on the within-cluster sum of squares, WSS, which is a measure of the within-cluster variation. The smaller this is, the more compact are the clusters. The WSS is plotted for several choices of k, and we look for an "elbow", just as we did when using a scree plot for PCA. That is, we look for the value of k such that increasing k further doesn't improve the WSS much. Let's have a look at an example, using `pam` for clustering:

```
library(factoextra)
fviz_nbclust(scale(seeds[, -8]), pam, method = "wss")

# Or, using a pipeline instead:
seeds[, -8] |> scale() |>
            fviz_nbclust(pam, method = "wss")
```

$k = 3$ seems like a good choice here.

`method = "silhouette"` produces a silhouette plot. The silhouette value measures how similar a point is compared to other points in its cluster. The closer to 1 this value is, the better. In a silhouette plot, the average silhouette value for points in the data are plotted against k:

```
fviz_nbclust(scale(seeds[, -8]), pam, method = "silhouette")
```

Judging by this plot, $k = 2$ appears to be the best choice.

Finally, `method = "gap_stat"` yields a plot of the gap statistic (Tibshirani et al., 2001), which is based on comparing the WSS to its expected value under a null distribution obtained using the bootstrap (Section 7.4). Higher values of the gap statistic are preferable:

```
fviz_nbclust(scale(seeds[, -8]), pam, method = "gap_stat")
```

In this case, $k = 3$ gives the best value.

In addition to plots for choosing k, `factoextra` provides the function `fviz_cluster` for creating PCA-based plots, with an option to add convex hulls or ellipses around the clusters:

```
# First, find the clusters:
seeds[, -8] |> scale() |>
            kmeans(centers = 3) -> seeds_cluster

# Plot clusters and their convex hulls:
library(factoextra)
fviz_cluster(seeds_cluster, data = seeds[, -8])

# Without row numbers:
fviz_cluster(seeds_cluster, data = seeds[, -8], geom = "point")

# With ellipses based on the multivariate normal distribution:
```

```
fviz_cluster(seeds_cluster, data = seeds[, -8],
             geom = "point", ellipse.type = "norm")
```

Note that in this plot, the shapes correspond to the clusters and not the varieties of seeds.

~

Exercise 4.30. The chorSub data from cluster contains measurements of 10 chemicals in 61 geological samples from the Kola Peninsula. Cluster this data using either kmeans or pam (does either seem to be a better choice here?). What is a good choice of k here? Visualise the results.

4.12.4   Fuzzy clustering

An alternative to k-means clustering is *fuzzy clustering*, in which each point is "spread out" over the k clusters instead of being placed in a single cluster. The more similar it is to other observations in a cluster, the higher is its membership in that cluster. Points can have a high degree of membership to several clusters, which is useful in applications where points should be allowed to belong to more than one cluster. An important example is genetics, where genes can encode proteins with more than one function. If each point corresponds to a gene, it then makes sense to allow the points to belong to several clusters, potentially associated with different functions. The opposite of fuzzy clustering is *hard clustering*, in which each point only belongs to one cluster.

fanny from cluster can be used to perform fuzzy clustering:

```
library(cluster)
seeds[, -8] |> scale() |>
             fanny(k = 3) -> seeds_cluster

# Check membership of each cluster for the different points:
seeds_cluster$membership

# Plot the closest hard clustering:
library(factoextra)
fviz_cluster(seeds_cluster, geom = "point")
```

As for kmeans and pam, we can use fviz_nbclust to determine how many clusters to use:

```
seeds[, -8] |> scale() |>
             fviz_nbclust(fanny, method = "wss")
seeds[, -8] |> scale() |>
             fviz_nbclust(fanny, method = "silhouette")
# Producing the gap statistic plot takes a while here, so
# you may want to skip it in this case:
seeds[, -8] |> scale() |>
             fviz_nbclust(fanny, method = "gap")
```

~

Exercise 4.31. Do a fuzzy clustering of the USArrests data. Is Maryland strongly associated with a single cluster, or with several clusters? What about New Jersey?

4.12.5 Model-based clustering

As a last option, we'll consider model-based clustering, in which each cluster is assumed to come from a multivariate normal distribution. This will yield ellipsoidal clusters. Mclust from the mclust package fits such a model, called a Gaussian finite mixture model, using the EM-algorithm (Scrucca et al., 2016). First, let's install the package:

```
install.packages("mclust")
```

Now, let's cluster the seeds data. The number of clusters is chosen as part of the clustering procedure. We'll use a function from the factoextra for plotting the clusters with ellipsoids.

```
library(mclust)
seeds_cluster <- Mclust(scale(seeds[, -8]))
summary(seeds_cluster)

# Plot results with ellipsoids:
library(factoextra)
fviz_cluster(seeds_cluster, geom = "point", ellipse.type = "norm")
```

Gaussian finite mixture models are based on the assumption that the data is numerical. For categorical data, we can use latent class analysis, which we'll discuss in Section 10.1.3, instead.

~

Exercise 4.32. Return to the chorSub data from Exercise 4.30. Cluster it using a Gaussian finite mixture model. How many clusters do you find?

4.12.6 Comparing clusters

Having found some interesting clusters, we are often interested in exploring differences between the clusters. To do so, we must first extract the cluster labels from our clustering (which are contained in the variables clustering for methods with Western female names, cluster for kmeans, and classification for Mclust). We can then add those labels to our data frame and use them when plotting.

For instance, using the seeds data, we can compare the area of seeds from different clusters:

```
# Cluster the seeds using k-means with k=3:
library(cluster)
seeds[, -8] |> scale() |>
            kmeans(centers = 3) -> seeds_cluster

# Add the results to the data frame:
seeds$clusters <- factor(seeds_cluster$cluster)
```

```
# Instead of $cluster, we'd use $clustering for agnes, pam, and fanny
# objects, and $classification for an Mclust object.

# Compare the areas of the 3 clusters using boxplots:
library(ggplot2)
ggplot(seeds, aes(x = Area, group = clusters, fill = clusters)) +
    geom_boxplot()

# Or using density estimates:
ggplot(seeds, aes(x = Area, group = clusters, fill = clusters)) +
    geom_density(alpha = 0.7)
```

We can also create a scatterplot matrix to look at all variables simultaneously:

```
library(GGally)
ggpairs(seeds[, -8], aes(colour = seeds$clusters, alpha = 0.2))
```

It may be tempting to run some statistical tests (e.g., a t-test) to see if there are differences between the clusters. Note, however, that in statistical hypothesis testing, it is typically assumed that the hypotheses that are being tested have been generated independently from the data. *Double-dipping*, where the data first is used to generate a hypothesis ("judging from this boxplot, there seems to be a difference in means between these two groups!" or "I found these clusters, and now I'll run a test to see if they are different") and then test that hypothesis, is generally frowned upon, as that substantially inflates the risk of a type I error. Recently, however, there have been some advances in valid techniques for testing differences in means between clusters found using hierarchical clustering; see Gao et al. (2020).

5

Dealing with messy data

...or, put differently, *welcome to the real world*. Real datasets are seldom as tidy and clean as those you have seen in the previous examples in this book. On the contrary, real data is messy. Things will be out of place and formatted in the wrong way. You'll need to filter the rows to remove those that aren't supposed to be used in the analysis. You'll need to remove some columns and merge others. You will need to wrestle, clean, coerce, and coax your data until it finally has the right format. Only then will you be able to actually analyse it.

This chapter contains a number of examples that serve as cookbook recipes for common data wrangling tasks. And as with any cookbook, you'll find yourself returning to some recipes more or less every day, until you know them by heart, while you never find the right time to use other recipes. You definitely do not have to know all of them by heart and can always go back and look up a recipe that you need.

After working with the material in this chapter, you will be able to use R to:

- Handle numeric and categorical data,
- Manipulate and find patterns in text strings,
- Work with dates and times,
- Filter, subset, sort, and reshape your data using `data.table`, `dplyr`, and `tidyr`,
- Split and merge datasets,
- Scrape data from the web, and
- Import data from different file formats.

5.1 Changing data types

In Exercise 2.21 you discovered that R implicitly coerces variables into other data types when needed. For instance, if you add a `numeric` to a `logical`, the result is a `numeric`. And if you place them together in a vector, the vector will contain two `numeric` values:

```
TRUE + 5
v1 <- c(TRUE, 5)
v1
```

However, if you add a `numeric` to a `character`, the operation fails. And if you put them together in a vector, both become `character` strings:

```
"One" + 5
v2 <- c("One", 5)
v2
```

There is a hierarchy for data types in R: logical < integer < numeric < character. When variables of different types are somehow combined (with addition, put in the same vector, and so on), R will coerce both to the higher ranking type. That is why v1 contained numeric variables (numeric is higher ranked than logical) and v2 contained character values (character is higher ranked than numeric).

Automatic coercion is often useful, but will sometimes cause problems. As an example, a vector of numbers may accidentally be converted to a character vector, which will confuse plotting functions. Luckily it is possible to convert objects to other data types. The functions most commonly used for this are as.logical, as.numeric and as.character. Here are some examples of how they can be used:

```
as.logical(1)             # Should be TRUE
as.logical("FALSE")       # Should be FALSE
as.numeric(TRUE)          # Should be 1
as.numeric("2.718282")    # Should be numeric 2.718282
as.character(2.718282)    # Should be the string "2.718282"
as.character(TRUE)        # Should be the string "TRUE"
```

A word of warning though – conversion only works if R can find a natural conversion between the types. Here are some examples where conversion fails. Note that only some of them cause warning messages:

```
as.numeric("two")                    # Should be 2
as.numeric("1+1")                    # Should be 2
as.numeric("2,718282")               # Should be numeric 2.718282
as.logical("Vaccines cause autism")  # Should be FALSE
```

~

Exercise 5.1. The following tasks are concerned with converting and checking data types:

1. What happens if you apply as.logical to the numeric values 0 and 1? What happens if you apply it to other numbers?

2. What happens if you apply as.character to a vector containing numeric values?

3. The functions is.logical, is.numeric and is.character can be used to check if a variable is a logical, numeric or character, respectively. What type of object do they return?

4. Is NA a logical, numeric or character?

5.2 Working with lists

A data structure that is very convenient for storing data of different types is list. You can think of a list as a data frame where you can put different types of objects in each column: like a numeric vector of length 5 in the first, a data frame in the second, and a single `character` in the third[1]. Here is an example of how to create a `list` using the function of the same name:

```
my_list <- list(my_numbers = c(86, 42, 57, 61, 22),
                my_data = data.frame(a = 1:3, b = 4:6),
                my_text = "Lists are the best.")
```

To access the elements in the list, we can use the same $ notation as for data frames:

```
my_list$my_numbers
my_list$my_data
my_list$my_text
```

In addition, we can access them using indices, but using *double* brackets:

```
my_list[[1]]
my_list[[2]]
my_list[[3]]
```

To access elements within the elements of lists, additional brackets can be added. For instance, if you wish to access the second element of the `my_numbers` vector, you can use either of these:

```
my_list[[1]][2]
my_list$my_numbers[2]
```

5.2.1   Splitting vectors into lists

Consider the `airquality` dataset, which among other things describes the temperature on each day during a five-month period. Suppose that we wish to split the `airquality$Temp` vector into five separate vectors: one for each month. We could do this by repeated filtering, e.g.,

```
temp_may <- airquality$Temp[airquality$Month == 5]
temp_june <- airquality$Temp[airquality$Month == 6]
# ...and so on.
```

Apart from the fact that this isn't a very good-looking solution, this would be infeasible if we needed to split our vector into a larger number of new vectors. Fortunately, there is a function that allows us to split the vector by month, storing the result as a list - `split`:

```
temps <- split(airquality$Temp, airquality$Month)
temps

# To access the temperatures for June:
temps$`6`
```

---

[1]In fact, the opposite is true: under the hood, a data frame is a list of vectors of equal length.

```
temps[[2]]

# To give more informative names to the elements in the list:
names(temps) <- c("May", "June", "July", "August", "September")
temps$June
```

Note that, in breach of the rules for variable names in R, the original variable names here were numbers (actually `character` variables that happened to contain numeric characters). When accessing them using $ notation, you need to put them between backticks (`` ` ``), e.g., ``temps$`6` ``, to make it clear that 6 is a variable name and not a number.

5.2.2 Collapsing lists into vectors

Conversely, there are times where you want to collapse a list into a vector. This can be done using `unlist`:

```
unlist(temps)
```

~

Exercise 5.2. Load the `vas.csv` data from Exercise 2.30. Split the VAS vector so that you get a list containing one vector for each patient. How can you then access the visual analogue scale (VAS) values for patient 212?

5.3 Working with numbers

A lot of data analyses involve numbers, which typically are represented as `numeric` values in R. We've already seen in Section 2.4.5 that there are numerous mathematical operators that can be applied to numbers in R. But there are also other functions that come in handy when working with numbers.

5.3.1 Rounding numbers

At times you may want to round numbers, either for presentation purposes or for some other reason. There are several functions that can be used for this:

```
a <- c(2.1241, 3.86234, 4.5, -4.5, 10000.1001)
round(a, 3)          # Rounds to 3 decimal places
signif(a, 3)         # Rounds to 3 significant digits
ceiling(a)           # Rounds up to the nearest integer
floor(a)             # Rounds down to the nearest integer
trunc(a)             # Rounds to the nearest integer, toward 0
                     # (note the difference in how 4.5
                     #  and -4.5 are treated!)
```

5.3.2 Sums and means in data frames

When working with numerical data, you'll frequently find yourself wanting to compute sums or means of either columns or rows of data frames. The `colSums`, `rowSums`, `colMeans` and `rowMeans` functions can be used to do this. Here is an example with an expanded version of the `bookstore` data, where three purchases have been recorded for each customer:

```r
bookstore2 <- data.frame(purchase1 = c(20, 59, 2, 12, 22, 160,
                                       34, 34, 29),
                         purchase2 = c(14, 67, 9, 20, 20, 81,
                                       19, 55, 8),
                         purchase3 = c(4, 62, 11, 18, 33, 57,
                                       24, 49, 29))

colSums(bookstore2)    # The total amount for customers' 1st, 2nd and
                       # 3rd purchases
rowSums(bookstore2)    # The total amount for each customer
colMeans(bookstore2)   # Mean purchase for 1st, 2nd and 3rd purchases
rowMeans(bookstore2)   # Mean purchase for each customer
```

Moving beyond sums and means, in Section 6.5 you'll learn how to apply any function to the rows or columns of a data frame.

5.3.3 Summaries of series of numbers

When a `numeric` vector contains a series of consecutive measurements, as is the case, e.g., in a time series, it is often of interest to compute various cumulative summaries. For instance, if the vector contains the daily revenue of a business during a month, it may be of value to know the total revenue up to each day, that is, the *cumulative sum* for each day.

Let's return to the `a10` data from Section 4.7, which described the monthly anti-diabetic drug sales in Australia during 1991-2008.

```r
library(fpp2)
a10
```

Elements 7 to 18 contain the sales for 1992. We can compute the total, highest, and smallest monthly sales up to and including each month using `cumsum`, `cummax`, and `cummin`:

```r
a10[7:18]
cumsum(a10[7:18])   # Total sales
cummax(a10[7:18])   # Highest monthly sales
cummin(a10[7:18])   # Lowest monthly sales

# Plot total sales up to and including each month:
plot(1:12, cumsum(a10[7:18]),
     xlab = "Month",
     ylab = "Total sales",
     type = "b")
```

In addition, the `cumprod` function can be used to compute cumulative products.

At other times, we are interested in studying *run lengths* in series, that is, the lengths of runs of equal values in a vector. Consider the `upp_temp` vector defined in the code chunk below, which contains the daily temperatures in Uppsala, Sweden, in February 2020[2].

```
upp_temp <- c(5.3, 3.2, -1.4, -3.4, -0.6, -0.6, -0.8, 2.7, 4.2, 5.7,
              3.1, 2.3, -0.6, -1.3, 2.9, 6.9, 6.2, 6.3, 3.2, 0.6, 5.5,
              6.1, 4.4, 1.0, -0.4, -0.5, -1.5, -1.2, 0.6)
```

It could be interesting to look at runs of sub-zero days, i.e., consecutive days with sub-zero temperatures. The `rle` function counts the lengths of runs of equal values in a vector. To find the length of runs of temperatures below or above zero we can use the vector defined by the condition `upp_temp < 0`, the values of which are `TRUE` on sub-zero days and `FALSE` when the temperature is 0 or higher. When we apply `rle` to this vector, it returns the length and value of the runs:

```
rle(upp_temp < 0)
```

We first have a 2-day run of above zero temperatures (`FALSE`), then a 5-day run of sub-zero temperatures (`TRUE`), then a 5-day run of above zero temperatures, and so on.

5.3.4   Scientific notation: `1e-03`

When printing very large or very small numbers, R uses *scientific notation*, meaning that 7,000,000 (7 followed by 6 zeroes) is displayed as (the mathematically equivalent) $7 \cdot 10^6$, and 0.0000007 is displayed as $7 \cdot 10^{-7}$. Well, almost, the *10 raised to the power of x* bit isn't really displayed as 10^x, but as `e+x`, a notation used in many programming languages and calculators. Here are some examples:

```
7000000
0.0000007
7e+07
exp(30)
```

Scientific notation is a convenient way to display very large and very small numbers, but it's not always desirable. If you just want to print the number, the `format` function can be used to convert it to a character, suppressing scientific notation:

```
format(7000000, scientific = FALSE)
```

If you still want your number to be a `numeric` (as you often do), a better choice is to change the option for when R uses scientific notation. This can be done using the `scipen` argument in the `options` function:

```
options(scipen = 1000)
7000000
0.0000007
```

---

[2]Courtesy of the Department of Earth Sciences at Uppsala University.

```
7e+07
exp(30)
```

To revert this option back to the default, you can use:

```
options(scipen = 0)
7000000
0.0000007
7e+07
exp(30)
```

Note that this option only affects how R *prints* numbers, and not how they are treated in computations.

5.3.5   Floating point arithmetics

Some numbers cannot be written in finite decimal forms. Take 1/3 for example, the decimal form of which is

$$0.3333333333333333333333333333333\ldots.$$

Clearly, the computer cannot store this number exactly, as that would require an infinite memory[3]. Because of this, numbers in computers are stored as *floating point numbers*, which aim to strike a balance between *range* (being able to store both very small and very large numbers) and *precision* (being able to represent numbers accurately). Most of the time, calculations with floating points yield exactly the results that we'd expect, but sometimes these non-exact representations of numbers will cause unexpected problems. If we wish to compute $1.5 - 0.2$ and $1.1 - 0.2$, say, we could of course use R for that. Let's see if it gets the answers right:

```
1.5 - 0.2
1.5 - 0.2 == 1.3   # Check if 1.5-0.2=1.3
1.1 - 0.2
1.1 - 0.2 == 0.9   # Check if 1.1-0.2=0.9
```

The limitations of floating point arithmetics cause the second calculation to fail. To see what has happened, we can use `sprintf` to print numbers with 30 decimals (by default, R prints a rounded version with fewer decimals):

```
sprintf("%.30f", 1.1 - 0.2)
sprintf("%.30f", 0.9)
```

The first 12 decimals are identical, but after that the two numbers `1.1 - 0.2` and `0.9` diverge. In our other example, $1.5 - 0.2$, we don't encounter this problem – both `1.5 - 0.2` and `0.3` have the same floating point representation:

---

[3]This is not strictly speaking true; if we use base 3, 1/3 is written as 0.1 which can be stored in a finite memory. But then other numbers become problematic instead.

```
sprintf("%.30f", 1.5 - 0.2)
sprintf("%.30f", 1.3)
```

The order of the operations also matters in this case. The following three calculations would all yield identical results if performed with real numbers, but in floating point arithmetics the results differ:

```
1.1 - 0.2 - 0.9
1.1 - 0.9 - 0.2
1.1 - (0.9 + 0.2)
```

In most cases, it won't make a difference whether a variable is represented as 0.90000000000000013... or 0.90000000000000002..., but in some cases tiny differences like that can propagate and cause massive problems. A famous example of this involves the US Patriot surface-to-air defence system, which at the end of the first Gulf War missed an incoming missile due to an error in floating point arithmetics[4]. It is important to be aware of the fact that floating point arithmetics occasionally will yield incorrect results. This can happen for numbers of any size, but it is more likely to occur when very large and very small numbers appear in the same computation.

So, `1.1 - 0.2` and `0.9` may not be the same thing in floating point arithmetics, but at least they are *nearly* the same thing. The `==` operator checks if two numbers are exactly equal, but there is an alternative that can be used to check if two numbers are nearly equal: `all.equal`. If the two numbers are (nearly) equal, it returns TRUE, and if they are not, it returns a description of how they differ. In order to avoid the latter, we can use the `isTRUE` function to return FALSE instead:

```
1.1 - 0.2 == 0.9
all.equal(1.1 - 0.2, 0.9)
all.equal(1, 2)
isTRUE(all.equal(1, 2))
```

~

Exercise 5.3. These tasks showcase some problems that are commonly faced when working with numeric data:

    1. The vector `props <- c(0.1010, 0.2546, 0.6009, 0.0400, 0.0035)` contains proportions (which, by definition, are between 0 and 1). Convert the proportions to percentages with one decimal place.

    2. Compute the highest and lowest temperatures up to and including each day in the `airquality` dataset.

    3. What is the longest run of days with temperatures above 80 in the `airquality` dataset?

Exercise 5.4. These tasks are concerned with floating point arithmetics:

---

[4]Not in R though.

1. Very large numbers, like `10e500`, are represented by `Inf` (infinity) in R. Try to find out what the largest number that can be represented as a floating point number in R is.

2. Due to an error in floating point arithmetics, `sqrt(2)^2 - 2` is not equal to `0`. Change the order of the operations so that the results is `0`.

5.4 Working with categorical data and factors

In Sections 2.6.2 and 2.8 we looked at how to analyse and visualise categorical data, i.e., data where the variables can take a fixed number of possible values that somehow correspond to groups or categories. But so far we haven't really gone into how to handle categorical variables in R.

Categorical data is stored in R as `factor` variables. You may ask why a special data structure is needed for categorical data, when we could just use `character` variables to represent the categories. Indeed, the latter is what R does by default, e.g., when creating a `data.frame` object or reading data from `.csv` and `.xlsx` files.

Let's say that you've conducted a survey on students' smoking habits. The possible responses are *Never*, *Occasionally*, *Regularly*, and *Heavy*. From 10 students, you get the following responses:

```
smoke <- c("Never", "Never", "Heavy", "Never", "Occasionally",
           "Never", "Never", "Regularly", "Regularly", "No")
```

Note that the last answer is invalid – `No` was not one of the four answers that were allowed for the question.

You could use `table` to get a summary of how many answers of each type that you got:

```
table(smoke)
```

But the categories are not presented in the correct order! There is a clear order between the different categories, *Never* < *Occasionally* < *Regularly* < *Heavy*, but `table` doesn't present the results in that way. Moreover, R didn't recognise that `No` was an invalid answer and treats it just the same as the other categories.

This is where `factor` variables come in. They allow you to specify which values your variable can take, and the ordering between them (if any).

5.4.1 Creating factors

When creating a `factor` variable, you typically start with a `character`, `numeric` or `logical` variable, the values of which are turned into categories. To turn the `smoke` vector that you created in the previous section into a `factor`, you can use the `factor` function:

```
smoke2 <- factor(smoke)
```

You can inspect the elements, and *levels*, i.e., the values that the categorical variable takes, as follows:

```
smoke2
levels(smoke2)
```

So far, we haven't solved neither the problem of the categories being in the wrong order nor that invalid No value. To fix both these problems, we can use the levels argument in factor:

```
smoke2 <- factor(smoke, levels = c("Never", "Occasionally",
                                   "Regularly", "Heavy"),
                 ordered = TRUE)

# Check the results:
smoke2
levels(smoke2)
table(smoke2)
```

You can control the order in which the levels are presented by choosing which order we write them in the levels argument. The ordered = TRUE argument specifies that the order of the variables is *meaningful*. It can be excluded in cases where you wish to specify the order in which the categories should be presented purely for presentation purposes (e.g., when specifying whether to use the order Male/Female/Other or Female/Male/Other). Also note that the No answer now became an NA, which in the case of factor variables represents both missing observations and invalid observations. To find the values of smoke that became NA in smoke2, you can use which and is.na:

```
smoke[which(is.na(smoke2))]
```

By checking the original values of the NA elements, you can see if they should be excluded from the analysis or recoded into a proper category (No could for instance be recoded into Never). In Section 5.5.3 you'll learn how to replace values in larger datasets automatically using regular expressions.

5.4.2 Changing factor levels

When we created smoke2, one of the elements became an NA. NA was however not included as a level of the factor. Sometimes, it is desirable to include NA as a level, for instance when you want to analyse rows with missing data. This is easily done using the addNA function:

```
smoke2 <- addNA(smoke2)
```

If you wish to change the name of one or more of the factor levels, you can do it directly via the levels function. For instance, we can change the name of the NA category, which is the 5th level of smoke2, as follows:

```
levels(smoke2)[5] <- "Invalid answer"
```

The above solution is a little brittle in that it relies on specifying the index of the level name, which can change if we're not careful. More robust solutions using the `data.table` and `dplyr` packages are presented in Section 5.7.6.

Finally, if you've added more levels than what are actually used, these can be dropped using the `droplevels` function:

```
smoke2 <- factor(smoke, levels = c("Never", "Occasionally",
                                   "Regularly", "Heavy",
                                   "Constantly"),
                        ordered = TRUE)
levels(smoke2)
smoke2 <- droplevels(smoke2)
levels(smoke2)
```

5.4.3 Changing the order of levels

Now suppose that we'd like the levels of the `smoke2` variable to be presented in the reverse order: *Heavy, Regularly, Occasionally*, and *Never*. This can be done by a new call to `factor`, where the new level order is specified in the `levels` argument:

```
smoke2 <- factor(smoke2, levels = c("Heavy", "Regularly",
                                    "Occasionally", "Never"))

# Check the results:
levels(smoke2)
```

5.4.4 Combining levels

Finally, `levels` can be used to merge categories by replacing their separate names with a single name. For instance, we can combine the smoking categories *Occasionally, Regularly*, and *Heavy* to a single category named *Yes*. Assuming that these are first, second, and third in the list of names (as will be the case if you've run the last code chunk above), here's how to do it:

```
levels(smoke2)[1:3] <- "Yes"

# Check the results:
levels(smoke2)
```

Alternative ways to do this are presented in Section 5.7.6.

~

Exercise 5.5. In Exercise 2.27 you learned how to create a `factor` variable from a `numeric` variable using `cut`. Return to your solution (or the solution at the back of the book) and do the following:

1. Change the category names to `Mild`, `Moderate`, and `Hot`.

2. Combine `Moderate` and `Hot` into a single level named `Hot`.

Exercise 5.6. Load the `msleep` data from the `ggplot2` package. Note that categorical variable `vore` is stored as a `character`. Convert it to a `factor` by running `msleep$vore <- factor(msleep$vore)`.

1. How are the resulting factor levels ordered? Why are they ordered in that way?

2. Compute the mean value of `sleep_total` for each `vore` group.

3. Sort the factor levels according to their `sleep_total` means. Hint: this can be done manually, or more elegantly using, e.g., a combination of the functions `rank` and `match` in an intermediate step.

5.5 Working with strings

Text in R is represented by `character` strings. These are created using double or single quotes. I recommend double quotes for three reasons. First, it is the default in R and is the recommended style (see, e.g., `?Quotes`). Second, it improves readability – code with double quotes is easier to read because double quotes are easier to spot than single quotes. Third, it will allow you to easily use apostrophes in your strings, which single quotes don't (because apostrophes will be interpreted as the end of the string). Single quotes can however be used if you need to include double quotes inside your string:

```
# This works:
text1 <- "An example of a string. Isn't this great?"
text2 <- 'Another example of a so-called "string".'

# This doesn't work:
text1_fail <- 'An example of a string. Isn't this great?'
text2_fail <- "Another example of a so-called "string"."
```

If you check what these two strings look like, you'll notice something funny about `text2`:

```
text1
text2
```

R has put backslash characters, \, before the double quotes. The backslash is called an *escape character*, which invokes a different interpretation of the character that follows it. In fact, you can use this to put double quotes inside a string that you define using double quotes:

```
text2_success <- "Another example of a so-called \"string\"."
```

There are a number of other special characters that can be included using a backslash: \n for a line break (a new line) and \t for a tab (a long whitespace) being the most important[5]:

---

[5]See `?Quotes` for a complete list.

```
text3 <- "Text...\n\tWith indented text on a new line!"
```

To print your string in the Console in a way that shows special characters instead of their escape character-versions, use the function `cat`:

```
cat(text3)
```

You can also use `cat` to print the string to a text file...

```
cat(text3, file = "new_findings.txt")
```

...and to append text at the end of a text file:

```
cat("Let's add even more text!", file = "new_findings.txt",
    append = TRUE)
```

(Check the output by opening `new_findings.txt`!)

5.5.1   Concatenating strings

If you wish to concatenate multiple strings, `cat` will do that for you:

```
first <- "This is the beginning of a sentence"
second <- "and this is the end."
cat(first, second)
```

By default, cat places a single white space between the two strings, so that `"This is the beginning of a sentence"` and `"and this is the end."` are concatenated to `"This is the beginning of a sentence and this is the end."`. You can change that using the `sep` argument in `cat`. You can also add as many strings as you like as input:

```
cat(first, second, sep = "; ")
cat(first, second, sep = "\n")
cat(first, second, sep = "")
cat(first, second, "\n", "And this is another sentence.")
```

At other times, you want to concatenate two or more strings without printing them. You can then use `paste` in exactly the same way as you'd use `cat`, the exception being that `paste` returns a string instead of printing it.

```
my_sentence <- paste(first, second, sep = "; ")
my_novel <- paste(first, second, "\n",
                  "And this is another sentence.")

# View results:
my_sentence
```

```
my_novel
cat(my_novel)
```

Finally, if you wish to create a number of similar strings based on information from other variables, you can use `sprintf`, which allows you to write a string using `%s` as a placeholder for the values that should be pulled from other variables:

```
names <- c("Irma", "Bea", "Lisa")
ages <- c(5, 59, 36)

sprintf("%s is %s years old.", names, ages)
```

There are many more uses of `sprintf` (we've already seen some in Section 5.3.5), but this is enough for us for now.

5.5.2   Changing case

If you need to translate characters from lowercase to uppercase or vice versa, that can be done using `toupper` and `tolower`:

```
my_string <- "SOMETIMES I SCREAM (and sometimes I whisper)."
toupper(my_string)
tolower(my_string)
```

If you only wish to change the case of some particular element in your string, you can use `substr`, which allows you to access substrings:

```
months <- c("january", "february", "march", "aripl")

# Replacing characters 2-4 of months[4] with "pri":
substr(months[4], 2, 4) <- "pri"
months

# Replacing characters 1-1 (i.e. character 1) of each element of month
# with its uppercase version:
substr(months, 1, 1) <- toupper(substr(months, 1, 1))
months
```

5.5.3   Finding patterns using regular expressions

Regular expressions (regexps) are special strings that describe patterns. They are extremely useful if you need to find, replace, or otherwise manipulate a number of strings depending on whether or not a certain pattern exists in each one of them. For instance, you may want to find all strings containing only numbers and convert them to `numeric`, or find all strings that contain an email address and remove said addresses (for censoring purposes, say). Regular expressions are incredibly useful but can be daunting. Not everyone will need them, and if this all seems a bit too much, you can safely skip this section or just skim through it and return to it at a later point.

To illustrate the use of regular expressions, we will use a sheet from the `projects-email.xlsx` file from the book's web page. In Exercise 2.32, you explored the second sheet in this file, but here we'll use the third instead. Set `file_path` to the path to the file, and then run the following code to import the data:

```
library(openxlsx)
contacts <- read.xlsx(file_path, sheet = 3)
str(contacts)
```

There are now three variables in `contacts`. We'll primarily be concerned with the third one: `Address`. Some people have email addresses attached to them, others have postal addresses, and some have no address at all:

```
contacts$Address
```

You can find loads of guides on regular expressions online, but few of them are easy to use with R, the reason being that regular expressions in R sometimes require escape characters that aren't needed in some other programming languages. In this section we'll take a look at regular expressions, *as they are written in R*.

The basic building blocks of regular expressions are patterns consisting of one or more characters. If, for instance, we wish to find all occurrences of the letter y in a vector of strings, the regular expression describing that "pattern" is simply `"y"`. The functions used to find occurrences of patterns are called `grep` and `grepl`. They differ only in the output they return: `grep` returns the indices of the strings containing the pattern, and `grepl` returns a `logical` vector with TRUE at indices matching the patterns and FALSE at other indices.

To find all addresses containing a lowercase y, we use `grep` and `grepl` as follows:

```
grep("y", contacts$Address)
grepl("y", contacts$Address)
```

Note how both outputs contain the same information presented in different ways.

In the same way, we can look for word or substrings. For instance, we can find all addresses containing the string `"Edin"`:

```
grep("Edin", contacts$Address)
grepl("Edin", contacts$Address)
```

Similarly, we can also look for special characters. Perhaps we can find all email addresses by looking for strings containing the @ symbol:

```
grep("@", contacts$Address)
grepl("@", contacts$Address)

# To display the addresses matching the pattern:
contacts$Address[grep("@", contacts$Address)]
```

Interestingly, this includes two rows that aren't email addresses. To separate the email addresses from the other rows, we'll need a more complicated regular expression, describing the pattern of an email address in more general terms. Here are four examples or regular expressions that'll do the trick:

```
grep(".+@.+[.].+", contacts$Address)
grep(".+@.+\\..+", contacts$Address)
grep("[[:graph:]]+@[[:graph:]]+[.][[:alpha:]]+", contacts$Address)
grep("[[:alnum:]._-]+@[[:alnum:]._-]+[.][[:alpha:]]+",
    contacts$Address)
```

To try to wrap our head around what these mean, we'll have a look at the building blocks of regular expressions. These are:

- Patterns describing a single character.
- Patterns describing a class of characters, e.g., letters or numbers.
- Repetition quantifiers describing how many repetitions of a pattern to look for.
- Other operators.

We've already looked at single character expressions, as well as the multicharacter expression "Edin" which simply is a combination of four single-character expressions. Patterns describing classes of characters, e.g., characters with certain properties, are denoted by brackets [] (for manually defined classes) or double brackets [[]] (for pre-defined classes). One example of the latter is "[[:digit:]], which is a pattern that matches all digits: 0 1 2 3 4 5 6 7 8 9. Let's use it to find all addresses containing a number:

```
grep("[[:digit:]]", contacts$Address)
contacts$Address[grep("[[:digit:]]", contacts$Address)]
```

Some important pre-defined classes are:

- [[:lower:]] matches lowercase letters,
- [[:upper:]] matches UPPERCASE letters,
- [[:alpha:]] matches both lowercase and UPPERCASE letters,
- [[:digit:]] matches digits: 0 1 2 3 4 5 6 7 8 9,
- [[:alnum:]] matches alphanumeric characters (alphabetic characters and digits),
- [[:punct:]] matches punctuation characters: ! " # $ % & ' ( ) * + , - . / : ; < = > ? @ [\ ] ^ _ { | } ~,
- [[:space:]] matches space characters: space, tab, newline, and so on,
- [[:graph:]] matches letters, digits, and punctuation characters,
- [[:print:]] matches letters, digits, punctuation characters, and space characters,
- . matches *any* character.

Examples of manually defined classes are:

- [abcd] matches a, b, c, and d,
- [a-d] matches a, b, c, and d,
- [aA12] matches a, A, 1 and 2,
- [.] matches .,
- [.,] matches . and ,,
- [^abcd] matches anything except a, b, c, or d.

So, for instance, we can find all addresses that don't contain at least one of the letters y and
z using:

```
grep("[^yz]", contacts$Address)
contacts$Address[grep("[^yz]", contacts$Address)]
```

All of these patterns can be combined with patterns describing a single character:

- `gr[ea]y` matches `grey` and `gray` (but not `greay`!),
- `b[^o]g` matches `bag`, `beg`, and similar strings, but not `bog`,
- `[.]com` matches `.com`.

When using the patterns above, you only look for a single occurrence of the pattern.
Sometimes you may want a pattern like *a word of 2-4 letters* or *any number of digits in a
row*. To create these, you add repetition patterns to your regular expression:

- `?` means that the preceding pattern is matched *at most once*, i.e., 0 or 1 time,
- `*` means that the preceding pattern is matched *0 or more* times,
- `+` means that the preceding pattern is matched *at least once*, i.e., 1 time or more,
- `{n}` means that the preceding pattern is matched *exactly* n times,
- `{n,}` means that the preceding pattern is matched *at least* n times, i.e., n times or more,
- `{n,m}` means that the preceding pattern is matched *at least* n times *but not more than* m
  times.

Here are some examples of how repetition patterns can be used:

```
# There are multiple ways of finding strings containing two n's
# in a row:
contacts$Address[grep("nn", contacts$Address)]
contacts$Address[grep("n{2}", contacts$Address)]

# Find strings with words beginning with an uppercase letter, followed
# by at least one lowercase letter:
contacts$Address[grep("[[:upper:]][[:lower:]]+", contacts$Address)]

# Find strings with words beginning with an uppercase letter, followed
# by at least six lowercase letters:
contacts$Address[grep("[[:upper:]][[:lower:]]{6,}", contacts$Address)]

# Find strings containing any number of letters, followed by any
# number of digits, followed by a space:
contacts$Address[grep("[[:alpha:]]+[[:digit:]]+[[:space:]]",
                      contacts$Address)]
```

Finally, there are some other operators that you can use to create even more complex
patterns:

- `|` alteration, picks one of multiple possible patterns. For example, `ab|bc` matches `ab` or `bc`.
- `()` parentheses are used to denote a subset of an expression that should be evaluated
  separately. For example, `colo|our` matches `colo` or `our` while `col(o|ou)r` matches `color` or
  `colour`.
- `^`, when used outside of brackets `[]`, means that the match should be found at the start of
  the string. For example, `^a` matches strings beginning with `a`, but not `"dad"`.

- $ means that the match should be found at the end of the string. For example, a$ matches strings ending with a, but not "dad".
- \\ escape character that can be used to match special characters like ., ^ and $ (\\., \\^, \\$).

This may seem like a lot (and it is!), but there are in fact many more possibilities when working with regular expression. For the sake of some sort of brevity, we'll leave it at this for now though.

Let's return to those email addresses. We saw three regular expressions that could be used to find them:

```
grep(".+@.+[.].+", contacts$Address)
grep(".+@.+\\..+", contacts$Address)
grep("[[:graph:]]+@[[:graph:]]+[.][[:alpha:]]+", contacts$Address)
grep("[[:alnum:]._-]+@[[:alnum:]._-]+[.][[:alpha:]]+",
     contacts$Address)
```

The first two of these specify the same pattern: *any number of any characters, followed by an @, followed by any number of any characters, followed by a period ., followed by any number of characters.* This will match email addresses but would also match strings like "?=)(/x@!.a??", which isn't a valid email address. In this case, that's not a big issue, as our goal was to find addresses that looked like email addresses and not to verify that the addresses were valid.

The third alternative has a slightly different pattern: *any number of letters, digits, and punctuation characters, followed by an @, followed by any number of letters, digits, and punctuation characters, followed by a period ., followed by any number of letters.* This too would match "?=)(/x@!.a??" as it allows punctuation characters that don't usually occur in email addresses. The fourth alternative, however, won't match "?=)(/x@!.a??" as it only allows letters, digits and the symbols ., _ and - in the name and domain name of the address.

5.5.4 Substitution

An important use of regular expressions is in substitutions, where the parts of strings that match the pattern in the expression are replaced by another string. There are two email addresses in our data that contain (a) instead of @:

```
contacts$Address[grep("[(]a[)]]", contacts$Address)]
```

If we wish to replace the (a) by @, we can do so using sub and gsub. The former replaces only the *first* occurrence of the pattern in the input vector, whereas the latter replaces *all* occurrences.

```
contacts$Address[grep("[(]a[)]]", contacts$Address)]
sub("[(]a[)]]", "@", contacts$Address)    # Replace first occurrence
gsub("[(]a[)]]", "@", contacts$Address)   # Replace all occurrences
```

5.5.5 Splitting strings

At times you want to extract only a part of a string, for example if measurements recorded in a column contain units, e.g. `66.8 kg` instead of `66.8`. To split a string into different parts, we can use `strsplit`.

As an example, consider the email addresses in our `contacts` data. Suppose that we want to extract the user names from all email addresses, i.e., remove the `@domain.topdomain` part. First, we store all email addresses from the data in a new vector, and then we split them at the `@` sign:

```r
emails <- contacts$Address[grepl(
            "[[:alnum:]._-]+@[[:alnum:]._-]+[.][[:alpha:]]+",
            contacts$Address)]
emails_split <- strsplit(emails, "@")
emails_split
```

`emails_split` is a *list*. In this case, it seems convenient to convert the split strings into a matrix using `unlist` and `matrix` (you may want to have a quick look at Exercise 2.23 to re-familiarise yourself with `matrix`):

```r
emails_split <- unlist(emails_split)

# Store in a matrix with length(emails_split)/2 rows and 2 columns:
emails_matrix <- matrix(emails_split,
                        nrow = length(emails_split)/2,
                        ncol = 2,
                        byrow = TRUE)

# Extract usernames:
emails_matrix[,1]
```

Similarly, when working with data stored in data frames, it is sometimes desirable to split a column containing strings into two columns. Some convenience functions for this are discussed in Section 5.11.3.

5.5.6 Variable names

Variable names can be very messy, particularly when they are imported from files. You can access and manipulate the variable names of a data frame using `names`:

```r
names(contacts)
names(contacts)[1] <- "ID number"
grep("[aA]", names(contacts))
```

~

Exercise 5.7. Download the file `handkerchief.csv` from the book's web page[6]. It contains a short list of prices of Italian handkerchiefs from the 1769 publication[7] *A list of prices in those branches of the weaving manufactory, called the black branch, and the fancy branch.* Load the data in a data frame in R and then do the following:

1. Read the documentation for the function `nchar`. What does it do? Apply it to the `Italian.handkerchief` column of your data frame.

2. Use `grep` to find out how many rows of the `Italian.handkerchief` column that contain numbers.

3. Find a way to extract the prices in shillings (S) and pence (D) from the `Price` column, storing these in two new `numeric` variables in your data frame.

Exercise 5.8. Download the `oslo-biomarkers.xlsx` data from the book's web page[8]. It contains data from a medical study about patients with disc herniations, performed at the Oslo University Hospital, Ullevål (this is a modified[9] version of the data analysed by Moen et al. (2016)). Blood samples were collected from a number of patients with disc herniations at three time points: 0 weeks (first visit at the hospital), 6 weeks, and 12 months. The levels of some biomarkers related to inflammation were measured in each blood sample. The first column in the spreadsheet contains information about the patient ID and the time point of sampling. Load the data and check its structure. Each patient is uniquely identified by their ID number. How many patients were included in the study?

Exercise 5.9. What patterns do the following regular expressions describe? Apply them to the `Address` vector of the `contacts` data to check that you interpreted them correctly.

1. `"$g"`
2. `"^[^[[:digit:]]]"`
3. `"a(s|l)"`
4. `"[[:lower:]]+[.][[:lower:]]+"`

Exercise 5.10. Write code that, given a string, creates a vector containing all words from the string, with one word in each element and no punctuation marks. Apply it to the following string to check that it works:

```
x <- "This is an example of a sentence, with 10 words. Here are 4 more!"
```

5.6 Working with dates and times

Data describing dates and times can be complex, not least because they can be written is so many different formats. The date 1 April 2020 can be written as `2020-04-01`, `20/04/01`,

[6]http://www.modernstatisticswithr.com/data.zip
[7]https://books.google.se/books?id=rUxiAAAAcAAJ
[8]http://www.modernstatisticswithr.com/data.zip
[9]For patient confidentiality purposes.

`200401`, `1/4 2020`, `4/1/20`, `1 Apr 20`, and a myriad of other ways. The time 5 past 6 in the evening can be written as `18:05` or `6.05 pm`. In addition to this ambiguity, time zones, daylight saving time, leap years, and even leap seconds make working with dates and times even more complicated.

The default in R is to use the ISO8601 standards, meaning that dates are written as YYYY-MM-DD, and that times are written using the 24-hour hh:mm:ss format. In order to avoid confusion, you should always use these, unless you have *very* strong reasons not to.

Dates in R are represented as `Date` objects, and dates with times as `POSIXct` objects. The examples below are concerned with `Date` objects, but you will explore `POSIXct` too, in Exercise 5.12.

5.6.1 Date formats

The `as.Date` function tries to coerce a `character` string to a date. For some formats, it will automatically succeed, whereas for others, you have to provide the format of the date manually. To complicate things further, what formats work automatically will depend on your system settings. Consequently, the safest option is always to specify the format of your dates, to make sure that the code still will run if you at some point have to execute it on a different machine. To help describe date formats, R has a number of tokens to describe days, months, and years:

- `%d` - day of the month as a number (01-31).
- `%m` - month of the year as a number (01-12).
- `%y` - year without century (00-99).
- `%Y` - year with century (e.g., 2020).

Here are some examples of date formats, all describing 1 April 2020 – try them both with and without specifying the format to see what happens:

```
as.Date("2020-04-01")
as.Date("2020-04-01", format = "%Y-%m-%d")
as.Date("4/1/20")
as.Date("4/1/20", format = "%m/%d/%y")


# Sometimes dates are expressed as the number of days since a
# certain date. For instance, 1 April 2020 is 43,920 days after
# 1 January 1900:
as.Date(43920, origin = "1900-01-01")
```

If the date includes month or weekday names, you can use tokens to describe that as well:

- `%b` - abbreviated month name, e.g., `Jan`, `Feb`.
- `%B` - full month name, e.g., `January`, `February`.
- `%a` - abbreviated weekday, e.g., `Mon`, `Tue`.
- `%A` - full weekday, e.g., `Monday`, `Tuesday`.

Things become a little more complicated now though, because R will interpret the names as if they were written in the language set in your *locale*, which contains a number of settings related to your language and region. To find out what language is in your locale, you can use:

```r
Sys.getlocale("LC_TIME")
```

I'm writing this on a machine with Swedish locale settings (my output from the above code chunk is `"sv_SE.UTF-8"`). The Swedish word for *Wednesday* is *onsdag*[10], and therefore the following code doesn't work on my machine:

```r
as.Date("Wednesday 1 April 2020", format = "%A %d %B %Y")
```

However, if I translate it to Swedish, it runs just fine:

```r
as.Date("Onsdag 1 april 2020", format = "%A %d %B %Y")
```

You may at times need to make similar translations of dates. One option is to use `gsub` to translate the names of months and weekdays into the correct language (see Section 5.5.4). Alternatively, you can change the locale settings. On most systems, the following setting will allow you to read English months and days properly:

```r
Sys.setlocale("LC_TIME", "C")
```

The locale settings will revert to the defaults the next time you start R.

Conversely, you may want to extract a substring from a `Date` object, for instance the day of the month. This can be done using `strftime`, using the same tokens as above. Here are some examples, including one with the token `%j`, which can be used to extract the day of the year:

```r
dates <- as.Date(c("2020-04-01", "2021-01-29", "2021-02-22"),
                 format = "%Y-%m-%d")

# Extract the day of the month:
strftime(dates, format = "%d")

# Extract the month:
strftime(dates, format = "%m")

# Extract the year:
strftime(dates, format = "%Y")

# Extract the day of the year:
strftime(dates, format = "%j")
```

Should you need to, you can of course convert these objects from `character` to `numeric` using `as.numeric`.

For a complete list of tokens that can be used to describe date patterns, see `?strftime`.

∼

---

[10]The Swedish *onsdag* and English *Wednesday* both derive from the proto-Germanic *Wodensdag*, Odin's day, in honour of the old Germanic god of that name.

Exercise 5.11. Consider the following `Date` vector:

```
dates <- as.Date(c("2015-01-01", "1984-03-12", "2012-09-08"),
                 format = "%Y-%m-%d")
```

1. Apply the functions `weekdays`, `months`, and `quarters` to the vector. What do they do?

2. Use the `julian` function to find out how many days passed between 1970-01-01 and the dates in `dates`.

Exercise 5.12. Consider the three `character` objects created below:

```
time1 <- "2020-04-01 13:20"
time2 <- "2020-04-01 14:30"
time3 <- "2020-04-03 18:58"
```

1. What happens if you convert the three variables to `Date` objects using `as.Date` without specifying the date format?

2. Convert `time1` to a `Date` object and add `1` to it. What is the result?

3. Convert `time3` and `time1` to `Date` objects and subtract them. What is the result?

4. Convert `time2` and `time1` to `Date` objects and subtract them. What is the result?

5. What happens if you convert the three variables to `POSIXct` date and time objects using `as.POSIXct` without specifying the date format?

6. Convert `time3` and `time1` to `POSIXct` objects and subtract them. What is the result?

7. Convert `time2` and `time1` to `POSIXct` objects and subtract them. What is the result?

8. Use the `difftime` to repeat the calculation in task 6, but with the result presented in hours.

Exercise 5.13. In some fields, e.g., economics, data is often aggregated on a quarter-year level, as in these examples:

```
qvec1 <- c("2020 Q4", "2021 Q1", "2021 Q2")
qvec2 <- c("Q4/20", "Q1/21", "Q2/21")
qvec3 <- c("Q4-2020", "Q1-2021", "Q2-2021")
```

To convert `qvec1` to a `Date` object, we can use `as.yearqtr` from the `zoo` package in two ways:

```
library(zoo)
as.Date(as.yearqtr(qvec1, format = "%Y Q%q"))
as.Date(as.yearqtr(qvec1, format = "%Y Q%q"), frac = 1)
```

1. Describe the results. What is the difference? Which do you think is preferable?

2. Convert `qvec2` and `qvec3` to `Date` objects in the same way. Make sure that you get the `format` argument, which describes the date format, right.

5.6.2 Plotting with dates

`ggplot2` automatically recognises `Date` objects and will usually plot them in a nice way. That only works if it actually has the dates though. Consider the following plot, which we created in Section 4.7.7 – it shows the daily electricity demand in Victoria, Australia in 2014:

```
library(plotly)
library(fpp2)

# Create the plot object
myPlot <- autoplot(elecdaily[,"Demand"])

# Create the interactive plot
ggplotly(myPlot)
```

When you hover over the points, the formatting of the dates looks odd. We'd like to have proper dates instead. In order to do so, we'll use `seq.Date` to create a sequence of dates, ranging from 2014-01-01 to 2014-12-31:

```
# Create a data frame with better formatted dates
elecdaily2 <- as.data.frame(elecdaily)
elecdaily2$Date <- seq.Date(as.Date("2014-01-01"),
                            as.Date("2014-12-31"),
                            by = "day")

# Create the plot object
myPlot <- ggplot(elecdaily2, aes(Date, Demand)) +
    geom_line()

# Create the interactive plot
ggplotly(myPlot)
```

`seq.Date` can be used analogously to create sequences where there is a week, month, quarter, or year between each element of the sequence, by changing the `by` argument.

∼

Exercise 5.14. Return to the plot from Exercise 4.12, which was created using

```
library(fpp2)
autoplot(elecdaily, facets = TRUE)
```

You'll notice that the x-axis shows week numbers rather than dates (the dates in the `elecdaily` time series object are formatted as weeks with decimal numbers). Make a time series plot of the `Demand` variable with dates (2014-01-01 to 2014-12-31) along the x-axis (your solution is likely to rely on standard R techniques rather than `autoplot`).

Exercise 5.15. Create an interactive version time series plot of the `a10` anti-diabetic drug sales data, as in Section 4.7.7. Make sure that the dates are correctly displayed.

5.7 Data manipulation with `data.table`, `dplyr`, and `tidyr`

In the remainder of this chapter, we will use three packages that contain functions for fast and efficient data manipulation: `data.table` and the tidyverse packages `dplyr` and `tidyr`. To begin with, it is therefore a good idea to install `data.table` and `tidyr`, which we haven't used before. And while you wait for the installation to finish, read on.

```
install.packages(c("tidyr", "data.table"))
```

There is almost always more than one way to solve a problem in R. We now know how to access vectors and elements in data frames, e.g., to compute means. We also know how to modify and add variables to data frames. Indeed, you can do just about anything using the functions in base R. Sometimes, however, those solutions become rather cumbersome, as they can require a fair amount of programming and verbose code. `data.table` and the tidyverse packages offer simpler solutions and speed up the workflow for these types of problems. Both can be used for the same tasks. You can learn one of them or both. The syntax used for `data.table` is often more concise and arguably more consistent than that in `dplyr` (it is in essence an extension of the `[i, j]` notation that we have already used for data frames). Second, it is fast and memory-efficient, which makes a huge difference if you are working with big data (you'll see this for yourself in Section 6.6). On the other hand, many people prefer the syntax in `dplyr` and `tidyr`, which lends itself exceptionally well for usage with pipes. If you work with small- or medium-sized datasets, the difference in performance between the two packages is negligible. `dplyr` is also much better suited for working directly with databases, which is a huge selling point if your data already is in a database[11].

In the sections below, we will see how to perform different operations using both `data.table` and the tidyverse packages. Perhaps you already know which one you want to use (`data.table` if performance is important to you, `dplyr`+`tidyr` if you like to use pipes or will be doing a lot of work with databases). If not, you can use these examples to guide your choice. Or not choose at all! I regularly use all three packages myself, to harness the strength of them all. There is no harm in knowing how to use both a hammer and a screwdriver.

5.7.1 `data.table` and tidyverse syntax basics

`data.table` relies heavily on the `[i, j]` notation that is used for data frames in R. It also adds a third element: `[i, j, by]`. Using this, R selects the rows indicated by `i`, the columns indicated by `j`, and groups them by `by`. This makes it easy, e.g., to compute grouped summaries.

---

[11]There is also a package called `dtplyr`, which allows you to use the fast functions from `data.table` with `dplyr` syntax. It is useful if you are working with big data, already know `dplyr`, and don't want to learn `data.table`. If that isn't an accurate description of you, you can safely ignore `dtplyr` for now.

With the tidyverse packages, you will instead use specialised functions with names like `filter` and `summarise` to perform operations on your data. These are typically combined using the pipe operator, `|>`, which makes the code flow nicely from left to right.

It's almost time to look at some examples of what this actually looks like in practice. First though, now that you've installed `data.table` and `dplyr`, it's time to load them (we'll get to `tidyr` a little later). We'll also create a `data.table` version of the `airquality` data, which we'll use in the examples below. This is required in order to use `data.table` syntax, as it only works on `data.table` objects. Luckily, `dplyr` works perfectly when used on `data.table` objects, so we can use the same object for the examples for both packages.

```
library(data.table)
library(dplyr)

aq <- as.data.table(airquality)
```

When importing data from csv files, you can import them as `data.table` objects instead of `data.frame` objects by replacing `read.csv` with `fread` from the `data.table` package. The latter function also has the benefit of being substantially faster when importing large (several megabytes) csv files.

Note that, similar to what we saw in Section 5.2.1, variables in imported data frames can have names that would not be allowed in base R, for instance including forbidden characters like `-`. `data.table` and `dplyr` allow you to work with such variables by wrapping their names in apostrophes: referring to the illegally named variable as `illegal-character-name` won't work, but `` `illegal-character-name` `` will.

5.7.2 Modifying a variable

As a first example, let's consider how to use `data.table` and `dplyr` to modify a variable in a data frame. The wind speed in `airquality` is measured in miles per hour (mph). We can convert that to metres per second (m/s) by multiplying the speed by 0.44704. Using only base R, we'd do this using `airquality$Wind <- airquality$Wind * 0.44704`. With `data.table` we can instead do this using `[i, j]` notation, and with `dplyr` we can do it by using a function called `mutate` (because it "mutates" your data).

Change wind speed to m/s instead of mph:

With `data.table`:

```
aq[, Wind := Wind * 0.44704]
```

With `dplyr`:

```
aq |> mutate(Wind =
        Wind * 0.44704) -> aq
```

Note that when using `data.table`, there is not an explicit assignment. We don't use `<-` to assign the new data frame to `aq` – instead, the assignment happens automatically. This means that you have to be a bit careful, so that you don't inadvertently make changes to your data when playing around with it.

In this case, using `data.table` or `dplyr` doesn't make anything easier. Where these packages really shine is when we attempt more complicated operations. Before that though, let's look at a few more simple examples.

5.7.3   Computing a new variable based on existing variables

What if we wish to create new variables based on other variables in the data frame? For instance, maybe we want to create a *dummy variable* called `Hot`, containing a `logical` that describes whether a day was hot (temperature above 90 - `TRUE`) or not (`FALSE`). That is, we wish to check the condition `Temp > 90` for each row and put the resulting `logical` in the new variable `Hot`.

Add a dummy variable describing whether it is hotter than 90:

With `data.table`:   With `dplyr`:

```
aq[, Hot := Temp > 90]
```
```
aq |> mutate(Hot = Temp > 90) -> aq
```

5.7.4   Renaming a variable

To change the name of a variable, we can use `setnames` from `data.table` or `rename` from `dplyr`. Let's change the name of the variable `Hot` that we created in the previous section, to `HotDay`:

With `data.table`:   With `dplyr`:

```
setnames(aq, "Hot", "HotDay")
```
```
aq |> rename(HotDay = Hot) -> aq
```

5.7.5   Removing a variable

Maybe adding `Hot` to the data frame wasn't such a great idea after all. How can we remove it?

Removing `Hot`:

With `data.table`:   With `dplyr`:

```
aq[, Hot := NULL]
```
```
aq |> select(-Hot) -> aq
```

If we wish to remove multiple columns at once, the syntax is similar:

Removing multiple columns:

With `data.table`:   With `dplyr`:

```
aq[, c("Month", "Day") := NULL]
```
```
aq |> select(-Month, -Day) -> aq
```

~

Exercise 5.16. Load the VAS pain data `vas.csv` from Exercise 2.30. Then do the following:

1. Remove the columns `X` and `X.1`.

2. Add a dummy variable called `highVAS` that indicates whether a patient's VAS is 7 or greater on any given day.

5.7.6 Recoding `factor` levels

Changing the names of `factor` levels in base R typically relies on using indices of level names, as in Section 5.4.2. This can be avoided using `data.table` or the `recode` function in `dplyr`. We return to the `smoke` example from Section 5.4 and put it in a `data.table`:

```
library(data.table)
library(dplyr)

smoke <- c("Never", "Never", "Heavy", "Never", "Occasionally",
           "Never", "Never", "Regularly", "Regularly", "No")

smoke2 <- factor(smoke, levels = c("Never", "Occasionally",
                                   "Regularly", "Heavy"),
                        ordered = TRUE)

smoke3 <- data.table(smoke2)
```

Suppose that we want to change the levels' names to abbreviated versions: *Nvr, Occ, Reg,* and *Hvy*. Here's how to do this:

With `data.table`:

```
new_names = c("Nvr", "Occ", "Reg", "Hvy")
smoke3[.(smoke2 = levels(smoke2), to = new_names),
       on = "smoke2",
       smoke2 := i.to]
smoke3[, smoke2 := droplevels(smoke2)]
```

With `dplyr`:

```
# Version 1:
smoke3 |> mutate(smoke2 = case_match(smoke2,
                "Never" ~ "Nvr",
                "Occasionally" ~ "Occ",
                "Regularly" ~ "Reg",
                "Heavy" ~ "Hvy")) -> smoke3

# Version 2:
smoke3 |> mutate(smoke2 = recode(smoke2,
                "Never" = "Nvr",
                "Occasionally" = "Occ",
```

```
                         "Regularly" = "Reg",
                         "Heavy" = "Hvy")) -> smoke3
```

Next, we can combine the *Occ*, *Reg*, and *Hvy* levels into a single level called *Yes*:

With `data.table`:

```
smoke3[.(smoke2 = c("Occ", "Reg", "Hvy"), to = "Yes"),
       on = "smoke2",
       smoke2 := i.to]
```

With `dplyr`:

```
# Version 1:
smoke3 |> mutate(smoke2 = case_match(smoke2,
                 "Nvr" ~ "Nvr",
                 c("Occ", "Reg", "Hvy") ~ "Yes")) -> smoke3

# Version 2:
smoke3 |> mutate(smoke2 = recode(smoke2,
                 "Occ" = "Yes",
                 "Reg" = "Yes",
                 "Hvy" = "Yes")) -> smoke3
```

~

Exercise 5.17. In Exercise 2.27 you learned how to create a `factor` variable from a `numeric` variable using `cut`. Return to your solution (or the solution at the back of the book) and do the following using `data.table` and/or `dplyr`:

  1. Change the category names to `Mild`, `Moderate`, and `Hot`.

  2. Combine `Moderate` and `Hot` into a single level named `Hot`.

5.7.7   Grouped summaries

We've already seen how we can use `aggregate` and `by` to create grouped summaries. However, in many cases it is as easy or easier to use `data.table` or `dplyr` for such summaries.

To begin with, let's load the packages again (in case you don't already have them loaded), and let's recreate the `aq` `data.table`, which we made a bit of a mess of by removing some important columns in the previous section:

```
library(data.table)
library(dplyr)

aq <- data.table(airquality)
```

Now, let's compute the mean temperature for each month. Both `data.table` and `dplyr` will return a data frame with the results. In the `data.table` approach, assigning a name to the summary statistic (`mean`, in this case) is optional, but not in `dplyr`.

With `data.table`:

```
aq[, mean(Temp), Month]
# or, to assign a name:
aq[, .(meanTemp = mean(Temp)),
   Month]
```

With `dplyr`:

```
aq |> group_by(Month) |>
      summarise(meanTemp =
                    mean(Temp))
```

You'll recall that if we apply `mean` to a vector containing `NA` values, it will return `NA`:

With `data.table`:

```
aq[, mean(Ozone), Month]
```

With `dplyr`:

```
aq |> group_by(Month) |>
      summarise(meanTemp =
                    mean(Ozone))
```

In order to avoid this, we can pass the argument `na.rm = TRUE` to `mean`, just as we would in other contexts. To compute the mean ozone concentration for each month, ignoring `NA` values:

With `data.table`:

```
aq[, mean(Ozone, na.rm = TRUE),
   Month]
```

With `dplyr`:

```
aq |> group_by(Month) |>
      summarise(meanTemp =
                    mean(Ozone,
                       na.rm = TRUE))
```

What if we want to compute a grouped summary statistic involving two variables? For instance, the correlation between temperature and wind speed for each month?

With `data.table`:

```
aq[, cor(Temp, Wind), Month]
```

With `dplyr`:

```
aq |> group_by(Month) |>
      summarise(cor =
                    cor(Temp, Wind))
```

The syntax for computing multiple grouped statistics is similar. We compute both the mean temperature and the correlation for each month:

With `data.table`:

```
aq[, .(meanTemp = mean(Temp),
   cor = cor(Temp, Wind)),
   Month]
```

With dplyr:

```
aq |> group_by(Month) |>
    summarise(meanTemp =
        mean(Temp),
        cor =
        cor(Temp, Wind))
```

At times, you'll want to compute summaries for all variables that share some property. As an example, you may want to compute the mean of all `numeric` variables in your data frame. In dplyr there is a convenience function called `across` that can be used for this: `summarise(across(where(is.numeric), mean))` will compute the mean of all numeric variables. In `data.table`, we can instead utilise the `apply` family of functions from base R, that we'll study in Section 6.5. To compute the mean of all `numeric` variables:

With `data.table`:

```
aq[, lapply(.SD, mean),
   Month,
   .SDcols = is.numeric]
```

With dplyr:

```
aq |> group_by(Month) |>
        summarise(across(
           where(is.numeric),
           mean, na.rm = TRUE))
```

Both packages have special functions for counting the number of observations in groups: `.N` for `data.table` and `n` for dplyr. For instance, we can count the number of days in each month:

With `data.table`:

```
aq[, .N, Month]
```

With dplyr:

```
aq |> group_by(Month) |>
        summarise(days = n())
```

Similarly, you can count the number of unique values of variables using `uniqueN` for `data.table` and `n_distinct` for dplyr:

With `data.table`:

```
aq[, uniqueN(Month)]
```

With dplyr:

```
aq |> summarise(months =
            n_distinct(Month))
```

~

Exercise 5.18. Load the VAS pain data `vas.csv` from Exercise 2.30. Then do the following using `data.table` and/or dplyr:

1. Compute the mean VAS for each patient.

    2. Compute the lowest and highest VAS recorded for each patient.

    3. Compute the number of high-VAS days, defined as days with where the VAS was at least 7, for each patient.

Exercise 5.19. We return to the `datasauRus` package and the `datasaurus_dozen` dataset from Exercise 2.28. Check its structure and then do the following using `data.table` and/or `dplyr`:

    1. Compute the mean of x, mean of y, standard deviation of x, standard deviation of y, and correlation between x and y, grouped by `dataset`. Are there any differences between the 12 datasets?

    2. Make a scatterplot of x against y for each dataset. Are there any differences between the 12 datasets?

5.7.8   Filling in missing values

In some cases, you may want to fill in missing values of a variable with the previous non-missing entry. To see an example of this, let's create a version of `aq` where the value of `Month` are missing for some days:

```
aq$Month[c(2:3, 36:39, 70)] <- NA

# Some values of Month are now missing:
head(aq)
```

To fill in the missing values with the last non-missing entry, we can now use `nafill` or `fill` as follows:

With `data.table`:

```
aq[, Month := nafill(
    Month, "locf")]
```

With `tidyr`:

```
aq |> fill(Month) -> aq
```

To instead fill in the missing values with the *next* non-missing entry:

With `data.table`:

```
aq[, Month := nafill(
    Month, "nocb")]
```

With `tidyr`:

```
aq |> fill(Month,
        .direction = "up") -> aq
```

~

Exercise 5.20. Load the VAS pain data `vas.csv` from Exercise 2.30. Fill in the missing values in the `Visit` column with the last non-missing value.

5.7.9 Chaining commands together

When working with tidyverse packages, commands are usually chained together using |>
pipes. When using data.table, commands are chained by repeated use of [] brackets on
the same line. This is probably best illustrated using an example. Assume again that there
are missing values in Month in aq:

```
aq$Month[c(2:3, 36:39, 70)] <- NA
```

To fill in the missing values with the last non-missing entry (Section 5.7.8) and then count
the number of days in each month (Section 5.7.7), we can do as follows:

With data.table:

```
aq[, Month := nafill(Month, "locf")][, .N, Month]
```

With tidyr and dplyr:

```
aq |> fill(Month) |>
      group_by(Month) |>
      summarise(days = n())
```

5.8 Filtering: select rows

You'll frequently want to filter away some rows from your data. Perhaps you only want
to select rows where a variable exceeds some value, or you want to exclude rows with NA
values. This can be done in several different ways: using row numbers, using conditions, at
random, or using regular expressions. Let's have a look at them, one by one. We'll use aq,
the data.table version of airquality that we created before, for the examples.

```
library(data.table)
library(dplyr)

aq <- data.table(airquality)
```

5.8.1 Filtering using row numbers

If you know the row numbers of the rows that you wish to remove (perhaps you've found
them using which, as in Section 2.11.3?), you can use those numbers for filtering. Here are
four examples.

To select row 3:

With data.table: With dplyr:

```
aq[3,]
```
```
aq |> slice(3)
```

To select rows 3 to 5:

With `data.table`:

```
aq[3:5,]
```

With `dplyr`:

```
aq |> slice(3:5)
```

To select rows 3, 7, and 15:

With `data.table`:

```
aq[c(3, 7, 15),]
```

With `dplyr`:

```
aq |> slice(c(3, 7, 15))
```

To select all rows except rows 3, 7, and 15:

With `data.table`:

```
aq[-c(3, 7, 15),]
```

With `dplyr`:

```
aq |> slice(-c(3, 7, 15))
```

5.8.2 Filtering using conditions

Filtering is often done using conditions, e.g., to select observations with certain properties. Here are some examples:

To select rows where `Temp` is greater than 90:

With `data.table`:

```
aq[Temp > 90,]
```

With `dplyr`:

```
aq |> filter(Temp > 90)
```

To select rows where `Month` is 6 (June):

With `data.table`:

```
aq[Month == 6,]
```

With `dplyr`:

```
aq |> filter(Month == 6)
```

To select rows where `Temp` is greater than 90 and `Month` is 6 (June):

With `data.table`:

```
aq[Temp > 90 & Month == 6,]
```

With `dplyr`:

```
aq |> filter(Temp > 90,
             Month == 6)
```

To select rows where `Temp` is between 80 and 90 (including 80 and 90):

With `data.table`:   With `dplyr`:

```
aq[Temp %between% c(80, 90),]
```

```
aq |> filter(between(Temp,
                     80, 90))
```

To select the five rows with the highest `Temp`:

With `data.table`:   With `dplyr`:

```
aq[frankv(-Temp,
     ties.method = "min") <= 5,
   ]
```

```
aq |> top_n(5, Temp)
```

In this case, the above code returns more than 5 rows because of ties. To remove duplicate rows:

With `data.table`:   With `dplyr`:

```
unique(aq)
```

```
aq |> distinct()
```

To remove rows with missing data (`NA` values) in at least one variable:

With `data.table`:   With `tidyr`:

```
na.omit(aq)
```

```
library(tidyr)
aq |> drop_na()
```

To remove rows with missing `Ozone` values:

With `data.table`:   With `tidyr`:

```
na.omit(aq, "Ozone")
```

```
library(tidyr)
aq |> drop_na("Ozone")
```

At times, you want to filter your data based on whether the observations are connected to observations in a different dataset. Such filters are known as semijoins and antijoins. They are discussed in Section 5.12.4.

5.8.3 Selecting rows at random

In some situations, for instance when training and evaluating machine learning models, you may wish to draw a random sample from your data. This is done using the `sample` (`data.table`) and `sample_n` (`dplyr`) functions.

To select five rows at random:

With `data.table`:

```
aq[sample(.N, 5),]
```

With `dplyr`:

```
aq |> sample_n(5)
```

If you run the code multiple times, you will get different results each time. See Section 7.1 for more on random sampling and how it can be used.

5.8.4 Using regular expressions to select rows

In some cases, particularly when working with text data, you'll want to filter using regular expressions (see Section 5.5.3). `data.table` has a convenience function called `%like%` that can be used to call `grepl` in an alternative (less opaque?) way. With `dplyr` we use `grepl` in the usual fashion. To have some text data to try this out on, we'll use this data frame, which contains descriptions of some dogs:

```
dogs <- data.table(Name = c("Bianca", "Bella", "Mimmi", "Daisy",
                            "Ernst", "Smulan"),
                Breed = c("Greyhound", "Greyhound", "Pug", "Poodle",
                          "Bedlington Terrier", "Boxer"),
                Desc = c("Fast, playful", "Fast, easily worried",
                         "Intense, small, loud",
                         "Majestic, protective, playful",
                         "Playful, relaxed",
                         "Loving, cuddly, playful"))
View(dogs)
```

To select all rows with names beginning with B:

With `data.table`:

```
dogs[Name %like% "^B",]
# or:
dogs[grepl("^B", Name),]
```

With `dplyr`:

```
dogs |> filter(grepl("B[a-z]",
                     Name))
```

To select all rows where `Desc` includes the word `playful`:

With `data.table`:

```
dogs[Desc %like% "[pP]layful",]
```

With `dplyr`:

```
dogs |> filter(grepl("[pP]layful",
                     Desc))
```

~

Exercise 5.21. Download the `ucdp-onesided-191.csv` data file from the book's web page[12]. It contains data about international attacks on civilians by governments and formally organised armed groups during the period 1989-2018, collected as part of the Uppsala Conflict Data Program (Eck & Hultman, 2007; Petterson et al., 2019). Among other things, it contains information about the actor (attacker), fatality rate, and attack location. Load the data and check its structure.

1. Filter the rows so that only conflicts that took place in Colombia are retained. How many different actors were responsible for attacks in Colombia during the period?

2. Using the `best_fatality_estimate` column to estimate fatalities, calculate the number of worldwide fatalities caused by government attacks on civilians during 1989-2018.

Exercise 5.22. Load the `oslo-biomarkers.xlsx` data from Exercise 5.8. Use `data.table` and/or `dplyr` to do the following:

1. Select only the measurements from blood samples taken at 12 months.

2. Select only the measurements from the patient with ID number 6.

5.9 Subsetting: select columns

Another common situation is that you want to remove some variables from your data. Perhaps the variables aren't of interest in a particular analysis that you're going to perform, or perhaps you've simply imported more variables than you need. As with rows, this can be done using numbers, names, or regular expressions. Let's look at some examples using the aq data:

```
library(data.table)
library(dplyr)

aq <- data.table(airquality)
```

5.9.1 Selecting a single column

When selecting a single column from a data frame, you sometimes want to extract the column as a vector and sometimes as a single-column data frame (for instance, if you are going to pass it to a function that takes a data frame as input). You should be a bit careful when doing this, to make sure that you get the column in the correct format:

---

[12]http://www.modernstatisticswithr.com/data.zip

With `data.table`:

```
# Return a vector:
aq$Temp
# or
aq[, Temp]

# Return a data.table:
aq[, "Temp"]
```

With `dplyr`:

```
# Return a vector:
aq$Temp
# or
aq |> pull(Temp)

# Return a tibble:
aq |> select(Temp)
```

5.9.2 Selecting multiple columns

Selecting multiple columns is more straightforward, as the object that is returned always will be a data frame. Here are some examples.

To select `Temp`, `Month`, and `Day`:

With `data.table`:

```
aq[, .(Temp, Month, Day)]
```

With `dplyr`:

```
aq |> select(Temp, Month, Day)
```

To select all columns between `Wind` and `Month`:

With `data.table`:

```
aq[, Wind:Month]
```

With `dplyr`:

```
aq |> select(Wind:Month)
```

To select all columns except `Month` and `Day`:

With `data.table`:

```
aq[, -c("Month", "Day")]
```

With `dplyr`:

```
aq |> select(-Month, -Day)
```

To select all numeric variables (which for the `aq` data is all variables!):

With `data.table`:

```
aq[, .SD, .SDcols = is.numeric]
```

With `dplyr`:

```
aq |> select_if(is.numeric)
```

To remove columns with missing (`NA`) values:

With `data.table`:   With `dplyr`:

```
aq[, .SD,                                       aq |> select_if(~all(!is.na(.)))
   .SDcols = !anyNA]
```

5.9.3 Using regular expressions to select columns

In `data.table`, using regular expressions to select columns is done using `grep`. `dplyr` differs in that it has several convenience functions for selecting columns, like `starts_with`, `ends_with`, `contains`. As an example, we can select variables the name of which contains the letter n:

With `data.table`:

```
vars <- grepl("n", names(aq))
aq[, ..vars]
```

With `dplyr`:

```
# contains is a convenience function for checking if a name contains a string:
aq |> select(contains("n"))
# matches can be used with any regular expression:
aq |> select(matches("n"))
```

5.9.4 Subsetting using column numbers

It is also possible to subset using column numbers, but you need to be careful if you want to use that approach. Column numbers can change, for instance if a variable is removed from the data frame. More importantly, however, using column numbers can yield different results depending on what type of data table you're using. Let's have a look at what happens if we use this approach with different types of data tables:

```
# data.frame:
aq <- as.data.frame(airquality)
str(aq[,2])

# data.table:
aq <- as.data.table(airquality)
str(aq[,2])

# tibble:
aq <- as_tibble(airquality)
str(aq[,2])
```

As you can see, `aq[, 2]` returns a vector, a data table or a tibble, depending on what type of object `aq` is. Unfortunately, this approach is used by several R packages and can cause problems because it may return the wrong type of object.

A better approach is to use `aq[[2]]`, which works the same for data frames, data tables, and tibbles, returning a vector:

```
# data.frame:
aq <- as.data.frame(airquality)
str(aq[[2]])

# data.table:
aq <- as.data.table(airquality)
str(aq[[2]])

# tibble:
aq <- as_tibble(airquality)
str(aq[[2]])
```

~

Exercise 5.23. Return to the `ucdp-onesided-191.csv` data from Exercise 5.21. To have a cleaner and less bloated dataset to work with, it can make sense to remove some columns. Select only the `actor_name`, `year`, `best_fatality_estimate` and `location` columns.

5.10 Sorting

Sometimes you don't want to filter rows but rearrange their order according to their values for some variable. Similarly, you may want to change the order of the columns in your data. I often do this after merging data from different tables (as we'll do in Section 5.12). This is often useful for presentation purposes but can at times also aid in analyses.

5.10.1 Changing the column order

It is straightforward to change column positions using `setcolorder` in `data.table` and `relocate` in `dplyr`.

To put `Month` and `Day` in the first two columns, without rearranging the other columns:

With `data.table`:   With `dplyr`:

```
setcolorder(aq, c("Month", "Day"))        aq |> relocate("Month", "Day")
```

5.10.2 Changing the row order

In `data.table`, `order` is used for sorting rows, and in `dplyr`, `arrange` is used (sometimes in combination with `desc`). The syntax differs depending on whether you wish to sort your rows in ascending or descending order. We will illustrate this using the `airquality` data.

```
library(data.table)
library(dplyr)
```

```
aq <- data.table(airquality)
```

First of all, if you're just looking to sort a single vector, rather than an entire data frame, the quickest way to do so is to use sort:

```
sort(aq$Wind)
sort(aq$Wind, decreasing = TRUE)
sort(c("C", "B", "A", "D"))
```

If you're looking to sort an entire data frame by one or more variables, you need to move beyond sort. To sort rows by Wind (*ascending* order):

With data.table:

```
aq[order(Wind),]
```

With dplyr:

```
aq |> arrange(Wind)
```

To sort rows by Wind (*descending* order):

With data.table:

```
aq[order(-Wind),]
```

With dplyr:

```
aq |> arrange(-Wind)
# or
aq |> arrange(desc(Wind))
```

To sort rows, first by Temp (ascending order) and then by Wind (descending order):

With data.table:

```
aq[order(Temp, -Wind),]
```

With dplyr:

```
aq |> arrange(Temp, desc(Wind))
```

~

Exercise 5.24. Load the oslo-biomarkers.xlsx data from Exercise 5.8. Note that it is not ordered in a natural way. Reorder it by patient ID instead.

5.11 Reshaping data

The gapminder dataset from the gapminder package contains information about life expectancy, population size and GDP per capita for 142 countries for 12 years from 1952 to 2007. To begin with, let's have a look at the data[13]:

---

[13]You may need to install the package first, using install.packages("gapminder").

```
library(gapminder)
?gapminder
View(gapminder)
```

Each row contains data for one country and one year, meaning that the data for each country is spread over 12 rows. This is known as *long data* or *long format*. As another option, we could store it in *wide format*, where the data is formatted so that all observations corresponding to a country are stored on the same row:

```
Country        Continent    lifeExp1952 lifeExp1957 lifeExp1962 ...
Afghanistan    Asia            28.8        30.2        32.0       ...
Albania        Europe          55.2        59.3        64.8       ...
```

Sometimes, it makes sense to spread an observation over multiple rows (long format), and sometimes it makes more sense to spread a variable across multiple columns (wide format). Some analyses require long data, whereas others require wide data. And if you're unlucky, data will arrive in the wrong format for the analysis you need to do. In this section, you'll learn how to transform your data from long to wide, and back again.

5.11.1   From long to wide

When going from a long format to a wide format, you choose columns to group the observations by (in the gapminder case: country and maybe also continent), columns to take value names from (lifeExp, pop, and gdpPercap), and columns to create variable names from (year).

In data.table, the transformation from long to wide is done using the dcast function. dplyr does not contain functions for such transformations, but its sibling, the tidyverse package tidyr, does.

The tidyr function used for long-to-wide formatting is pivot_wider. First, we convert the gapminder data frame to a data.table object:

```
library(data.table)
library(tidyr)

gm <- as.data.table(gapminder)
```

To transform the gm data from long to wide and store it as gmw:

With data.table:

```
gmw <- dcast(gm, country + continent ~ year,
             value.var = c("pop", "lifeExp", "gdpPercap"))
```

With tidyr:

```
gm |> pivot_wider(id_cols = c(country, continent),
                  names_from = year,
```

```
values_from =
  c(pop, lifeExp, gdpPercap)) -> gmw
```

5.11.2   From wide to long

We've now seen how to transform the long format `gapminder` data to the wide format `gmw` data. But what if we want to go from wide format to long? Let's see if we can transform `gmw` back to the long format.

In `data.table`, wide-to-long formatting is done using `melt`, and in `dplyr` it is done using `pivot_longer`.

To transform the `gmw` data from long to wide:

With `data.table`:

```
gm <- melt(gmw, id.vars = c("country", "continent"),
           measure.vars = 2:37)
```

With `tidyr`:

```
gmw |> pivot_longer(names(gmw)[2:37],
                    names_to = "variable",
                    values_to = "value") -> gm
```

The resulting data frames are perhaps *too* long, with each variable (`pop`, `lifeExp`, and `gdpPercapita`) being put on a different row. To make it look like the original dataset, we must first split the `variable` variable (into a column with variable names and column with years) and then make the data frame a little wider again. That is the topic of the next section.

5.11.3   Splitting columns

In the too long `gm` data that you created at the end of the last section, the observations in the `variable` column look like `pop_1952` and `gdpPercap_2007`, i.e., are of the form `variable-Name_year`. We'd like to split them into two columns: one with variable names and one with years. `dplyr` has a function called `tstrsplit` for this purpose, and `tidyr` has `separate`.

To split the `variable` column at the underscore `_`, and then reformat `gm` to look like the original gapminder data:

With `data.table`:

```
gm[, c("variable", "year") := tstrsplit(variable,
                              "_", fixed = TRUE)]
gm <- dcast(gm, country + year ~ variable,
            value.var = c("value"))
```

With `tidyr`:

```
gm |> separate(variable,
              into = c("variable", "year"),
              sep = "_") |>
    pivot_wider(id_cols = c(country, continent, year),
                names_from = variable,
                values_from = value) -> gm
```

5.11.4 Merging columns

Similarly, you may at times want to merge two columns, for instance if one contains the day+month part of a date and the other contains the year. An example of such a situation can be found in the `airquality` dataset, where we may want to merge the `Day` and `Month` columns into a new `Date` column. Let's re-create the `aq` `data.table` object one last time:

```
library(data.table)
library(tidyr)

aq <- as.data.table(airquality)
```

If we wanted to create a `Date` column containing the year (1973), month and day for each observation, we could use `paste` and `as.Date`:

```
as.Date(paste(1973, aq$Month, aq$Day, sep = "-"))
```

The natural `data.table` approach is just this, whereas `tidyr` offers a function called `unite` to merge columns, which can be combined with `mutate` to paste the year to the date. To merge the `Month` and `Day` columns with a year and convert it to a `Date` object:

With `data.table`:

```
aq[, Date := as.Date(paste(1973,
                aq$Month,
                aq$Day,
                sep = "-"))]
```

With `tidyr` and `dplyr`:

```
aq |> unite("Date", Month, Day,
            sep = "-") |>
    mutate(Date = as.Date(
        paste(1973,
              Date,
              sep = "-")))
```

~

Exercise 5.25. Load the `oslo-biomarkers.xlsx` data from Exercise 5.8. Then do the following using `data.table` and/or `dplyr`/`tidyr`:

    1. Split the `PatientID.timepoint` column in two parts: one with the patient ID and one with the timepoint.

    2. Sort the table by patient ID, in numeric order.

3. Reformat the data from long to wide, keeping only the IL-8 and VEGF-A measurements.

Save the resulting data frame – you will need it again in Exercise 5.26!

5.12   Merging data from multiple tables

It is common that data is spread over multiple tables: different sheets in Excel files, different .csv files, or different tables in databases. Consequently, it is important to be able to merge data from different tables.

As a first example, let's study the sales datasets available from the book's web page: `sales-rev.csv` and `sales-weather.csv`. The first dataset describes the daily revenue for a business in the first quarter of 2020, and the second describes the weather in the region (somewhere in Sweden) during the same period[14]. Store their respective paths as `file_path1` and `file_path2` and then load them:

```
rev_data <- read.csv(file_path1, sep = ";")
weather_data <- read.csv(file_path2, sep = ";")

str(rev_data)
View(rev_data)

str(weather_data)
View(weather_data)
```

5.12.1   Binds

The simplest types of merges are *binds*, which can be used when you have two tables where either the rows or the columns *match each other exactly*. To illustrate what this may look like, we will use `data.table/dplyr` to create subsets of the business revenue data. First, we format the tables as `data.table` objects and the `DATE` columns as `Date` objects:

```
library(data.table)
library(dplyr)

rev_data <- as.data.table(rev_data)
rev_data$DATE <- as.Date(rev_data$DATE)

weather_data <- as.data.table(weather_data)
weather_data$DATE <- as.Date(weather_data$DATE)
```

Next, we wish to subtract three subsets: the revenue in January (`rev_jan`), the revenue in February (`rev_feb`), and the weather in January (`weather_jan`).

---

[14]I've intentionally left out the details regarding the business – these are real sales data from a client, which can be sensitive information.

With `data.table`:

```
rev_jan <- rev_data[DATE %between%
                c("2020-01-01",
                "2020-01-31"),]
rev_feb <- rev_data[DATE %between%
                c("2020-02-01",
                "2020-02-29"),]
weather_jan <- weather_data[DATE
                %between%
                c("2020-01-01",
                "2020-01-31"),]
```

With `dplyr`:

```
rev_data |> filter(between(DATE,
                as.Date("2020-01-01"),
                as.Date("2020-01-31"))
                ) -> rev_jan
rev_data |> filter(between(DATE,
                as.Date("2020-02-01"),
                as.Date("2020-02-29"))
                ) -> rev_feb
weather_data |> filter(between(
                DATE,
                as.Date("2020-01-01"),
                as.Date("2020-01-31"))
                ) -> weather_jan
```

A quick look at the structure of the data reveals some similarities:

```
str(rev_jan)
str(rev_feb)
str(weather_jan)
```

The rows in `rev_jan` correspond one-to-one to the *rows* in `weather_jan`, with both tables being sorted in exactly the same way. We could therefore *bind their columns*, i.e., add the columns of `weather_jan` to `rev_jan`.

`rev_jan` and `rev_feb` contain the same *columns*. We could therefore *bind their rows*, i.e., add the rows of `rev_feb` to `rev_jan`. To perform these operations, we can use either base R or `dplyr`:

With base R:

```
# Join columns of datasets that have the same rows:
cbind(rev_jan, weather_jan)

# Join rows of datasets that have the same columns:
rbind(rev_jan, rev_feb)
```

With `dplyr`:

```
# Join columns of datasets that have the same rows:
bind_cols(rev_jan, weather_jan)

# Join rows of datasets that have the same columns:
bind_rows(rev_jan, rev_feb)
```

5.12.2   Merging tables using keys

A closer look at the business revenue data reveals that `rev_data` contains observations from 90 days, whereas `weather_data` only contains data for 87 days; revenue data for 2020-03-01 is missing, and weather data for 2020-02-05, 2020-02-06, 2020-03-10, and 2020-03-29 are missing.

Suppose that we want to study how weather affects the revenue of the business. In order to do so, we must merge the two tables. We cannot use a simple column bind, because the two tables have different numbers of rows. If we attempt a bind, R will produce a merged table by recycling the first few rows from `rev_data` – note that the two `DATE` columns aren't properly aligned:

```
tail(cbind(rev_data, weather_data))
```

Clearly, this is not the desired output! We need a way to connect the rows in `rev_data` with the right rows in `weather_data`. Put differently, we need something that allows us to connect the observations in one table to those in another. Variables used to connect tables are known as *keys*, and must in some way uniquely identify observations. In this case the `DATE` column gives us the key – each observation is uniquely determined by its `DATE`. So to combine the two tables, we can combine rows from `rev_data` with the rows from `weather_data` that have the same `DATE` values. In the following sections, we'll look at different ways of merging tables using `data.table` and `dplyr`.

But first, a word of warning: finding the right keys for merging tables is not always straightforward. For a more complex example, consider the `nycflights13` package, which contains five separate but connected datasets:

```
library(nycflights13)
?airlines   # Names and carrier codes of airlines.
?airports   # Information about airports.
?flights    # Departure and arrival times and delay information for
            # flights.
?planes     # Information about planes.
?weather    # Hourly meteorological data for airports.
```

Perhaps you want to include weather information with the flight data, to study how weather affects delays. Or perhaps you wish to include information about the longitude and latitude of airports (from `airports`) in the `weather` dataset. In `airports`, each observation can be uniquely identified in three different ways: either by its airport code `faa`, its name `name`, or its latitude and longitude, `lat` and `lon`:

```
?airports
head(airports)
```

If we want to use either of these options as a key when merging with `airports` data with another table, that table should also contain the same key.

The `weather` data requires no less than four variables to identify each observation: `origin`, `month`, `day` and `hour`:

```
?weather
head(weather)
```

It is not perfectly clear from the documentation, but the `origin` variable is actually the FAA airport code of the airport corresponding to the weather measurements. If we wish to add longitude and latitude to the weather data, we could therefore use `faa` from `airports` as a key.

5.12.3 Inner and outer joins

An operation that combines columns from two tables is called a *join*. There are two main types of joins: *inner joins* and *outer joins.*

*   *Inner joins*: create a table containing all observations for which the key appeared in both tables. So if we perform an inner join on the `rev_data` and `weather_data` tables using `DATE` as the key, it won't contain data for the days that are missing from either the revenue table or the weather table.

In contrast, outer joins create a table retaining rows, even if there is no match in the other table. There are three types of outer joins:

*   *Left join*: retains all rows from the first table. In the revenue example, this means all dates present in `rev_data`.
*   *Right join*: retains all rows from the second table. In the revenue example, this means all dates present in `weather_data`.
*   *Full join*: retains all rows present in at least one of the tables. In the revenue example, this means all dates present in at least one of `rev_data` and `weather_data`.

We will use the `rev_data` and `weather_data` datasets to exemplify the different types of joins. To begin with, we convert them to `data.table` objects (which is optional if you wish to use `dplyr`):

```
library(data.table)
library(dplyr)

rev_data <- as.data.table(rev_data)
weather_data <- as.data.table(weather_data)
```

Remember that revenue data for 2020-03-01 is missing, and weather data for 2020-02-05, 2020-02-06, 2020-03-10, and 2020-03-29 are missing. This means that out of the 91 days in the period, only 86 have complete data. If we perform an inner join, the resulting table should therefore have 86 rows.

To perform an inner join of `rev_data` and `weather_data` using `DATE` as key:

With `data.table`:

```
merge(rev_data, weather_data,
      by = "DATE")

# Or:
setkey(rev_data, DATE)
rev_data[weather_data, nomatch = 0]
```

With `dplyr`:

```
rev_data |> inner_join(
                weather_data,
                by = "DATE")
```

A left join will retain the 90 dates present in `rev_data`. To perform a(n outer) left join of `rev_data` and `weather_data` using `DATE` as key:

With `data.table`:   With `dplyr`:

```
merge(rev_data, weather_data,
      all.x = TRUE, by = "DATE")
# Or:
setkey(weather_data, DATE)
weather_data[rev_data]
```

```
rev_data |> left_join(
                weather_data,
                by = "DATE")
```

A right join will retain the 87 dates present in `weather_data`. To perform a(n outer) right join of `rev_data` and `weather_data` using `DATE` as key:

With `data.table`:   With `dplyr`:

```
merge(rev_data, weather_data,
      all.y = TRUE, by = "DATE")
# Or:
setkey(rev_data, DATE)
rev_data[weather_data]
```

```
rev_data |> right_join(
                weather_data,
                by = "DATE")
```

A full join will retain the 91 dates present in at least one of `rev_data` and `weather_data`. To perform a(n outer) full join of `rev_data` and `weather_data` using `DATE` as key:

With `data.table`:   With `dplyr`:

```
merge(rev_data, weather_data,
      all = TRUE, by = "DATE")
```

```
rev_data |> full_join(
                weather_data,
                by = "DATE")
```

5.12.4 Semijoins and antijoins

Semijoins and antijoins are similar to joins but work on observations rather than variables. That is, they are used for filtering one table using data from another table:

- *Semijoin*: retains all observations in the first table that have a match in the second table.
- *Antijoin*: retains all observations in the first table that *do not* have a match in the second table.

The same thing can be achieved using the filtering techniques of Section 5.8, but semijoins and antijoins are simpler to use when the filtering relies on conditions from another table.

Suppose that we are interested in the revenue of our business for days in February with subzero temperatures. First, we can create a table called `filter_data` listing all such days:

With `data.table`:

```
rev_data$DATE <- as.Date(rev_data$DATE)
weather_data$DATE <- as.Date(weather_data$DATE)
filter_data <- weather_data[TEMPERATURE < 0 &
                     DATE %between%
                     c("2020-02-01",
                       "2020-02-29"),]
```

With `dplyr`:

```
rev_data$DATE <- as.Date(rev_data$DATE)
weather_data$DATE <- as.Date(weather_data$DATE)
weather_data |> filter(TEMPERATURE < 0,
                   between(DATE,
                   as.Date("2020-02-01"),
                   as.Date("2020-02-29"))
                       ) -> filter_data
```

Next, we can use a semijoin to extract the rows of `rev_data` corresponding to the days of `filter_data`:

With `data.table`:

```
setkey(rev_data, DATE)
rev_data[rev_data[filter_data,
             which = TRUE]]
```

With `dplyr`:

```
rev_data |> semi_join(
             filter_data,
             by = "DATE")
```

If instead we wanted to find all days *except* the days in February with subzero temperatures, we could perform an antijoin:

With `data.table`:

```
setkey(rev_data, DATE)
rev_data[!filter_data]
```

With `dplyr`:

```
rev_data |> anti_join(
             filter_data,
             by = "DATE")
```

∼

Exercise 5.26. We return to the `oslo-biomarkers.xlsx` data from Exercises 5.8 and 5.25. Load the data frame that you created in Exercise 5.25 (or copy the code from its solution).

You should also load the `oslo-covariates.xlsx` data from the book's web page; it contains information about the patients, such as age, gender, and smoking habits.

Then do the following using `data.table` and/or `dplyr/tidyr`:

1. Merge the wide data frame from Exercise 5.25 with the `oslo-covariates.xlsx` data, using patient ID as key.

2. Use the `oslo-covariates.xlsx` data to select data for smokers from the wide data frame Exercise 5.25.

5.13   Scraping data from websites

Web scraping is the process of extracting data from a webpage. For instance, let's say that we'd like to download the list of Nobel laureates from the Wikipedia page https://en.wikipedia.org/wiki/List_of_Nobel_laureates. As with most sites, the text and formatting of the page is stored in an HTML file. In most browsers, you can view the HTML code by right-clicking on the page and choosing *View page source*. As you can see, all the information from the table can be found there, albeit in a format that is only just human-readable:

```
...
<tbody><tr>
<th>Year
</th>
<th width="18%"><a href="/wiki/List_of_Nobel_laureates_in_Physics"
  title="List of Nobel laureates in Physics">Physics</a>
</th>
<th width="16%"><a href="/wiki/List_of_Nobel_laureates_in_Chemistry"
title="List of Nobel laureates in Chemistry">Chemistry</a>
</th>
<th width="18%"><a href="/wiki/List_of_Nobel_laureates_in_Physiology_
or_Medicine" title="List of Nobel laureates in Physiology or Medicine
">Physiology<br />or Medicine</a>
</th>
<th width="16%"><a href="/wiki/List_of_Nobel_laureates_in_Literature"
title="List of Nobel laureates in Literature">Literature</a>
</th>
<th width="16%"><a href="/wiki/List_of_Nobel_Peace_Prize_laureates"
title="List of Nobel Peace Prize laureates">Peace</a>
</th>
<th width="15%"><a href="/wiki/List_of_Nobel_laureates_in_Economics"
class="mw-redirect" title="List of Nobel laureates in Economics">
  Economics</a><br />(The Sveriges Riksbank Prize)<sup id="cite_ref-
11" class="reference"><a href="#cite_note-11">&#91;11&#93;</a></sup>
</th></tr>
<tr>
<td align="center">1901
```

```
</td>
<td><span data-sort-value="Röntgen, Wilhelm"><span class="vcard"><span
class="fn"><a href="/wiki/Wilhelm_R%C3%B6ntgen" title="Wilhelm
Röntgen">  Wilhelm Röntgen</a></span></span></span>
</td>
<td><span data-sort-value="Hoff, Jacobus Henricus van &#39;t"><span
class="vcard"><span class="fn"><a href="/wiki/Jacobus_Henricus_van_
%27t_Hoff" title="Jacobus Henricus van &#39;t Hoff">Jacobus Henricus
van 't Hoff</a></span></span></span>
</td>
<td><span data-sort-value="von Behring, Emil Adolf"><span class=
"vcard">
<span class="fn"><a href="/wiki/Emil_Adolf_von_Behring" class="mw-
redirect" title="Emil Adolf von Behring">Emil Adolf von Behring</a>
</span></span></span>
</td>
...
```

To get hold of the data from the table, we could perhaps select all rows, copy them, and paste them into a spreadsheet software such as Excel. But it would be much more convenient to be able to just import the table to R straight from the HTML file. Because tables written in HTML follow specific formats, it is possible to write code that automatically converts them to data frames in R. The `rvest` package contains a number of functions for that. Let's install it:

```
install.packages("rvest")
```

To read the entire Wikipedia page, we use:

```
library(rvest)
url <- "https://en.wikipedia.org/wiki/List_of_Nobel_laureates"
wiki <- read_html(url)
```

The object `wiki` now contains all the information from the page – you can have a quick look at it by using `html_text`:

```
html_text(wiki)
```

That is more information than we need. To extract all tables from `wiki`, we can use `html_nodes`:

```
tables <- html_nodes(wiki, "table")
tables
```

The first table, starting with the HTML code

```
<table class="wikitable sortable"><tbody>\n<tr>\n<th>Year\n</th>
```

is the one we are looking for. To transform it to a data frame, we use `html_table` as follows:

```
laureates <- html_table(tables[[1]], fill = TRUE)
View(laureates)
```

The `rvest` package can also be used for extracting data from more complex website structures using the SelectorGadget[15] tool in the web browser Chrome[16]. The SelectorGadget
lets you select the page elements that you wish to scrape in your browser, and helps
you create the code needed to import them to R. For an example of how to use it, run
`vignette("selectorgadget")`.

~

Exercise 5.27. Scrape the table containing different keytar models from https://en.wik
ipedia.org/wiki/List_of_keytars. Perform the necessary operations to convert the `Dates`
column to `numeric`.

5.14   Other commons tasks

5.14.1   Deleting variables

If you no longer need a variable, you can delete it using `rm`:

```
my_variable <- c(1, 8, pi)
my_variable
rm(my_variable)
my_variable
```

This can be useful for instance if you have loaded a data frame that is no longer needed and
takes up a lot of memory. If you, for some reason, want to wipe all your variables, you can
use `ls`, which returns a vector containing the names of all variables, in combination with `rm`:

```
# Use this at your own risk! This deletes all currently loaded
# variables.

# Uncomment to run:
# rm(list = ls())
```

Variables are automatically deleted when you exit R (unless you choose to save your
workspace). On the rare occasions where I want to wipe all variables from memory, I usually
do a restart instead of using `rm`.

---

[15]https://selectorgadget.com/
[16]https://www.google.com/chrome/

5.14.2   Importing data from other statistical packages

The `foreign` library contains functions for importing data from other statistical packages, such as Stata (`read.dta`), Minitab (`read.mtp`), SPSS (`read.spss`), and SAS (XPORT files, `read.xport`). They work just like `read.csv` (see Section 2.15), with additional arguments specific to the file format used for the statistical package in question.

5.14.3   Importing data from databases

R and RStudio have excellent support for connecting to databases. However, this requires some knowledge about databases and topics like ODBC drivers and is therefore beyond the scope of the book. More information about using databases with R can be found at https://db.rstudio.com/.

5.14.4   Importing data from JSON files

JSON is a common file format for transmitting data between different systems. It is often used in web server systems where users can request data. One example of this is found in the JSON file at: https://opendata-download-metobs.smhi.se/api/version/1.0/parameter/2 /station/98210/period/latest-months/data.json. It contains daily mean temperatures from Stockholm, Sweden, during the last few months, accessible from the Swedish Meteorological and Hydrological Institute's server. Have a look at it in your web browser, and then install the `jsonlite` package:

```
install.packages("jsonlite")
```

We'll use the `fromJSON` function from `jsonlite` to import the data:

```
library(jsonlite)
url <- paste("https://opendata-download-metobs.smhi.se/api/version/",
             "1.0/parameter/2/station/98210/period/latest-months/",
             "data.json",
             sep = "")
stockholm <- fromJSON(url)
stockholm
```

By design, JSON files contain lists, and so `stockholm` is a `list` object. The temperature data that we were looking for is (in this particular case) contained in the list element called `value`:

```
stockholm$value
```

6

R programming

The tools in Chapters 2-5 will allow you to manipulate, summarise, and visualise your data in all sorts of ways. But what if you need to compute some statistic that there isn't a function for? What if you need automatic checks of your data and results? What if you need to repeat the same analysis for a large number of files? This is where the programming tools you'll learn about in this chapter, like loops and conditional statements, come in handy. And this is where you take the step from being able to use R for routine analyses to being able to use R for *any* analysis.

After working with the material in this chapter, you will be able to use R to:

- Write your own R functions,
- Use several new pipe operators,
- Use conditional statements to perform different operations depending on whether or not a condition is satisfied,
- Iterate code operations multiple times using loops,
- Iterate code operations multiple times using functionals, and
- Measure the performance of your R code.

6.1 Functions

Suppose that we wish to compute the mean of a vector x. One way to do this would be to use sum and length:

```
x <- 1:100
# Compute mean:
sum(x)/length(x)
```

Now suppose that we wish to compute the mean of several vectors. We could do this by repeated use of sum and length:

```
x <- 1:100
y <- 1:200
z <- 1:300

# Compute means:
sum(x)/length(x)
sum(y)/length(y)
sum(z)/length(x)
```

But wait! I made a mistake when I copied the code to compute the mean of z – I forgot to change `length(x)` to `length(z)`! This is an easy mistake to make when you repeatedly copy and paste code. In addition, repeating the same code multiple times just doesn't look good. It would be much more convenient to have a single function for computing the means. Fortunately, such a function exists – `mean`:

```
# Compute means
mean(x)
mean(y)
mean(z)
```

As you can see, using `mean` makes the code shorter and easier to read and reduces the risk of errors induced by copying and pasting code (we only have to change the argument of one function instead of two).

You've already used a ton of different functions in R: functions for computing means, manipulating data, plotting graphics, and more. All these functions have been written by somebody who thought that they needed to repeat a task (e.g., computing a mean or plotting a bar chart) over and over again. And in such cases, it is much more convenient to have a function that does that task than to have to write or copy code every time you want to do it. This is true also for your own work – whenever you need to repeat the same task several times, it is probably a good idea to write a function for it. It will reduce the amount of code you have to write and lessen the risk of errors caused by copying and pasting old code. In this section, you will learn how to write your own functions.

6.1.1   Creating functions

For the sake of the example, let's say that we wish to compute the mean of several vectors but that the function `mean` doesn't exist. We would therefore like to write our own function for computing the mean of a vector. An R function takes some variables as input (arguments or parameters) and returns an object. Functions are defined using `function`. The definition follows a particular format:

```
function_name <- function(argument1, argument2, ...)
{
     # ...
     # Some rows with code that creates some_object
     # ...
     return(some_object)
}
```

In the case of our function for computing a mean, this could look like:

```
average <- function(x)
{
     avg <- sum(x)/length(x)
     return(avg)
}
```

This defines a function called `average` that takes an object called `x` as input. It computes the sum of the elements of `x`, divides that by the number of elements in `x`, and returns the resulting mean.

If we now make a call to `average(x)`, our function will compute the mean value of the vector `x`. Let's try it out, to see that it works:

```
x <- 1:100
y <- 1:200
average(x)
average(y)
```

6.1.2   Local and global variables

Note that despite the fact that the vector was called `x` in the code we used to define the function, `average` works regardless of whether the input is called `x` or `y`. This is because R distinguishes between *global variables* and *local variables*. A global variable is created in the *global environment* outside a function. It is available to all functions (these are the variables that you can see in the Environment panel in RStudio). A local variable is created in the *local environment* inside a function. It is only available to that particular function. For instance, our `average` function creates a variable called `avg`, yet when we attempt to access `avg` after running `average` this variable doesn't seem to exist:

```
average(x)
avg
```

Because `avg` is a local variable, it is only available inside of the `average` function. Local variables take precedence over global variables inside the functions to which they belong. Because we named the argument used in the function `x`, `x` becomes the name of a local variable in `average`. As far as `average` is concerned, there is only one variable named `x`, and that is whatever object that was given as input to the function, regardless of what its original name was. Any operations performed on the local variable `x` won't affect the global variable `x` at all.

Functions can access global variables:

```
y_squared <- function()
{
      return(y^2)
}

y <- 2
y_squared()
```

But operations performed on global variables inside functions won't affect the global variable:

```
add_to_y <- function(n)
{
      y <- y + n
```

```
}
```

```
y <- 1
add_to_y(1)
y
```

Suppose you really need to change a global variable inside a function[1]. In that case, you can use an alternative assignment operator, <<-, which assigns a value to the variable in the *parent environment* to the current environment. If you use <<- for assignment inside a function that is called from the global environment, this means that the assignment takes place in the global environment. But if you use <<- in a function (function 1) that is called by another function (function 2), the assignment will take place in the environment for function 2, thus affecting a local variable in function 2. Here is an example of a global assignment using <<-:

```
add_to_y_global <- function(n)
{
    y <<- y + n
}
```

```
y <- 1
add_to_y_global(1)
y
```

6.1.3   Will your function work?

It is always a good idea to test if your function works as intended, and to try to figure out what can cause it to break. Let's return to our average function:

```
average <- function(x)
{
    avg <- sum(x)/length(x)
    return(avg)
}
```

We've already seen that it seems to work when the input x is a numeric vector. But what happens if we input something else instead?

```
average(c(1, 5, 8)) # Numeric input
average(c(TRUE, TRUE, FALSE)) # Logical input
average(c("Lady Gaga", "Tool", "Dry the River")) # Character input
average(data.frame(x = c(1, 1, 1), y = c(2, 2, 1))) # Numeric df
average(data.frame(x = c(1, 5, 8), y = c("A", "B", "C"))) # Mixed type
```

The first two of these render the desired output (the logical values being represented by 0s and 1s), but the rest don't. Many R functions include checks that the input is of the correct

---

[1]Do you *really*?

type, or checks to see which method should be applied depending on what data type the input is. We'll learn how to perform such checks in Section 6.3.

As a side note, it is possible to write functions that don't end with `return`. In that case, the output (i.e., what would be written in the Console if you'd run the code there) from the last line of the function will automatically be returned. I prefer to (almost) always use `return` though, as it is easy to accidentally make the function return nothing by finishing it with a line that yields no output. Below are two examples of how we could have written `average` without a call to `return`. The first doesn't work as intended, because the function's final (and only) line doesn't give any output.

```r
average_bad <- function(x)
{
      avg <- sum(x)/length(x)
}

average_ok <- function(x)
{
      sum(x)/length(x)
}

average_bad(c(1, 5, 8))
average_ok(c(1, 5, 8))
```

6.1.4 More on arguments

It is possible to create functions with as many arguments as you like, but it will become quite unwieldy if the user has to supply too many arguments to your function. It is therefore common to provide default values to arguments, which is done by setting a value in the function call. Here is an example of a function that computes x^n, using $n = 2$ as the default:

```r
power_n <- function(x, n = 2)
{
      return(x^n)
}
```

If we don't supply n, power_n uses the default n = 2:

```r
power_n(3)
```

But if we supply an n, power_n will use that instead:

```r
power_n(3, 1)
power_n(3, 3)
```

For clarity, you can specify which value corresponds to which argument:

```r
power_n(x = 2, n = 5)
```

... and can then even put the arguments in the wrong order:

```
power_n(n = 5, x = 2)
```

However, if we only supply n we get an error, because there is no default value for x:

```
power_n(n = 5)
```

It is possible to pass a function as an argument. Here is a function that takes a vector and a function as input and applies the function to the first two elements of the vector:

```
apply_to_first2 <- function(x, func)
{
    result <- func(x[1:2])
    return(result)
}
```

By supplying different functions to apply_to_first2, we can make it perform different tasks:

```
x <- c(4, 5, 6)
apply_to_first2(x, sqrt)
apply_to_first2(x, is.character)
apply_to_first2(x, power_n)
```

But what if the function that we supply requires additional arguments? Using apply_to_first2 with sum and the vector c(4, 5, 6) works fine:

```
apply_to_first2(x, sum)
```

But if we instead use the vector c(4, NA, 6), the function returns NA:

```
x <- c(4, NA, 6)
apply_to_first2(x, sum)
```

Perhaps we'd like to pass na.rm = TRUE to sum to ensure that we get a numeric result, if at all possible. This can be done by adding ... to the list of arguments for both functions, which indicates additional parameters (to be supplied by the user) that will be passed to func:

```
apply_to_first2 <- function(x, func, ...)
{
    result <- func(x[1:2], ...)
    return(result)
}

x <- c(4, NA, 6)
apply_to_first2(x, sum)
apply_to_first2(x, sum, na.rm = TRUE)
```

~

Exercise 6.1. Write a function that converts temperature measurements in degrees Fahrenheit to degrees Celsius, and apply it to the `Temp` column of the `airquality` data.

Exercise 6.2. Practice writing functions by doing the following:

1. Write a function that takes a vector as input and returns a vector containing its minimum and maximum, without using `min` and `max`.

2. Write a function that computes the mean of the squared values of a vector using `mean`, and that takes additional arguments that it passes on to `mean` (e.g., `na.rm`).

6.1.5 Namespaces

It is possible, and even likely, that you will encounter functions in packages with the same name as functions in other packages. Or, similarly, that there are functions in packages with the same names as those you have written yourself. This is of course a bit of a headache, but it's actually something that can be overcome without changing the names of the functions. Just like variables can live in different environments, R functions live in *namespaces*, usually corresponding to either the global environment or the package they belong to. By specifying which namespace to look for the function in, you can use multiple functions that all have the same name.

For example, let's create a function called `sqrt`. There is already such a function in the `base` package[2] (see `?sqrt`), but let's do it anyway:

```
sqrt <- function(x)
{
      return(x^10)
}
```

If we now apply `sqrt` to an object, the function that we just defined will be used:

```
sqrt(4)
```

But if we want to use the `sqrt` from `base`, we can specify that by writing the namespace (which almost always is the package name) followed by `::` and the function name:

```
base::sqrt(4)
```

The `::` notation can also be used to call a function or object from a package without loading the package's namespace:

```
msleep # Doesn't work if ggplot2 isn't loaded
ggplot2::msleep # Works, without loading the ggplot2 namespace!
```

---

[2]`base` is automatically loaded when you start R and contains core functions such as `sqrt`.

When you call a function, R will look for it in all active namespaces, following a particular order. To see the order of the namespaces, you can use `search`:

```
search()
```

Note that the global environment is first in this list – meaning that the functions that you define always will be preferred to functions in packages.

All this being said, note that it is bad practice to give your functions and variables the same names as common functions. Don't name them `mean`, `c`, or `sqrt`. Nothing good can ever come from that sort of behaviour.

Nothing.

6.1.6 Sourcing other scripts

If you want to reuse a function that you have written in a new script, you can of course copy it into that script. But if you then make changes to your function, you will quickly end up with several different versions of it. A better idea can therefore be to put the function in a separate script, which you then can call in each script where you need the function. This is done using `source`. If, for instance, you have code that defines some functions in a file called `helper-functions.R` in your working directory, you can run it (thus defining the functions) when the rest of your code is run by adding `source("helper-functions.R")` to your code.

Another option is to create an R package containing the function, but that is beyond the scope of this book. Should you choose to go down that route, I highly recommend reading *R Packages*[3] by Wickham and Bryan.

6.2 More on pipes

We have seen how the pipe operator `|>` can be used to chain functions together. But there are also other pipe operators that are useful. In this section we'll look at some of them, and see how you can create functions using pipes.

6.2.1 *Ce ne sont pas non plus des pipes*

Although `|>` is the pipe operator that we use the most, there are situations where other pipe operators are more appropriate. The `magrittr` package provides a number of other pipes that are useful in certain situations.

One example is when you want to pass variables rather than an entire dataset to the next function. This is needed for instance if you want to use `cor` to compute the correlation between two variables, because `cor` takes two vectors as input instead of a data frame. Previously, we solved this by using `with`:

```
library(dplyr)
airquality |>
```

---

[3]https://r-pkgs.org/index.html

```
    filter(Temp > 80) |>
    with(cor(Temp, Wind))
```

Another option is to use the %$% pipe, which passes on the names of all variables in your data frame instead of the actual data frame:

```
library(magrittr)
airquality |>
      filter(Temp > 80) %$%
      cor(Temp, Wind)
```

If you want to modify a variable using a pipe, you can use the *compound assignment* pipe %<>%. The following three lines all yield exactly the same result:

```
x <- 1:8;    x <- sqrt(x);      x
x <- 1:8;    x |> sqrt() -> x;  x
x <- 1:8;    x %<>% sqrt;       x
```

As long as the first pipe in the pipeline is the compound assignment operator %<>%, you can combine it with other pipes:

```
x <- 1:8
x %<>% subset(x > 5) |> sqrt()
x
```

Sometimes, you want to do something in the middle of a pipeline, like creating a plot, before continuing to the next step in the chain. The *tee* operator %T>% can be used to execute a function without passing on its output (if any). Instead, it passes on the output to its left. Here is an example:

```
airquality |>
      filter(Temp > 80) %T>%
      plot() %$%
      cor(Temp, Wind)
```

Note that if we'd used an ordinary pipe |> instead, we'd get an error:

```
airquality |>
      filter(Temp > 80) |>
      plot() %$%
      cor(Temp, Wind)
```

The reason is that cor looks for the variables Temp and Wind in the plot object, and not in the data frame. The tee operator takes care of this by passing on the data from its left side.

When using the tee operator with ggplot, you need to wrap ggplot in curly brackets and in a call to print:

```
library(ggplot2)
airquality |>
      filter(Temp > 80) %T>%
      {print(ggplot(., aes(Temp, Wind)) + geom_point())} %$%
      cor(Temp, Wind)
```

6.2.2 Writing functions with pipes

If you will be reusing the same pipeline multiple times, you may want to create a function for it. Let's say that you have a data frame containing only numeric variables, and that you want to create a scatterplot matrix (which can be done using plot) and compute the correlations between all variables (using cor). As an example, you could do this for airquality as follows:

```
airquality %T>% plot |>  cor()
```

To define a function for this combination of operators, we write:

```
plot_and_cor <- function(x) { x %T>% plot |>  cor() }
```

If you only use magrittr pipes in your function, you can write this in a shorter form, without the function(...) part:

```
plot_and_cor <- . %T>% plot %>% cor
```

We can now use this function just like any other:

```
# With the airquality data:
airquality |> plot_and_cor
plot_and_cor(airquality)

# With the bookstore data:
age <- c(28, 48, 47, 71, 22, 80, 48, 30, 31)
purchase <- c(20, 59, 2, 12, 22, 160, 34, 34, 29)
bookstore <- data.frame(age, purchase)
bookstore |> plot_and_cor
```

$\sim$

Exercise 6.3. Write a function that takes a data frame as input and uses pipes to print the number of NA values in the data, remove all rows with NA values and return a summary of the remaining data.

Exercise 6.4. Pipes are operators, that is, functions that take two variables as input and can be written without parentheses (other examples of operators are + and *). You can define your own operators just as you would any other function. For instance, we can define an operator called quadratic that takes two numbers a and b as input and computes the quadratic expression $(a + b)^2$:

```
`%quadratic%` <- function(a, b) { (a + b)^2 }
2 %quadratic% 3
```

Create an operator called %against% that takes two vectors as input and draws a scatterplot of them.

6.3 Checking conditions

Sometimes you'd like your code to perform different operations depending on whether or not a certain condition is fulfilled. Perhaps you want it to do something different if there is missing data, if the input is a character vector, or if the largest value in a numeric vector is greater than some number. In Section 2.11.3 you learned how to filter data using conditions. In this section, you'll learn how to use conditional statements for a number of other tasks.

6.3.1 if and else

The most important functions for checking whether a condition is fulfilled are if and else. The basic syntax is

```
if(condition) { do something } else { do something else }
```

The condition should return a single logical value, so that it evaluates to either TRUE or FALSE. If the condition is fulfilled, i.e., if it is TRUE, the code inside the first pair of curly brackets will run, and if it's not (FALSE), the code within the second pair of curly brackets will run instead.

As a first example, assume that you want to compute the reciprocal of x, $1/x$, unless $x = 0$, in which case you wish to print an error message:

```
x <- 2
if(x == 0) { cat("Error! Division by zero.") } else { 1/x }
```

Now try running the same code with x set to 0:

```
x <- 0
if(x == 0) { cat("Error! Division by zero.") } else { 1/x }
```

Alternatively, we could check if $x \neq 0$ and then change the order of the segments within the curly brackets:

```
x <- 0
if(x != 0) { 1/x } else { cat("Error! Division by zero.") }
```

You don't have to write all of the code on the same line, but you must make sure that the else part is on the same line as the first }:

```
if(x == 0)
{
    cat("Error! Division by zero.")
} else
{
    1/x
}
```

You can also choose not to have an `else` part at all. In that case, the code inside the curly brackets will run if the condition is satisfied, and if not, nothing will happen:

```
x <- 0
if(x == 0) { cat("x is 0.") }

x <- 2
if(x == 0) { cat("x is 0.") }
```

Finally, if you need to check a number of conditions one after another, in order to list different possibilities, you can do so by repeated use of `if` and `else`:

```
if(x == 0)
{
    cat("Error! Division by zero.")
} else if(is.infinite((x)))
{
    cat("Error! Division by infinity.")
} else if(is.na((x)))
{
    cat("Error! Division by NA.")
} else
{
    1/x
}
```

6.3.2 & & &&

Just as when we used conditions for filtering in Sections 2.11.3 and 5.8.2, it is possible to combine several conditions into one using & (AND) and | (OR). However, the & and | operators are vectorised, meaning that they will return a vector of `logical` values whenever possible. This is not desirable in conditional statements, where the condition must evaluate to a single value. Using a condition that returns a vector results in an error message:

```
if(c(1, 2) == 2) { cat("The vector contains the number 2.\n") }
if(c(2, 1) == 2) { cat("The vector contains the number 2.\n") }
```

Usually, if a condition evaluates to a vector, it is because you've made an error in your code. Remember, if you really need to evaluate a condition regarding the elements in a vector, you can collapse the resulting `logical` vector to a single value using `any` or `all`.

Some texts recommend using the operators && and || instead of & and | in conditional statements. These work almost like & and |, but check the conditions on the left-hand and right-hand sides sequentially. & and | always evaluate all the conditions that you're combining, while && and || don't: && stops as soon as it encounters a FALSE, and || stops as soon as it encounters a TRUE. Consequently, you can put the conditions you wish to combine in a particular order to make sure that they can be evaluated. For instance, you may want first to check that a variable exists, and then check a property. This can be done using exists to check whether or not it exists – note that the variable name must be written within quotes:

```r
# "a" is a variable that doesn't exist

# Using && works:
if(exists("a") && a > 0)
{
    cat("The variable exists and is positive.")
} else { cat("a doesn't exist or is negative.") }

# But using & doesn't, because it attempts to evaluate a>0
# even though a doesn't exist:
if(exists("a") & a > 0)
{
    cat("The variable exists and is positive.")
} else { cat("a doesn't exist or is negative.") }
```

6.3.3 ifelse

It is common that you want to assign different values to a variable depending on whether or not a condition is satisfied:

```r
x <- 2

if(x == 0)
{
    reciprocal <- "Error! Division by zero."
} else
{
    reciprocal <- 1/x
}

reciprocal
```

In fact, this situation is so common that there is a special function for it: ifelse:

```r
reciprocal <- ifelse(x == 0, "Error! Division by zero.", 1/x)
```

ifelse evaluates a condition and then returns different answers depending on whether the condition is TRUE or FALSE. It can also be applied to vectors, in which case it checks the condition for each element of the vector and returns an answer for each element:

```
x <- c(-1, 1, 2, -2, 3)
ifelse(x > 0, "Positive", "Negative")
```

6.3.4 `switch`

For the sake of readability, it is usually a good idea to try to avoid chains of the type `if()`
`{} else if() {} else if() {} else {}`. One function that can be useful for this is `switch`,
which lets you list a number of possible results, either by position (a number) or by name:

```
position <- 2
switch(position,
      "First position",
      "Second position",
      "Third position")
```

```
name <- "First"
switch(name,
      First = "First name",
      Second = "Second name",
      Third = "Third name")
```

You can for instance use this to decide what function should be applied to your data:

```
x <- 1:3
y <- c(3, 5, 4)
method <- "nonparametric2"
cor_xy <- switch(method,
      parametric = cor(x, y, method = "pearson"),
      nonparametric1 = cor(x, y, method = "spearman"),
      nonparametric2 = cor(x, y, method = "kendall"))
cor_xy
```

6.3.5 Failing gracefully

Conditional statements are useful for ensuring that the input to a function you've written is
of the correct type. In Section 6.1.3 we saw that our `average` function failed if we applied it
to a `character` vector:

```
average <- function(x)
{
      avg <- sum(x)/length(x)
      return(avg)
}
```

```
average(c("Lady Gaga", "Tool", "Dry the River"))
```

By using a conditional statement, we can provide a more informative error message. We can check that the input is numeric and, if it's not, stop the function and print an error message, using stop:

```
average <- function(x)
{
      if(is.numeric(x))
      {
            avg <- sum(x)/length(x)
            return(avg)
      } else
      {
            stop("The input must be a numeric vector.")
      }
}

average(c(1, 5, 8))
average(c("Lady Gaga", "Tool", "Dry the River"))
```

~

Exercise 6.5. Which of the following conditions are TRUE? First think about the answer, and then check it using R.

```
x <- 2
y <- 3
z <- -3
```

   1. x > 2

   2. x > y | x > z

   3. x > y & x > z

   4. abs(x*z) >= y

Exercise 6.6. Fix the errors in the following code:

```
x <- c(1, 2, pi, 8)

# Only compute square roots if x exists
# and contains positive values:
if(exists(x)) { if(x > 0) { sqrt(x) } }
```

6.4 Iteration using loops

We have already seen how you can use functions to make it easier to repeat the same task over and over. But there is still a part of the puzzle missing – what if, for example, you wish to apply a function to each column of a data frame? What if you want to apply it to data from a number of files, one at a time? The solution to these problems is to use *iteration*. In this section, we'll explore how to perform iteration using *loops*.

6.4.1 `for` loops

`for` loops can be used to run the same code several times, with different settings, e.g., different data, in each iteration. Their use is perhaps best explained by some examples. We create the loop using `for`, give the name of a *control variable* and a vector containing its values (the control variable controls how many iterations to run), and then write the code that should be repeated in each iteration of the loop. In each iteration, a new value of the control variable is used in the code, and the loop stops when all values have been used.

As a first example, let's write a `for` loop that runs a block of code five times, where the block prints the current iteration number:

```
for(i in 1:5)
{
    cat("Iteration", i, "\n")
}
```

This is equivalent to writing:

```
cat("Iteration", 1, "\n")
cat("Iteration", 2, "\n")
cat("Iteration", 3, "\n")
cat("Iteration", 4, "\n")
cat("Iteration", 5, "\n")
```

The upside is that we didn't have to copy and edit the same code multiple times – and as you can imagine, this benefit becomes even more pronounced if you have more complicated code blocks.

The values for the control variable are given in a vector, and the code block will be run once for each element in the vector – we say that we *loop over the values in the vector*. The vector doesn't have to be `numeric` – here is an example with a `character` vector:

```
for(word in c("one", "two", "five hundred and fifty five"))
{
    cat("Iteration", word, "\n")
}
```

Of course, loops are used for so much more than merely printing text on the screen. A common use is to perform some computation and then store the result in a vector. In this case, we must first create an empty vector to store the result in, e.g., using `vector`, which creates an empty vector of a specific type and length:

```
squares <- vector("numeric", 5)

for(i in 1:5)
{
    squares[i] <- i^2
}
squares
```

In this case, it would have been both simpler and computationally faster to compute the squared values by running `(1:5)^2`. This is known as a *vectorised* solution and is very important in R. We'll discuss vectorised solutions in detail in Section 6.5.

When creating the values used for the control variable, we often wish to create different sequences of numbers. Two functions that are very useful for this are `seq`, which creates sequences, and `rep`, which repeats patterns:

```
seq(0, 100)
seq(0, 100, by = 10)
seq(0, 100, length.out = 21)

rep(1, 4)
rep(c(1, 2), 4)
rep(c(1, 2), c(4, 2))
```

Finally, `seq_along` can be used to create a sequence of indices for a vector of a data frame, which is useful if you wish to iterate some code for each element of a vector or each column of a data frame:

```
seq_along(airquality) # Gives the indices of all columns of the data
                      # frame
seq_along(airquality$Temp) # Gives the indices of all elements of the
                           # vector
```

Here is an example of how to use `seq_along` to compute the mean of each column of a data frame:

```
# Compute the mean for each column of the airquality data:
means <- vector("double", ncol(airquality))

# Loop over the variables in airquality:
for(i in seq_along(airquality))
{
    means[i] <- mean(airquality[[i]], na.rm = TRUE)
}

# Check that the results agree with those from the colMeans function:
```

means
```
colMeans(airquality, na.rm = TRUE)
```

The line inside the loop could have read means[i] <- mean(airquality[,i], na.rm = TRUE), but that would have caused problems if we'd used it with a data.table or tibble object; see Section 5.9.4.

Finally, we can also change the values of the data in each iteration of the loop. Some machine learning methods require that the data is *standardised*, i.e., that all columns have mean 0 and standard deviation 1. This is achieved by subtracting the mean from each variable and then dividing each variable by its standard deviation. We can write a function for this that uses a loop, changing the values of a column in each iteration:

```
standardise <- function(df, ...)
{
    for(i in seq_along(df))
    {
        df[[i]] <- (df[[i]] - mean(df[[i]], ...))/sd(df[[i]], ...)
    }
    return(df)
}

# Try it out:
aqs <- standardise(airquality, na.rm = TRUE)
colMeans(aqs, na.rm = TRUE) # Non-zero due to floating point
                            # arithmetics!
sd(aqs$Wind)
```

~

Exercise 6.7. Practice writing for loops by doing the following:

1. Compute the mean temperature for each month in the airquality dataset using a loop rather than an existing function.

2. Use a for loop to compute the maximum and minimum value of each column of the airquality data frame, storing the results in a data frame.

3. Make your solution to the previous task reusable by writing a function that returns the maximum and minimum value of each column of a data frame.

Exercise 6.8. Use rep or seq to create the following vectors:

1. 0.25 0.5 0.75 1

2. 1 1 1 2 2 5

Exercise 6.9. As an alternative to seq_along(airquality) and seq_along (airquality$Temp), we could create the same sequences using 1:ncol(airquality) and 1:length(airquality$Temp). Use x <- c() to create a vector of length zero. Then create loops that use `seq_along(x)` and `1:length(x)` as values for the control variable. How many iterations are the two loops run? Which solution is preferable?

Exercise 6.10. An alternative to standardisation is *normalisation*, where all `numeric` variables are rescaled so that their smallest value is 0 and their largest value is 1. Write a function that normalises the variables in a data frame containing `numeric` columns.

Exercise 6.11. The function `list.files` can be used to create a vector containing the names of all files in a folder. The `pattern` argument can be used to supply a regular expression describing a file name pattern. For instance, if `pattern = "\\.csv$"` is used, only .csv files will be listed.

Create a loop that goes through all .csv files in a folder and prints the names of the variables for each file.

6.4.2 Loops within loops

In some situations, you'll want to put a loop inside another loop. Such loops are said to be *nested*. An example is if we want to compute the correlation between all pairs of variables in `airquality` and store the result in a matrix:

```
cor_mat <- matrix(NA, nrow = ncol(airquality),
                  ncol = ncol(airquality))
for(i in seq_along(airquality))
{
    for(j in seq_along(airquality))
    {
        cor_mat[i, j] <- cor(airquality[[i]], airquality[[j]],
                             use = "pairwise.complete")
    }
}

# Element [i, j] of the matrix now contains the correlation between
# variables i and j:
cor_mat
```

Once again, there is a vectorised solution to this problem, given by `cor(airquality, use = "pairwise.complete")`. As we will see in Section 6.6, vectorised solutions like this can be several times faster than solutions that use nested loops. In general, solutions involving nested loops tend to be fairly slow; but, on the other hand, they are often easy and straightforward to implement.

6.4.3 Keeping track of what's happening

Sometimes each iteration of your loop takes a long time to run, and you'll want to monitor its progress. This can be done using printed messages or a progress bar in the Console panel, or sound notifications. We'll showcase each of these using a loop containing a call to `Sys.sleep`, which pauses the execution of R commands for a short time (determined by the user).

First, we can use `cat` to print a message describing the progress. Adding \r to the end of a string allows us to print all messages on the same line, with each new message replacing the old one:

```r
# Print each message on a new same line:
for(i in 1:5)
{
    cat("Step", i, "out of 5\n")
    Sys.sleep(1) # Sleep for 1 second
}

# Replace the previous message with the new one:
for(i in 1:5)
{
    cat("Step", i, "out of 5\r")
    Sys.sleep(1) # Sleep for one second
}
```

Adding a progress bar is a little more complicated, because we must first start the bar by using `txtProgressBar` and then update it using `setTxtProgressBar`:

```r
sequence <- 1:5
pbar <- txtProgressBar(min = 0, max = max(sequence), style = 3)
for(i in sequence)
{
    Sys.sleep(1) # Sleep for 1 second
    setTxtProgressBar(pbar, i)
}
close(pbar)
```

Finally, the `beepr` package[4] can be used to play sounds, with the function `beep`:

```r
install.packages("beepr")

library(beepr)
# Play all 11 sounds available in beepr:
for(i in 1:11)
{
    beep(sound = i)
    Sys.sleep(2) # Sleep for 2 seconds
}
```

6.4.4 Loops and lists

In our previous examples of loops, it has always been clear from the start how many iterations the loop should run and what the length of the output vector (or data frame) should be. This isn't always the case. To begin with, let's consider the case where the length of the

---

[4]Arguably the best add-on package for R.

output is unknown or difficult to know in advance. Let's say that we want to go through the `airquality` data to find days that are extreme in the sense that at least one variable attains its maximum on those days. That is, we wish to find the indices of the maximum of each variable, and store them in a vector. Because several days can have the same temperature or wind speed, there may be more than one such maximal index for each variable. For that reason, we don't know the length of the output vector in advance.

In such cases, it is usually a good idea to store the result from each iteration in a `list` (Section 5.2), and then collect the elements from the list once the loop has finished. We can create an empty list with one element for each variable in `airquality` using `vector`:

```
# Create an empty list with one element for each variable in
# airquality:
max_list <- vector("list", ncol(airquality))

# Naming the list elements will help us see which variable the maximal
# indices belong to:
names(max_list) <- names(airquality)

# Loop over the variables to find the maxima:
for(i in seq_along(airquality))
{
      # Find indices of maximum values:
      max_index <- which(airquality[[i]] == max(airquality[[i]],
                                               na.rm = TRUE))

      # Add indices to list:
      max_list[[i]] <- max_index
}

# Check results:
max_list

# Collapse to a vector:
extreme_days <- unlist(max_list)
```

(In this case, only the variables `Month` and `Days` have duplicate maximum values.)

6.4.5   `while` loops

In some situations, we want to run a loop until a certain condition is met, meaning that we don't know in advance how many iterations we'll need. This is more common in numerical optimisation and simulation, but sometimes it also occurs in data analyses.

When we don't know in advance how many iterations are needed, we can use `while` loops. Unlike `for` loops that iterate a fixed number of times, `while` loops keep iterating as long as some specified condition is met. Here is an example where the loop keeps iterating until `i` squared is greater than 100:

```
i <- 1

while(i^2 <= 100)
{
     cat(i,"squared is", i^2, "\n")
     i <- i +1
}


i
```

The code block inside the loop keeps repeating until the condition `i^2 <= 100` no longer is satisfied. We have to be a bit careful with this condition – if we set it in such a way that it is possible that the condition *always* will be satisfied, the loop will just keep running and running, creating what is known as an *infinite loop*. If you've accidentally created an infinite loop, you can break it by pressing the Stop button at the top of the Console panel in RStudio.

In Section 5.3.3 we saw how `rle` can be used to find and compute the lengths of runs of equal values in a vector. We can use nested `while` loops to create something similar. `while` loops are a good choice here, because we don't know how many runs are in the vector in advance. Here is an example, which you'll study in more detail in Exercise 6.12:

```
# Create a vector of 0's and 1's:
x <- rep(c(0, 1, 0, 1, 0), c(5, 1, 4, 2, 7))

# Create empty vectors where the results will be stored:
run_values <- run_lengths <- c()

# Set the initial condition:
i <- 1

# Iterate over the entire vector:
while(i < length(x))
{
    # A new run starts:
    run_length <- 1
    cat("A run starts at i =", i, "\n")

    # Check how long the run continues:
    while(x[i+1] == x[i] & i < length(x))
    {
         run_length <- run_length + 1
         i <- i + 1
    }

    i <- i + 1

    # Save results:
    run_values <- c(run_values, x[i-1])
    run_lengths <- c(run_lengths, run_length)
```

```
}
```

```
# Present the results:
data.frame(run_values, run_lengths)
```

~

Exercise 6.12. Consider the nested `while` loops in the run length example above. Go through the code and think about what happens in each step. What happens when `i` is `1`? When it is `5`? When it is `6`? Answer the following questions:

1. What does the condition for the outer while loop check? Why is it needed?

2. What does the condition for the inner while loop check? Why is it needed?

3. What does the line `run_values <- c(run_values, x[i-1])` do?

Exercise 6.13. The *control statements* `break` and `next` can be used inside both `for` and `while` loops to control their behaviour further. `break` stops a loop, and `next` skips to the next iteration of it. Use these functions to modify the following piece of code so that the loop skips to the next iteration if `x[i]` is `0`, and breaks if `x[i]` is `NA`:

```
x <- c(1, 5, 8, 0, 20, 0, 3, NA, 18, 2)

for(i in seq_along(x))
{
     cat("Step", i, "- reciprocal is", 1/x[i], "\n")
}
```

Exercise 6.14. Using the `cor_mat` computation from Section 6.4.2, write a function that computes all pairwise correlations in a data frame and uses `next` to only compute correlations for `numeric` variables. Test your function by applying it to the `msleep` data from `ggplot2`. Could you achieve the same thing without using `next`?

6.5 Iteration using vectorisation and functionals

Many operators and functions in R take vectors as input and handle them in a highly efficient way, usually by passing the vector on to an optimised function written in the C programming language[5]. So if we want to compute the squares of the numbers in a vector, we don't need to write a loop:

---

[5]Unlike R, C is a low-level language that allows the user to write highly specialised (and complex) code to perform operations very quickly.

```
squares <- vector("numeric", 5)

for(i in 1:5)
{
    squares[i] <- i^2
}
squares
```

Instead, we can simply apply the ^ operator, which uses fast C code to compute the squares:

```
squares <- (1:5)^2
```

These types of functions and operators are called *vectorised*. They take a vector as input and apply a function to all its elements, meaning that we can avoid slower solutions utilising loops in R[6]. Try to use vectorised solutions rather than loops whenever possible – it makes your code both easier to read and faster to run.

A related concept is *functionals*, which are functions that contain a for loop. Instead of writing a for loop, you can use a functional, supplying data, a function that should be applied in each iteration of the loop, and a vector to loop over. This won't necessarily make your loop run faster, but it does have other benefits:

- *Shorter code*: functionals allow you to write more concise code. Some would argue that they also allow you to write code that is easier to read, but that is obviously a matter of taste.
- *Efficient*: functionals handle memory allocation and other small tasks efficiently, meaning that you don't have to worry about creating a vector of an appropriate size to store the result.
- *No changes to your environment*: because all operations now take place in the local environment of the functional, you don't run the risk of accidentally changing variables in your global environment.
- *No left-overs*: for leaves the control variable (e.g., i) in the environment, functionals do not.
- *Easy to use with pipes*: because the loop has been wrapped in a function, it lends itself well to being used in a |> pipeline.

Explicit loops are preferable when:

- You think that they are easier to read and write.
- Your functions take data frames or other non-vector objects as input.
- Each iteration of your loop depends on the results from previous iterations.

In this section, we'll see how we can apply functionals to obtain elegant alternatives to (explicit) loops.

---

[6]The vectorised functions often use loops, but loops written in C, which are much faster.

6.5.1 A first example with `apply`

The prototypical functional is `apply`, which loops over either the rows or the columns of a data frame[7]. The arguments are a dataset, the margin to loop over (1 for rows, 2 for columns) and then the function to be applied.

In Section 6.4.1 we wrote a `for` loop for computing the mean value of each column in a data frame:

```
# Compute the mean for each column of the airquality data:
means <- vector("double", ncol(airquality))

# Loop over the variables in airquality:
for(i in seq_along(airquality))
{
    means[i] <- mean(airquality[[i]], na.rm = TRUE)
}
```

Using apply, we can reduce this to a single line. We wish to use the `airquality` data, loop over the columns (margin 2), and apply the function `mean` to each column:

```
apply(airquality, 2, mean)
```

Rather elegant, don't you think?

Additional arguments can be passed to the function inside `apply` by adding them to the end of the function call:

```
apply(airquality, 2, mean, na.rm = TRUE)
```

~

Exercise 6.15. Use `apply` to compute the maximum and minimum value of each column of the `airquality` data frame. Can you write a function that allows you to compute both with a single call to `apply`?

6.5.2 Variations on a theme

There are several variations of `apply` that are tailored to specific problems:

- `lapply`: takes a function and vector/list as input and returns a list.
- `sapply`: takes a function and vector/list as input and returns a vector or matrix.
- `vapply`: a version of `sapply` with additional checks of the format of the output.
- `tapply`: for looping over groups, e.g., when computing grouped summaries.
- `rapply`: a recursive version of `tapply`.
- `mapply`: for applying a function to multiple arguments; see Section 6.5.7.
- `eapply`: for applying a function to all objects in an environment.

---

[7]Actually, over the rows or columns of a matrix - `apply` converts the data frame to a `matrix` object.

We have already seen several ways to compute the mean temperature for different months in the `airquality` data (Sections 2.12 and 5.7.7, and Exercise 6.7). The *apply family offer several more:

```
# Create a list:
temps <- split(airquality$Temp, airquality$Month)

lapply(temps, mean)
sapply(temps, mean)
vapply(temps, mean, vector("numeric", 1))
tapply(airquality$Temp, airquality$Month, mean)
```

There is, as that delightful proverb goes, more than one way to skin a cat.

~

Exercise 6.16. Use an *apply function to simultaneously compute the monthly maximum and minimum temperature in the `airquality` data frame.

Exercise 6.17. Use an *apply function to simultaneously compute the monthly maximum and minimum temperature and windspeed in the `airquality` data frame.

Hint: start by writing a function that simultaneously computes the maximum and minimum temperature and windspeed for a data frame containing data from a single month.

6.5.3 purrr

If you feel enthusiastic about ~~skinning cats~~ using functionals instead of loops, the tidyverse package `purrr` is a great addition to your toolbox. It contains a number of specialised alternatives to the *apply functions. More importantly, it also contains certain shortcuts that come in handy when working with functionals. For instance, it is fairly common to define a short function inside your functional, which is useful for instance when you don't want the function to take up space in your environment. This can be done a little more elegantly with `purrr` functions using a shortcut denoted by ~. Let's say that we want to standardise all variables in `airquality`. The `map` function is the `purrr` equivalent of `lapply`. We can use it with or without the shortcut, and with or without pipes (we mention the use of pipes now because it will be important in what comes next):

```
# Base solution:
lapply(airquality, function(x) { (x-mean(x))/sd(x) })

# Base solution with pipe:
library(magrittr)
airquality |> lapply(function(x) { (x-mean(x))/sd(x) })

# purrr solution:
library(purrr)
map(airquality, function(x) { (x-mean(x))/sd(x) })

# We can make the purrr solution less verbose using a shortcut:
```

```
map(airquality, ~(.-mean(.))/sd(.))

# purr solution with pipe and shortcut:
airquality |> map(~(.-mean(.))/sd(.))
```

Where this shortcut really shines is if you need to use multiple functionals. Let's say that we want to standardise the `airquality` variables, compute a `summary` and then extract columns 2 and 5 from the summary (which contains the 1st and 3rd quartile of the data):

```
# Impenetrable base solution:
lapply(lapply(lapply(airquality,
                     function(x) { (x-mean(x))/sd(x) }),
             summary),
      function(x) { x[c(2, 5)] })

# Base solution with pipe:
airquality |>
      lapply(function(x) { (x-mean(x))/sd(x) }) |>
      lapply(summary) |>
      lapply(function(x) { x[c(2, 5)] })

# purrr solution:
airquality |>
      map(~(.-mean(.))/sd(.)) |>
      map(summary) |>
      map(~.[c(2, 5)])
```

Once you know the meaning of ~, the `purrr` solution is a lot cleaner than the base solutions.

6.5.4   Specialised functions

So far, it may seem like `map` is just like `lapply` but with a shortcut for defining functions, which is more or less true. But `purrr` contains a lot more functionals that you can use, each tailored to specific problems.

For instance, if you need to specify that the output should be a vector of a specific type, you can use:

- `map_dbl(data, function)` instead of `vapply(data, function, vector("numeric", length))`,
- `map_int(data, function)` instead of `vapply(data, function, vector("integer", length))`,
- `map_chr(data, function)` instead of `vapply(data, function, vector("character", length))`,
- `map_lgl(data, function)` instead of `vapply(data, function, vector("logical", length))`.

If you need to specify that the output should be a data frame, you can use:

- `map_dfr(data, function)` instead of `sapply(data, function)`.

The ~ shortcut for functions is available for all these map_* functions. In case you need to pass additional arguments to the function inside the functional, just add them at the end of the functional call:

```
airquality |> map_dbl(max)
airquality |> map_dbl(max, na.rm = TRUE)
```

Another specialised function is the walk function. It works just like map, but it doesn't return anything. This is useful if you want to apply a function with no output, such as cat or read.csv:

```
# Returns a list of NULL values:
airquality |> map(~cat("Maximum:", max(.), "\n"))

# Returns nothing:
airquality |> walk(~cat("Maximum:", max(.), "\n"))
```

~

Exercise 6.18. Use a map_* function to simultaneously compute the monthly maximum and minimum temperature in the airquality data frame, returning a vector.

6.5.5 Exploring data with functionals

Functionals are great for creating custom summaries of your data. For instance, if you want to check the data type and number of unique values of each variable in your dataset, you can do that with a functional:

```
library(palmerpenguins)
penguins |> map_dfr(~(data.frame(unique_values = length(unique(.)),
                                 class = class(.))))
```

You can of course combine purrr functionals with functions from other packages, e.g., to replace length(unique(.)) with a function from your favourite data manipulation package:

```
# Using uniqueN from data.table:
library(data.table)
peng <- as.data.table(penguins)
peng |> map_dfr(~(data.frame(unique_values = uniqueN(.),
                             class = class(.))))

# Using n_distinct from dplyr:
library(dplyr)
penguins |> map_dfr(~(data.frame(unique_values = n_distinct(.),
                                 class = class(.))))
```

When creating summaries, it can often be useful to be able to loop over both the elements of a vector and their indices. In purrr, this is done using the usual map* functions, but with an i (for index) in the beginning of their names, e.g., imap and iwalk:

```
# Returns a list of NULL values:
imap(airquality, ~ cat(.y, ": ", median(.x), "\n", sep = ""))

# Returns nothing:
iwalk(airquality, ~ cat(.y, ": ", median(.x), "\n", sep = ""))
```

Note that `.x` is used to denote the variable, and that `.y` is used to denote the *name* of the variable. If `i*` functions are used on vectors without element names, indices are used instead. The names of elements of vectors can be set using `set_names`:

```
# Without element names:
x <- 1:5
iwalk(x, ~ cat(.y, ": ", exp(.x), "\n", sep = ""))

# Set element names:
x <- set_names(x, c("exp(1)", "exp(2)", "exp(3)", "exp(4)", "exp(5)"))
iwalk(x, ~ cat(.y, ": ", exp(.x), "\n", sep = ""))
```

~

Exercise 6.19. Write a function that takes a data frame as input and returns the following information about each variable in the data frame: variable name, number of unique values, data type, and number of missing values. The function should, as you will have guessed, use a functional.

Exercise 6.20. In Exercise 6.11 you wrote a function that printed the names and variables for all `.csv` files in a folder given by `folder_path`. Use `purrr` functionals to do the same thing.

6.5.6 Keep calm and carry on

Another neat feature of `purrr` is the `safely` function, which can be used to wrap a function that will be used inside a functional and makes sure that the functional returns a result even if there is an error. For instance, let's say that we want to compute the logarithm of all variables in the `msleep` data:

```
library(ggplot2)
msleep
```

Note that some columns are `character` vectors, which will cause `log` to throw an error:

```
log(msleep$name)
log(msleep)
lapply(msleep, log)
map(msleep, log)
```

Note that the error messages we get from `lapply` and `map` here don't give any information about which variable caused the error, making it more difficult to figure out what's gone wrong.

If first we wrap `log` with `safely`, we get a list containing the correct output for the numeric variables, and error messages for the non-numeric variables:

```
safe_log <- safely(log)
lapply(msleep, safe_log)
map(msleep, safe_log)
```

Not only does this tell us where the errors occur, but it also returns the logarithms for all variables that `log` actually could be applied to.

If you'd like your functional to return some default value, e.g., `NA`, instead of an error message, you can use `possibly` instead of `safely`:

```
pos_log <- possibly(log, otherwise = NA)
map(msleep, pos_log)
```

6.5.7 Iterating over multiple variables

A final important case is when you want to iterate over more than one variable. This is often the case when fitting statistical models that should be used for prediction, as you'll see in Section 8.2.5. Another example is when you wish to create plots for several subsets in your data. For instance, we could create a plot of `body_mass_g` versus `flipper_length_mm` for each combination of `species` and `sex` in the `penguins` data. To do this for a single combination, we'd use something like this:

```
library(ggplot2)
library(dplyr)
library(palmerpenguins)

penguins |> filter(species == "Chinstrap",
                   sex == "male") |>
         ggplot(aes(body_mass_g, flipper_length_mm)) +
             geom_point() +
             ggtitle("Chinstrap, male")
```

To create such a plot for all combinations of `species` and `sex`, we must first create a data frame containing all unique combinations, which can be done using the `distinct` function from `dplyr`:

```
combos <- penguins |> distinct(species, sex) |> na.omit()
all_species <- combos$species
all_sexes <- combos$sex
```

`map2` and `walk2` from `purrr` loop over the elements of two vectors, `x` and `y`, say. They combine the first element of `x` with the first element of `y`, the second element of `x` with the second element of `y`, and so on – meaning that they won't automatically loop over all combinations of elements. That is the reason why we use `distinct` above to create two vectors where each pair (`x[i]`, `y[i]`) corresponds to a combination. Apart from the fact that we add a second vector to the call, `map2` and `walk2` work just like `map` and `walk`:

```
# Print all pairs:
walk2(all_species, all_sexes, ~cat(.x, .y, "\n"))

# Create a plot for each pair:
combos_plots <- map2(all_species, all_sexes, ~{
                    penguins |> filter(species == .x,
                                       sex == .y) |>
                    ggplot(aes(body_mass_g, flipper_length_mm)) +
                        geom_point() +
                        ggtitle(paste(.x, .y, sep =", "))})

# View some plots:
combos_plots[[1]]
combos_plots[[6]]

# Save all six plots in a pdf file, with one plot per page:
pdf("all_combos_plots.pdf", width = 8, height = 8)
combos_plots
dev.off()
```

The base function `mapply` could also have been used here. If you need to iterate over more than two vectors, you can use `pmap` or `pwalk`, which work analogously to `map2` and `walk2`.

∼

Exercise 6.21. Using the `gapminder` data from the `gapminder` package, create scatterplots of `pop` and `lifeExp` for each combination of `continent` and `year`. Save each plot as a separate `.png` file.

6.6   Measuring code performance

There are probably as many ideas about what good code is as there are programmers. Some prefer readable code; others prefer concise code. Some prefer to work with separate functions for each task, while others would rather continue to combine a few basic functions in new ways. Regardless of what you consider to be *good code*, there are a few objective measures that can be used to assess the quality of your code. In addition to writing code that works and is bug-free, you'd like your code to be:

- *Fast*: meaning that it runs quickly. Some tasks can take seconds or weeks, depending on what code you write for them. Speed is particularly important if you're going to run your code many times.
- *Memory efficient*: meaning that it uses as little of your computer's memory as possible. Software running on your computer uses its memory – its RAM – to store data. If you're not careful with RAM, you may end up with a full memory and a sluggish or frozen

computer. Memory efficiency is critical if you're working with big datasets that take up a lot of RAM to begin with.

In this section we'll have a look at how you can measure the speed and memory efficiency of R functions. A caveat is that while speed and memory efficiency are important, the most important thing is to come up with a solution that works in the first place. You should almost always start by solving a problem, and then worry about speed and memory efficiency, not the other way around. The reason for this is that efficient code often is more difficult to write, read, and debug, which can slow down the process of writing it considerably.

Note also that speed and memory usage is system-dependent. The clock frequency and architecture of your processor and speed and size of your RAM will affect how your code performs, as will what operating system you use and what other programs you are running at the same time. That means that if you wish to compare how two functions perform, you need to compare them on the same system under the same conditions.

As a side note, a great way to speed up functions that use either loops or functionals is parallelisation. We cover that topic in Section 12.2.

6.6.1 Timing functions

To measure how long a piece of code takes to run, we can use `system.time` as follows:

```r
rtime <- system.time({
    x <- rnorm(1e6)
    mean(x)
    sd(x)
})

# elapsed is the total time it took to execute the code:
rtime
```

This isn't the best way of measuring computational time though and doesn't allow us to compare different functions easily. Instead, we'll use the bench package, which contains a function called mark that is very useful for measuring the execution time of functions and blocks of code. Let's start by installing it:

```r
install.packages("bench")
```

In Section 6.1.1 we wrote a function for computing the mean of a vector:

```r
average <- function(x)
{
    return(sum(x)/length(x))
}
```

Is this faster or slower than mean? We can use mark to apply both functions to a vector multiple times, and measure how long each execution takes:

```
library(bench)
x <- 1:100
bm <- mark(mean(x), average(x))
bm # Or use View(bm) if you don't want to print the results in the
   # Console panel.
```

mark has executed both function n_itr times each, and measured how long each execution took to perform. The execution time varies – in the output, you can see the shortest (min) and median (median) execution times, as well as the number of iterations per second (itr/sec). Be a little wary of the units for the execution times so that you don't get them confused – a millisecond (ms, 10^{-3} seconds) is 1,000 microseconds (μs, 1 μs is 10^{-6} seconds), and 1 microsecond is 1,000 nanoseconds (ns, 1 ns is 10^{-9} seconds).

The result here may surprise you – it appears that average is faster than mean! The reason is that mean does a lot of things that average does not: it checks the data type and gives error messages if the data is of the wrong type (e.g., character), and then traverses the vector twice to lower the risk of errors due to floating point arithmetics. All of this takes time and makes the function slower (but safer to use).

We can plot the results using the ggbeeswarm package:

```
install.packages("ggbeeswarm")

plot(bm)
```

It is also possible to place blocks of code inside curly brackets, { }, in mark. Here is an example comparing a vectorised solution for computing the squares of a vector with a solution using a loop:

```
x <- 1:100
bm <- mark(x^2,
    {
        y <- x
        for(i in seq_along(x))
        {
            y[i] <- x[i]*x[i]
        }
        y
    })
bm
plot(bm)
```

Although the above code works, it isn't the prettiest, and the bm table looks a bit confusing because of the long expression for the code block. I prefer to put the code block inside a function instead:

```
squares <- function(x)
{
        y <- x
```

```
        for(i in seq_along(x))
        {
            y[i] <- x[i]*x[i]
        }
        return(y)
}

x <- 1:100
bm <- mark(x^2, squares(x))
bm
plot(bm)
```

Note that `squares(x)` is faster than the original code block:

```
bm <- mark(squares(x),
    {
        y <- x
        for(i in seq_along(x))
        {
            y[i] <- x[i]*x[i]
        }
        y
    })
bm
```

Functions in R are *compiled* the first time they are run, which often makes them run faster than the same code would have outside of the function. We'll discuss this further next.

6.6.2   Measuring memory usage (and a note on compilation)

`mark` also shows us how much memory is allocated when running different code blocks, in the `mem_alloc` column of the output[8].

Unfortunately, measuring memory usage is a little tricky. To see why, restart R (yes, really, this is important!), and then run the following code to benchmark `x^2` versus `squares(x)`:

```
library(bench)

squares <- function(x)
{
        y <- x
        for(i in seq_along(x))
        {
            y[i] <- x[i]*x[i]
        }
```

---

[8]But only if your version of R has been *compiled with memory profiling*. If you are using a standard build of R, i.e., have downloaded the base R binary from R-project.org, you should be good to go. You can check that memory profiling is enabled by checking that `capabilities("profmem")` returns `TRUE`. If not, you may need to reinstall R if you wish to enable memory profiling.

```
        return(y)
}

x <- 1:100
bm <- mark(x^2, squares(x))
bm
```

Judging from the mem_alloc column, it appears that the squares(x) loop not only is slower, but it also uses more memory. But wait! Let's run the code again, just to be sure of the result:

```
bm <- mark(x^2, squares(x))
bm
```

This time, both functions use less memory, and squares now uses *less* memory than x^2. What's going on?

Computers can't read code written in R or most other programming languages directly. Instead, the code must be translated to *machine code* that the computer's processor uses, in a process known as *compilation*. R uses *just-in-time compilation* of functions and loops[9], meaning that it translates the R code for new functions and loops to machine code *during execution*. Other languages, such as C, use ahead-of-time compilation, translating the code *prior to execution*. The latter can make the execution much faster, but some flexibility is lost, and the code needs to be run through a compiler ahead of execution, which also takes time. When doing the just-in-time compilation, R needs to use some of the computer's memory, which causes the memory usage to be greater the first time the function is run. However, if an R function is run again, it has already been compiled, meaning R doesn't have to allocate memory for compilation.

In conclusion, if you want to benchmark the memory usage of functions, make sure to run them once before benchmarking. Alternatively, if your function takes a long time to run, you can compile it without running it using the cmpfun function from the compiler package:

```
library(compiler)
squares <- cmpfun(squares)
squares(1:10)
```

~

Exercise 6.22. Write a function for computing the mean of a vector using a for loop. How much slower than mean is it? Which function uses more memory?

Exercise 6.23. We have seen three different ways of filtering a data frame to only keep rows that fulfil a condition: using base R, data.table, and dplyr. Suppose that we want to extract all flights from 1 January from the flights data in the nycflights13 package:

---

[9]Since R 3.4.

```
library(data.table)
library(dplyr)
library(nycflights13)
# Read about the data:
?flights

# Make a data.table copy of the data:
flights.dt <- as.data.table(flights)

# Filtering using base R:
flights0101 <- flights[flights$month == 1 & flights$day == 1,]
# Filtering using data.table:
flights0101 <- flights.dt[month == 1 & day == 1,]
# Filtering using dplyr:
flights0101 <- flights |> filter(month ==1, day == 1)
```

Compare the speed and memory usage of these three approaches. Which has the best performance?

7

The role of simulation in modern statistics

Simulation is at the heart of the computer-intensive methods used in modern statistics. This chapter will introduce simulation and some of its many uses. Particular focus is put on how simulation can be used for analyses and for evaluating the properties of statistical procedures. Most of the content in subsequent chapters can be understood without reading this chapter, but if you're looking for a deeper understanding of modern statistical methods, the topics presented herein are essential.

After reading this chapter, you will be able to use R to:

- Generate random numbers,
- Perform simulations to assess the performance of statistical methods,
- Perform simulation-based hypothesis tests,
- Compute simulation-based confidence intervals, and
- Make sample size computations.

7.1 Simulation and distributions

A *random variable* is a variable whose value describes the outcome of a random phenomenon. A (probability) *distribution* is a mathematical function that describes the probability of different outcomes for a random variable. Random variables and distributions are at the heart of probability theory and most, if not all, statistical models.

As we shall soon see, they are also invaluable tools when evaluating statistical methods. A key component of modern statistical work is *simulation*, in which we generate artificial data that can be used both in the analysis of real data (e.g., in permutation tests and bootstrap confidence intervals, topics that we'll explore in this chapter) and for assessing different methods. Simulation is possible only because we can generate random numbers, so let's begin by having a look at how we can generate random numbers in R.

7.1.1 Generating random numbers

The function `sample` can be used to randomly draw a number of elements from a vector. For instance, we can use it to draw two random numbers from the first 10 integers: $1, 2, \ldots, 9, 10$:

```
sample(1:10, 2)
```

Try running the above code multiple times. You'll get different results each time, because each time it runs, the random number generator is in a different *state*. In most cases, this is desirable (if the results were the same each time we used `sample`, it wouldn't be random), but not if we want to replicate a result at some later stage.

When we are concerned about reproducibility, we can use `set.seed` to fix the state of the random number generator:

```
# Each run generates different results:
sample(1:10, 2); sample(1:10, 2)

# To get the same result each time, set the seed to a
# number of your choice:
set.seed(314); sample(1:10, 2)
set.seed(314); sample(1:10, 2)
```

We often want to use simulated data from a probability distribution, such as the normal distribution. The normal distribution is defined by its mean μ and its variance σ^2 (or, equivalently, its standard deviation σ). There are special functions for generating data from different distributions – for the normal distribution it is called `rnorm`. We specify the number of observations that we want to generate (`n`) and the parameters of the distribution (the mean `mu` and the standard deviation `sigma`):

```
rnorm(n = 10, mu = 2, sigma = 1)

# A shorter version:
rnorm(10, 2, 1)
```

Similarly, there are functions that can be used to compute the quantile function, density function, and cumulative distribution function (CDF) of the normal distribution. Here are some examples for a normal distribution with mean 2 and standard deviation 1:

```
qnorm(0.9, 2, 1)    # Upper 90% quantile of distribution
dnorm(2.5, 2, 1)    # Density function f(2.5)
pnorm(2.5, 2, 1)    # Cumulative distribution function F(2.5)
```

~

Exercise 7.1. Sampling can be done with or without *replacement*. If replacement is used, an observation can be drawn more than once. Check the documentation for `sample`. How can you change the settings to sample with replacement? Draw five random numbers from the first 10 integers, with replacement.

7.1.2 Some common distributions

Next, we provide the syntax for random number generation, quantile functions, density/probability functions, and cumulative distribution functions for some of the most commonly used distributions. This section is mainly intended as a reference for you to look up when you need to use one of these distributions – so there is no need to run all the code chunks below right now.

Normal distribution $N(\mu, \sigma^2)$ with mean μ and variance σ^2:

```
rnorm(n, mu, sigma)     # Generate n random numbers
qnorm(0.95, mu, sigma)  # Upper 95 %quantile of distribution
dnorm(x, mu, sigma)     # Density function f(x)
pnorm(x, mu, sigma)     # Cumulative distribution function F(X)
```

Continuous uniform distribution $U(a, b)$ on the interval (a, b), with mean $\frac{a+b}{2}$ and variance $\frac{(b-a)^2}{12}$:

```
runif(n, a, b)      # Generate n random numbers
qunif(0.95, a, b)   # Upper 95% quantile of distribution
dunif(x, a, b)      # Density function f(x)
punif(x, a, b)      # Cumulative distribution function F(X)
```

Exponential distribution $Exp(m)$ with mean m and variance m^2:

```
rexp(n, 1/m)     # Generate n random numbers
qexp(0.95, 1/m)  # Upper 95% quantile of distribution
dexp(x, 1/m)     # Density function f(x)
pexp(x, 1/m)     # Cumulative distribution function F(X)
```

Gamma distribution $\Gamma(\alpha, \beta)$ with mean $\frac{\alpha}{\beta}$ and variance $\frac{\alpha}{\beta^2}$:

```
rgamma(n, alpha, beta)      # Generate n random numbers
qgamma(0.95, alpha, beta)   # Upper 95% quantile of distribution
dgamma(x, alpha, beta)      # Density function f(x)
pgamma(x, alpha, beta)      # Cumulative distribution function F(X)
```

Lognormal distribution $LN(\mu, \sigma^2)$ with mean $\exp(\mu + \sigma^2/2)$ and variance $(\exp(\sigma^2) - 1)\exp(2\mu + \sigma^2)$:

```
rlnorm(n, mu, sigma)     # Generate n random numbers
qlnorm(0.95, mu, sigma)  # Upper 95% quantile of distribution
dlnorm(x, mu, sigma)     # Density function f(x)
plnorm(x, mu, sigma)     # Cumulative distribution function F(X)
```

t-distribution $t(\nu)$ with mean 0 (for $\nu > 1$) and variance $\frac{\nu}{\nu-2}$ (for $\nu > 2$):

```
rt(n, nu)       # Generate n random numbers
qt(0.95, nu)    # Upper 95% quantile of distribution
dt(x, nu)       # Density function f(x)
pt(x, nu)       # Cumulative distribution function F(X)
```

Chi-squared distribution $\chi^2(k)$ with mean k and variance $2k$:

```
rchisq(n, k)     # Generate n random numbers
qchisq(0.95, k)  # Upper 95% quantile of distribution
```

```
dchisq(x, k)     # Density function f(x)
pchisq(x, k)     # Cumulative distribution function F(X)
```

F-distribution $F(d_1, d_2)$ with mean $\frac{d_2}{d_2-2}$ (for $d_2 > 2$) and variance $\frac{2d_2^2(d_1+d_2-2)}{d_1(d_2-2)^2(d_2-4)}$ (for $d_2 > 4$):

```
rf(n, d1, d2)     # Generate n random numbers
qf(0.95, d1, d2)  # Upper 95% quantile of distribution
df(x, d1, d2)     # Density function f(x)
pf(x, d1, d2)     # Cumulative distribution function F(X)
```

Beta distribution $Beta(\alpha, \beta)$ with mean $\frac{\alpha}{\alpha+\beta}$ and variance $\frac{\alpha\beta}{(\alpha+\beta)^2(\alpha+\beta+1)}$:

```
rbeta(n, alpha, beta)      # Generate n random numbers
qbeta(0.95, alpha, beta)   # Upper 95% quantile of distribution
dbeta(x, alpha, beta)      # Density function f(x)
pbeta(x, alpha, beta)      # Cumulative distribution function F(X)
```

Binomial distribution $Bin(n, p)$ with mean np and variance $np(1-p)$:

```
rbinom(n, n, p)     # Generate n random numbers
qbinom(0.95, n, p)  # Upper 95% quantile of distribution
dbinom(x, n, p)     # Probability function f(x)
pbinom(x, n, p)     # Cumulative distribution function F(X)
```

Poisson distribution $Po(\lambda)$ with mean λ and variance λ:

```
rpois(n, lambda)     # Generate n random numbers
qpois(0.95, lambda)  # Upper 95% quantile of distribution
dpois(x, lambda)     # Probability function f(x)
ppois(x, lambda)     # Cumulative distribution function F(X)
```

Negative binomial distribution $NegBin(r, p)$ with mean $\frac{rp}{1-p}$ and variance $\frac{rp}{(1-p)^2}$:

```
rnbinom(n, r, p)     # Generate n random numbers
qnbinom(0.95, r, p)  # Upper 95% quantile of distribution
dnbinom(x, r, p)     # Probability function f(x)
pnbinom(x, r, p)     # Cumulative distribution function F(X)
```

Multivariate normal distribution with mean vector μ and covariance matrix Σ:

```
library(MASS)
mvrnorm(n, mu, Sigma) # Generate n random numbers
```

∼

Exercise 7.2. Use `runif` and (at least) one of `round`, `ceiling` and `floor` to generate observations from a discrete random variable on the integers $1, 2, 3, 4, 5, 6, 7, 8, 9, 10$.

7.1.3 Assessing distributional assumptions

So how can we know that the functions for generating random observations from distributions work? And when working with real data, how can we know what distribution fits the data? One answer is that we can visually compare the distribution of the generated (or real) data to the target distribution. This can for instance be done by comparing a histogram of the data to the target distribution's density function.

To do so, we must add `aes(y = ..density..)` to the call to `geom_histogram`, which rescales the histogram to have area 1 (just like a density function has). We can then add the density function using `geom_function`:

```
# Generate data from a normal distribution with mean 10 and
# standard deviation 1
generated_data <- data.frame(normal_data = rnorm(1000, 10, 1))


library(ggplot2)
# Compare to histogram:
ggplot(generated_data, aes(x = normal_data)) +
      geom_histogram(colour = "black", aes(y = ..density..)) +
      geom_function(fun = dnorm, colour = "red", size = 2,
                  args = list(mean = mean(generated_data$normal_data),
                                  sd = sd(generated_data$normal_data)))
```

Try increasing the number of observations generated. As the number of observations increases, the histogram should start to look more and more like the density function.

We could also add a density estimate for the generated data, to further aid the eye here – we'd expect this to be close to the theoretical density function:

```
# Compare to density estimate:
ggplot(generated_data, aes(x = normal_data)) +
      geom_histogram(colour = "black", aes(y = ..density..)) +
      geom_density(colour = "blue", size = 2) +
      geom_function(fun = dnorm, colour = "red", size = 2,
                  args = list(mean = mean(generated_data$normal_data),
                                  sd = sd(generated_data$normal_data)))
```

If instead we wished to compare the distribution of the data to a χ^2 distribution, we would change the value of `fun` and `args` in `geom_function` accordingly:

```
# Compare to density estimate:
ggplot(generated_data, aes(x = normal_data)) +
      geom_histogram(colour = "black", aes(y = ..density..)) +
      geom_density(colour = "blue", size = 2) +
      geom_function(fun = dchisq, colour = "red", size = 2,
                  args = list(df = mean(generated_data$normal_data)))
```

Note that the values of `args` have changed. `args` should always be a list containing values for the parameters of the distribution: `mu` and `sigma` for the normal distribution and `df` for the χ^2 distribution (the same as in Section 7.1.2).

Another option is to draw a quantile-quantile plot, or Q-Q plot for short, which compares the theoretical quantiles of a distribution to the empirical quantiles of the data, showing each observation as a point. If the data follows the theorised distribution, then the points should lie more or less along a straight line.

To draw a Q-Q plot for a normal distribution, we use the geoms `geom_qq` and `geom_qq_line`:

```
# Q-Q plot for normality:
ggplot(generated_data, aes(sample = normal_data)) +
        geom_qq() + geom_qq_line()
```

For all other distributions, we must provide the quantile function of the distribution (many of which can be found in Section 7.1.2):

```
# Q-Q plot for the lognormal distribution:
ggplot(generated_data, aes(sample = normal_data)) +
        geom_qq(distribution = qlnorm) +
        geom_qq_line(distribution = qlnorm)
```

Q-Q-plots can be a little difficult to read. There will always be points deviating from the line – in fact, that's expected. So how much must they deviate before we rule out a distributional assumption? Particularly when working with real data, I like to compare the Q-Q-plot of my data to Q-Q-plots of simulated samples from the assumed distribution, to get a feel for what kind of deviations can appear if the distributional assumption holds. Here's an example of how to do this, for the normal distribution:

```
# Look at solar radiation data for May from the airquality
# dataset:
May <- airquality[airquality$Month == 5,]

# Create a Q-Q-plot for the solar radiation data, and store
# it in a list:
qqplots <- list(ggplot(May, aes(sample = Solar.R)) +
  geom_qq() + geom_qq_line() + ggtitle("Actual data"))

# Compute the sample size n:
n <- sum(!is.na(May$Temp))

# Generate 8 new datasets of size n from a normal distribution.
# Then draw Q-Q-plots for these and store them in the list:
for(i in 2:9)
{
    generated_data <- data.frame(normal_data = rnorm(n, 10, 1))
    qqplots[[i]] <- ggplot(generated_data, aes(sample = normal_data)) +
      geom_qq() + geom_qq_line() + ggtitle("Simulated data")
}
```

```
# Plot the resulting Q-Q-plots side-by-side:
library(patchwork)
(qqplots[[1]] + qqplots[[2]] + qqplots[[3]]) /
  (qqplots[[4]] + qqplots[[5]] + qqplots[[6]]) /
  (qqplots[[7]] + qqplots[[8]] + qqplots[[9]])
```

You can run the code several times to get more examples of what Q-Q-plots can look like when the distributional assumption holds. In this case, the tail points in the Q-Q-plot for the solar radiation data deviate from the line more than the tail points in most simulated examples do, and, personally, I'd be reluctant to assume that the data comes from a normal distribution.

$$\sim$$

Exercise 7.3. Investigate the sleep times in the `msleep` data from the `ggplot2` package. Do they appear to follow a normal distribution? A lognormal distribution?

Exercise 7.4. Another approach to assessing distributional assumptions for real data is to use formal hypothesis tests. One example is the Shapiro-Wilk test for normality, available in `shapiro.test`. The null hypothesis is that the data comes from a normal distribution, and the alternative is that it doesn't (meaning that a low p-value is supposed to imply non-normality).

    1. Apply `shapiro.test` to the sleep times in the `msleep` dataset. According to the Shapiro-Wilk test, is the data normally distributed?

    2. Generate 2,000 observations from a $\chi^2(100)$ distribution. Compare the histogram of the generated data to the density function of a normal distribution. Are they similar? What are the results when you apply the Shapiro-Wilk test to the data?

7.1.4 Monte Carlo integration

In this chapter, we will use simulation to compute p-values and confidence intervals to compare different statistical methods, and to perform sample size computations. Another important use of simulation is in *Monte Carlo integration*, in which random numbers are used for numerical integration. It plays an important role in, for instance, statistical physics, computational biology, computational linguistics, and Bayesian statistics; fields that require the computation of complicated integrals.

To create an example of Monte Carlo integration, let's start by writing a function, `circle`, that defines a quarter-circle on the unit square. We will then plot it using the geom `geom_function`:

```
circle <- function(x)
{
    return(sqrt(1-x^2))
}

ggplot(data.frame(x = c(0, 1)), aes(x)) +
    geom_function(fun = circle)
```

Let's say that we are interested in computing the area under a quarter-circle. We can highlight the area in our plot using `geom_area`:

```
ggplot(data.frame(x = seq(0, 1, 1e-4)), aes(x)) +
    geom_area(aes(x = x,
                  y = ifelse(x^2 + circle(x)^2 <= 1, circle(x), 0)),
              fill = "pink") +
    geom_function(fun = circle)
```

To find the area, we will generate a large number of random points uniformly in the unit square. By the law of large numbers, the proportion of points that end up under the quarter-circle should be close to the area under the quarter-circle[1]. To do this, we generate 10,000 random values for the x and y coordinates of each point using the $U(0,1)$ distribution, that is, using `runif`:

```
B <- 1e4
unif_points <- data.frame(x = runif(B), y = runif(B))
```

Next, we add the points to our plot:

```
ggplot(unif_points, aes(x, y)) +
    geom_area(aes(x = x,
                  y = ifelse(x^2 + circle(x)^2 <= 1, circle(x), 0)),
              fill = "pink") +
    geom_point(size = 0.5, alpha = 0.25,
               colour = ifelse(unif_points$x^2 + unif_points$y^2 <= 1,
                               "red", "black")) +
    geom_function(fun = circle)
```

Note the order in which we placed the geoms – we plot the points after the area so that the pink colour won't cover the points, and the function after the points so that the points won't cover the curve.

To estimate the area, we compute the proportion of points that are below the curve:

```
mean(unif_points$x^2 + unif_points$y^2 <= 1)
```

In this case, we can also compute the area exactly: $\int_0^1 \sqrt{1 - x^2}dx = \pi/4 = 0.7853\ldots$. For more complicated integrals, however, numerical integration methods like Monte Carlo integration may be required. That being said, there are better numerical integration methods for low-dimensional integrals like this one. Monte Carlo integration is primarily used for higher-dimensional integrals, where other techniques fail.

---

[1] In general, the proportion of points that fall below the curve will be proportional to the area under the curve *relative* to the area of the sample space. In this case, the sample space is the unit square, which has area 1, meaning that the relative area is the same as the absolute area.

7.2   Evaluating statistical methods using simulation

An important use of simulation is in the evaluation of statistical methods. In this section, we will see how simulation can be used to compare the performance of two estimators, as well as the type I error rate and power of hypothesis tests.

7.2.1   Comparing estimators

Let's say that we want to estimate the mean μ of a normal distribution. We could come up with several different estimators for μ:

- The sample mean $\bar{x}$,
- The sample median $\tilde{x}$,
- The average of the largest and smallest value in the sample: $\frac{x_{max}+x_{min}}{2}$.

In this particular case (under normality), statistical theory tells us that the sample mean is the best estimator[2]. But how much better is it, really? And what if we didn't know statistical theory? Could we use simulation to find out which estimator to use?

To begin with, let's write a function that computes the estimate $\frac{x_{max}+x_{min}}{2}$:

```
max_min_avg <- function(x)
{
      return((max(x)+min(x))/2)
}
```

Next, we'll generate some data from a $N(0,1)$ distribution and compute the three estimates:

```
x <- rnorm(25)

x_mean <- mean(x)
x_median <- median(x)
x_mma <- max_min_avg(x)
x_mean; x_median; x_mma
```

As you can see, the estimates given by the different approaches differ, so clearly the choice of estimator matters. We can't determine which to use based on a single sample though. Instead, we typically compare the long-run properties of estimators, such as their *bias* and *variance*. The bias is the difference between the mean of the estimator and the parameter it seeks to estimate. An estimator is *unbiased* if its bias is 0, which is considered desirable at least in this setting. Among unbiased estimators, we prefer the one that has the smallest variance. So how can we use simulation to compute the bias and variance of estimators?

The key to using simulation here is to realise that x_mean is an observation of the random variable $\bar{X} = \frac{1}{25}(X_1 + X_2 + \cdots + X_{25})$ where each X_i is $N(0,1)$-distributed. We can generate observations of X_i (using rnorm), and can therefore also generate observations of $\bar{X}$. That means that we can obtain an arbitrarily large sample of observations of $\bar{X}$, which we can use to estimate its mean and variance. Here is an example:

---

[2]At least in terms of mean squared error.

```r
# Set the parameters for the normal distribution:
mu <- 0
sigma <- 1

# We will generate 10,000 observations of the estimators:
B <- 1e4
res <- data.frame(x_mean = vector("numeric", B),
                  x_median = vector("numeric", B),
                  x_mma = vector("numeric", B))

# Start progress bar:
pbar <- txtProgressBar(min = 0, max = B, style = 3)

for(i in seq_along(res$x_mean))
{
    x <- rnorm(25, mu, sigma)
    res$x_mean[i] <- mean(x)
    res$x_median[i] <- median(x)
    res$x_mma[i] <- max_min_avg(x)

    # Update progress bar
    setTxtProgressBar(pbar, i)
}
close(pbar)

# Compare the estimators:
colMeans(res-mu) # Bias
apply(res, 2, var) # Variances
```

All three estimators appear to be unbiased (even if the simulation results aren't exactly 0, they are very close). The sample mean has the smallest variance (and is therefore preferable!), followed by the median. The $\frac{x_{max}+x_{min}}{2}$ estimator has the worst performance, which is unsurprising, as it ignores all information not contained in the extremes of the dataset.

In Section 7.2.5 we'll discuss how to choose the number of simulated samples to use in your simulations. For now, we'll just note that the estimate of the estimators' biases becomes more stable as the number of simulated samples increases, as can be seen from this plot, which utilises `cumsum`, described in Section 5.3.3:

```r
# Compute estimates of the bias of the sample mean for each
# iteration:
res$iterations <- 1:B
res$x_mean_bias <- cumsum(res$x_mean)/1:B - mu

# Plot the results:
library(ggplot2)
ggplot(res, aes(iterations, x_mean_bias)) +
    geom_line() +
    xlab("Number of iterations") +
    ylab("Estimated bias")
```

```
# Cut the x-axis to better see the oscillations for smaller
# numbers of iterations:
ggplot(res, aes(iterations, x_mean_bias)) +
    geom_line() +
    xlab("Number of iterations") +
    ylab("Estimated bias") +
    xlim(0, 1000)
```

~

Exercise 7.5. Repeat the above simulation for different sample sizes n between 10 and 100. Plot the resulting variances as a function of n.

Exercise 7.6. Repeat the simulation in Exercise 7.5, but with a $t(3)$ distribution instead of the normal distribution. Which estimator is better in this case?

7.2.2   Type I error rate of hypothesis tests

In the same vein that we just compared estimators, we can also compare hypothesis tests or confidence intervals. Let's have a look at the former and evaluate how well the old-school two-sample t-test fares compared to a permutation t-test and the Wilcoxon-Mann-Whitney test.

For our first comparison, we will compare the type I error rate of the three tests, i.e., the risk of rejecting the null hypothesis if the null hypothesis is true. Nominally, this is the significance level α, which we set to be 0.05.

We write a function for such a simulation, to which we can pass the sizes n1 and n2 of the two samples, as well as a function distr to generate data:

```
# Load package used for permutation t-test:
library(MKinfer)

# Create a function for running the simulation:
simulate_type_I <- function(n1, n2, distr, level = 0.05, B = 999,
                            alternative = "two.sided", ...)
{
    # Create a data frame to store the results in:
    p_values <- data.frame(p_t_test = vector("numeric", B),
                    p_perm_t_test = vector("numeric", B),
                    p_wilcoxon = vector("numeric", B))

    # Start progress bar:
    pbar <- txtProgressBar(min = 0, max = B, style = 3)

    for(i in 1:B)
    {
        # Generate data:
        x <- distr(n1, ...)
```

```
        y <- distr(n2, ...)

        # Compute p-values:
        p_values[i, 1] <- t.test(x, y,
                           alternative = alternative)$p.value
        p_values[i, 2] <- perm.t.test(x, y,
                           alternative = alternative,
                           R = 999)$perm.p.value
        p_values[i, 3] <- wilcox.test(x, y,
                           alternative = alternative)$p.value

        # Update progress bar:
        setTxtProgressBar(pbar, i)
    }

    close(pbar)

    # Return the type I error rates:
    return(colMeans(p_values < level))
}
```

First, let's try it with normal data. The simulation takes a little while to run, primarily because of the permutation t-test, so you may want to take a short break while you wait.

```
simulate_type_I(20, 20, rnorm, B = 9999)
```

Next, let's try it with a lognormal distribution, both with balanced and imbalanced sample sizes. Increasing the parameter σ (sdlog) increases the skewness of the lognormal distribution (i.e., makes it *more* asymmetric and therefore less similar to the normal distribution), so let's try that too. In case you are in a rush, the results from my run of this code block can be found below it.

```
simulate_type_I(20, 20, rlnorm, B = 9999, sdlog = 1)
simulate_type_I(20, 20, rlnorm, B = 9999, sdlog = 3)
simulate_type_I(20, 30, rlnorm, B = 9999, sdlog = 1)
simulate_type_I(20, 30, rlnorm, B = 9999, sdlog = 3)
```

My results were:

```
# Normal distribution, n1 = n2 = 20:
    p_t_test p_perm_t_test    p_wilcoxon
  0.04760476    0.04780478    0.04680468

# Lognormal distribution, n1 = n2 = 20, sigma = 1:
    p_t_test p_perm_t_test    p_wilcoxon
  0.03320332    0.04620462    0.04910491

# Lognormal distribution, n1 = n2 = 20, sigma = 3:
```

```
      p_t_test p_perm_t_test    p_wilcoxon
   0.00830083    0.05240524    0.04590459

# Lognormal distribution, n1 = 20, n2 = 30, sigma = 1:
      p_t_test p_perm_t_test    p_wilcoxon
   0.04080408    0.04970497    0.05300530

# Lognormal distribution, n1 = 20, n2 = 30, sigma = 3:
      p_t_test p_perm_t_test    p_wilcoxon
   0.01180118    0.04850485    0.05240524
```

What's noticeable here is that the permutation t-test and the Wilcoxon-Mann-Whitney test have type I error rates that are close to the nominal 0.05 in all five scenarios, whereas the t-test has too low a type I error rate when the data comes from a lognormal distribution. This makes the test too conservative in this setting. Next, let's compare the power of the tests.

7.2.3   Power of hypothesis tests

The power of a test is the probability of rejecting the null hypothesis if it is false. To estimate that, we need to generate data under the alternative hypothesis. For two-sample tests of the mean, the code is similar to what we used for the type I error simulation above, but we now need two functions for generating data – one for each group, because the groups differ under the alternative hypothesis. Bear in mind that the alternative hypothesis for the two-sample test is that the two distributions differ in location, so the two functions for generating data should reflect that.

```
# Load package used for permutation t-test:
library(MKinfer)

# Create a function for running the simulation:
simulate_power <- function(n1, n2, distr1, distr2, level = 0.05,
                           B = 999, alternative = "two.sided")
{
    # Create a data frame to store the results in:
    p_values <- data.frame(p_t_test = vector("numeric", B),
                   p_perm_t_test = vector("numeric", B),
                   p_wilcoxon = vector("numeric", B))

    # Start progress bar:
    pbar <- txtProgressBar(min = 0, max = B, style = 3)

    for(i in 1:B)
    {
        # Generate data:
        x <- distr1(n1)
        y <- distr2(n2)

        # Compute p-values:
```

```
           p_values[i, 1] <- t.test(x, y,
                              alternative = alternative)$p.value
           p_values[i, 2] <- perm.t.test(x, y,
                              alternative = alternative,
                              R = 999)$perm.p.value
           p_values[i, 3] <- wilcox.test(x, y,
                              alternative = alternative)$p.value


           # Update progress bar:
           setTxtProgressBar(pbar, i)
      }

   close(pbar)

   # Return power:
   return(colMeans(p_values < level))
}
```

Let's try this out with lognormal data, where the difference in the log means is 1:

```
# Balanced sample sizes:
simulate_power(20, 20, function(n) { rlnorm(n,
                                 meanlog = 2, sdlog = 1) },
                  function(n) { rlnorm(n,
                                 meanlog = 1, sdlog = 1) },
                  B = 9999)

# Imbalanced sample sizes:
simulate_power(20, 30, function(n) { rlnorm(n,
                                 meanlog = 2, sdlog = 1) },
                  function(n) { rlnorm(n,
                                 meanlog = 1, sdlog = 1) },
                  B = 9999)
```

Here are the results from my runs:

```
# Balanced sample sizes:
     p_t_test p_perm_t_test     p_wilcoxon
    0.6708671     0.7596760      0.8508851

# Imbalanced sample sizes:
     p_t_test p_perm_t_test     p_wilcoxon
    0.6915692     0.7747775      0.9041904
```

Among the three, the Wilcoxon-Mann-Whitney test appears to be preferable for lognormal data, as it manages to obtain the correct type I error rate (unlike the old-school t-test) and has the highest power (although we would have to consider more scenarios, including different samples sizes, other differences of means, and different values of σ to say for sure!).

Remember that both our estimates of power and type I error rates are proportions, meaning that we can use binomial confidence intervals to quantify the uncertainty in the estimates from our simulation studies. Let's do that for the lognormal setting with balanced sample sizes, using the results from my runs. The number of simulated samples were 9,999. For the t-test, the estimated type I error rate was `0.03320332`, which corresponds to $0.03320332 \cdot 9,999 = 332$ "successes". Similarly, there were 6,708 "successes" in the power study. The confidence intervals become:

```
library(MKinfer)
binomCI(332, 9999, conf.level = 0.95, method = "clopper-pearson")
binomCI(6708, 9999, conf.level = 0.95, method = "wilson")
```

~

Exercise 7.7. Repeat the simulation study of type I error rate and power for the old school t-test, permutation t-test, and the Wilcoxon-Mann-Whitney test with $t(3)$-distributed data. Which test has the best performance? How much lower is the type I error rate of the old-school t-test compared to the permutation t-test in the case of balanced sample sizes?

7.2.4 Power of some tests of location

The `MKpower` package contains functions for quickly performing power simulations for the old-school t-test and Wilcoxon-Mann-Whitney test in different settings. The arguments `rx` and `ry` are used to pass functions used to generate the random numbers, in line with the `simulate_power` function that we created above.

For the t-test, we can use `sim.power.t.test`:

```
library(MKpower)
sim.power.t.test(nx = 25, rx = rnorm, rx.H0 = rnorm,
                 ny = 25, ry = function(x) { rnorm(x, mean = 0.8) },
                 ry.H0 = rnorm)
```

For the Wilcoxon-Mann-Whitney test, we can use `sim.power.wilcox.test` for power simulations:

```
library(MKpower)
sim.power.wilcox.test(nx = 10, rx = rnorm, rx.H0 = rnorm,
                      ny = 15,
                      ry = function(x) { rnorm(x, mean = 2) },
                      ry.H0 = rnorm)
```

7.2.5 Some advice on simulation studies

There are two decisions that you need to make when performing a simulation study:

- How many *scenarios* to include, i.e., how many different settings for the model parameters to study, and
- How many *iterations* to use, i.e., how many simulated samples to create for each scenario.

The number of scenarios is typically determined by what the purpose of the study is. If you only are looking to compare two tests for a particular sample size and a particular difference in means, then maybe you only need that one scenario. On the other hand, if you want to know which of the two tests is preferable in general, or for different sample sizes, or for different types of distributions, then you need to cover more scenarios. In that case, the number of scenarios may well be determined by how much time you have available or how many you can fit into your report.

As for the number of iterations to run, that also partially comes down to computational power. If each iteration takes a long while to run, it may not be feasible to run tens of thousands of iterations (some advice for speeding up simulations by using parallelisation can be found in Section 12.2). In the best of all possible worlds, you have enough computational power available and can choose the number of iterations freely. In such cases, it is often a good idea to use confidence intervals to quantify the uncertainty in your estimate of power, bias, or whatever it is you are studying. For instance, the power of a test is estimated as the proportion of simulations in which the null hypothesis was rejected. This is a binomial experiment, and a confidence interval for the power can be obtained using the methods described in Section 3.5.1. Moreover, the `ssize.propCI` function described in said section can be used to determine the number of simulations that you need to obtain a confidence interval that is short enough for you to feel that you have a good idea about the actual power of the test.

As an example, if a small pilot simulation indicates that the power is about 0.8 and you want a confidence interval with width 0.01, the number of simulations needed can be computed as follows:

```
library(MKpower)
ssize.propCI(prop = 0.8, width = 0.01, method = "wilson")
```

In this case, you'd need 24,592 iterations to obtain the desired accuracy.

7.3 Sample size computations using simulation

Using simulation to compare statistical methods is a key tool in methodological statistical research and when assessing new methods. In applied statistics, a use of simulation that is just as important is sample size computations. In this section we'll have a look at how simulations can be useful in determining sample sizes.

7.3.1 Writing your own simulation

Suppose that we want to perform a correlation test and want to know how many observations we need to collect. As in the previous section, we can write a function to compute the power of the test:

```
simulate_power <- function(n, distr, level = 0.05, B = 999, ...)
{
    p_values <- vector("numeric", B)
```

```r
      # Start progress bar:
      pbar <- txtProgressBar(min = 0, max = B, style = 3)


      for(i in 1:B)
      {
              # Generate bivariate data:
              x <- distr(n)


              # Compute p-values:
              p_values[i] <- cor.test(x[,1], x[,2], ...)$p.value


              # Update progress bar:
              setTxtProgressBar(pbar, i)
      }


      close(pbar)


      return(mean(p_values < level))
}
```

Under the null hypothesis of no correlation, the correlation coefficient is 0. We want to find a sample size that will give us 90% power at the 5% significance level, for different hypothesised correlations. We will generate data from a bivariate normal distribution, because it allows us to easily set the correlation of the generated data. Note that the mean and variance of the marginal normal distributions are nuisance variables, which can be set to 0 and 1, respectively, without loss of generality (because the correlation test is invariant under scaling and shifts in location).

First, let's try our power simulation function:

```r
library(MASS) # Contains mvrnorm function for generating data
rho <- 0.5 # The correlation between the variables
mu <- c(0, 0)
Sigma <- matrix(c(1, rho, rho, 1), 2, 2)

simulate_power(50, function(n) { mvrnorm(n, mu, Sigma) }, B = 999)
```

To find the sample size we need, we will write a new function containing a while loop (see Section 6.4.5) that performs the simulation for increasing values of n until the test has achieved the desired power:

```r
library(MASS)


power.cor.test <- function(n_start = 10, rho, n_incr = 5, power = 0.9,
                           B = 999, ...)
{
    # Set parameters for the multivariate normal distribution:
    mu <- c(0, 0)
    Sigma <- matrix(c(1, rho, rho, 1), 2, 2)
```

```
# Set initial values
n <- n_start
power_cor <- 0

# Check power for different sample sizes:
while(power_cor < power)
{
    power_cor <- simulate_power(n,
                        function(n) { mvrnorm(n, mu, Sigma) },
                        B = B, ...)
    cat("n =", n, " - Power:", power_cor, "\n")
    n <- n + n_incr
}

# Return the result:
cat("\nWhen n =", n, "the power is", round(power_cor, 2), "\n")
return(n)
}
```

Let's try it out with different settings:

```
power.cor.test(n_start = 10, rho = 0.5, power = 0.9)
power.cor.test(n_start = 10, rho = 0.2, power = 0.8)
```

As expected, larger sample sizes are required to detect smaller correlations.

7.3.2   The Wilcoxon-Mann-Whitney test

The `sim.ssize.wilcox.test` in `MKpower` can be used to quickly perform sample size computations for the Wilcoxon-Mann-Whitney test, analogously to how we used `sim.power.wilcox.test` in Section 7.2.4:

```
library(MKpower)
sim.ssize.wilcox.test(rx = rnorm, ry = function(x) rnorm(x, mean = 2),
                    power = 0.8, n.min = 3, n.max = 10,
                    step.size = 1)
```

~

Exercise 7.8. Modify the functions we used to compute the sample sizes for the Pearson correlation test to instead compute sample sizes for the Spearman correlation tests. For bivariate normal data, are the required sample sizes lower or higher than those of the Pearson correlation test?

Exercise 7.9. In Section 3.5.1 we had a look at some confidence intervals for proportions, and saw how `ssize.propCI` can be used to compute sample sizes for such intervals using asymptotic approximations. Write a function to compute the exact sample size needed for the Clopper-Pearson interval to achieve a desired expected (average) width. Compare your results to those from the asymptotic approximations. Are the approximations good enough to be useful?

7.4   Bootstrapping

The bootstrap can be used for many things, most notably for constructing confidence intervals and running hypothesis tests. These tend to perform better than traditional parametric methods, such as the old-school t-test and its associated confidence interval, when the distributional assumptions of the parametric methods aren't met.

Confidence intervals and hypothesis tests are always based on a *statistic*, i.e., a quantity that we compute from the samples. The statistic could be the sample mean, a proportion, the Pearson correlation coefficient, or something else. In traditional parametric methods, we start by assuming that our data follows some distribution. For different reasons, including mathematical tractability, a common assumption is that the data is normally distributed. Under that assumption, we can then derive the distribution of the statistic that we are interested in analytically, like Gosset did for the t-test. That distribution can then be used to compute confidence intervals and p-values.

When using a bootstrap method, we follow the same steps, but we use the observed data and simulation instead. Rather than making assumptions about the distribution[3], we use the empirical distribution of the data. Instead of analytically deriving a formula that describes the statistic's distribution, we find a good approximation of the distribution of the statistic by using simulation. We can then use that distribution to obtain confidence intervals and p-values, just as in the parametric case.

The simulation step is important. We use a process known as *resampling*, where we repeatedly draw new observations *with replacement* from the original sample. We draw B samples this way, each with the same size n as the original sample. Each randomly drawn sample – called a *bootstrap sample* – will include different observations. Some observations from the original sample may appear more than once in a specific bootstrap sample, and some not at all. For each bootstrap sample, we compute the statistic in which we are interested. This gives us B observations of this statistic, which together form what is called the *bootstrap distribution* of the statistic. I recommend using $B = 9,999$ or greater, but we'll use smaller B in some examples to speed up the computations.

7.4.1   A general approach

The Pearson correlation test is known to be sensitive to deviations from normality. We can construct a more robust version of it using the bootstrap. To illustrate the procedure, we will use the `sleep_total` and `brainwt` variables from the `msleep` data. Here is the result from the traditional parametric Pearson correlation test:

```
library(ggplot2)

msleep %$% cor.test(sleep_total, brainwt, use = "pairwise.complete")
```

---

[3]Well, sometimes we make assumptions about the distribution *and* use the bootstrap. This is known as the parametric bootstrap and is discussed in Section 7.4.4.

To find the bootstrap distribution of the Pearson correlation coefficient, we can use resampling with a `for` loop (Section 6.4.1):

```r
# Extract the data that we are interested in:
mydata <- na.omit(msleep[,c("sleep_total", "brainwt")])

# Resampling using a for loop:
B <- 999 # Number of bootstrap samples
statistic <- vector("numeric", B)
for(i in 1:B)
{
        # Draw row numbers for the bootstrap sample:
        row_numbers <- sample(1:nrow(mydata), nrow(mydata),
                            replace = TRUE)

        # Obtain the bootstrap sample:
        sample <- mydata[row_numbers,]

        # Compute the statistic for the bootstrap sample:
        statistic[i] <- cor(sample[, 1], sample[, 2])
}

# Plot the bootstrap distribution of the statistic:
ggplot(data.frame(statistic), aes(statistic)) +
        geom_histogram(colour = "black")
```

Because this is such a common procedure, there are R packages that let us do resampling without having to write a `for` loop. In the remainder of the section, we will use the `boot` package to draw bootstrap samples. It also contains convenience functions that allows us to get confidence intervals from the bootstrap distribution quickly. Let's install it:

```r
install.packages("boot")
```

The most important function in this package is `boot`, which does the resampling. As input, it takes the original data, the number B of bootstrap samples to draw (called `R` here), and a function that computes the statistic of interest. This function should take the original data (`mydata` in our example above) and the row numbers of the sampled observation for a particular bootstrap sample (`row_numbers` in our example) as input.

For the correlation coefficient, the function that we input can look like this:

```r
cor_boot <- function(data, row_numbers, method = "pearson")
{
    # Obtain the bootstrap sample:
    sample <- data[row_numbers,]

    # Compute and return the statistic for the bootstrap sample:
    return(cor(sample[, 1], sample[, 2], method = method))
}
```

To get the bootstrap distribution of the Pearson correlation coefficient for our data, we can now use `boot` as follows:

```
library(boot)
```

```
# Base solution:
boot_res <- boot(na.omit(msleep[,c("sleep_total", "brainwt")]),
                 cor_boot,
                 999)
```

Next, we can plot the bootstrap distribution of the statistic computed in `cor_boot`:

```
plot(boot_res)
```

If you prefer, you can of course use a pipeline for the resampling instead:

```
library(boot)
library(dplyr)
```

```
msleep |> select(sleep_total, brainwt) |>
    drop_na() |>
    boot(cor_boot, 999) -> boot_res
```

7.4.2   Bootstrap confidence intervals

The next step is to use `boot.ci` to compute bootstrap confidence intervals. This is as simple as running:

```
boot.ci(boot_res)
```

Four intervals are presented: normal, basic, percentile, and BCa. The details concerning how these are computed based on the bootstrap distribution are presented in Section 14.1. It is generally agreed that the percentile and BCa intervals are preferable to the normal and basic intervals; see, e.g., Davison & Hinkley (1997) and Hall (1992); but, which one performs the best varies.

We also receive a warning message:

```
Warning message:
In boot.ci(boot_res) : bootstrap variances needed for studentized
intervals
```

A fifth type of confidence interval, the studentised interval, requires bootstrap estimates of the standard error of the test statistic. These are obtained by running an *inner bootstrap*, i.e., by bootstrapping each bootstrap sample to get estimates of the variance of the test statistic. Let's create a new function that does this, and then compute the bootstrap confidence intervals:

```
cor_boot_student <- function(data, i, method = "pearson")
{
    sample <- data[i,]

    correlation <- cor(sample[, 1], sample[, 2], method = method)

    inner_boot <- boot(sample, cor_boot, 100)
    variance <- var(inner_boot$t)

    return(c(correlation, variance))
}

library(ggplot2)
library(boot)

boot_res <- boot(na.omit(msleep[,c("sleep_total", "brainwt")]),
                 cor_boot_student,
                 999)

# Show bootstrap distribution:
plot(boot_res)

# Compute confidence intervals - including studentised:
boot.ci(boot_res)
```

While theoretically appealing (Hall, 1992), studentised intervals can be a little erratic in practice. I prefer to use percentile and BCa intervals instead.

For two-sample problems, we need to make sure that the number of observations drawn from each sample is the same as in the original data. The `strata` argument in `boot` is used to achieve this. Let's return to the example studied in Section 3.6, concerning the difference in how long carnivores and herbivores sleep. Let's say that we want a confidence interval for the difference of two means, using the `msleep` data. The simplest approach is to create a Welch-type interval, where we allow the two populations to have different variances. We can then resample from each population separately:

```
# Function that computes the mean for each group:
mean_diff_msleep <- function(data, i)
{
    sample1 <- subset(data[i, 1], data[i, 2] == "carni")
    sample2 <- subset(data[i, 1], data[i, 2] == "herbi")
    return(mean(sample1[[1]]) - mean(sample2[[1]]))
}

library(ggplot2) # Load the data
library(boot)    # Load bootstrap functions

# Create the data set to resample from:
boot_data <- na.omit(subset(msleep,
             vore == "carni" | vore == "herbi")[,c("sleep_total",
```

```
                                                          "vore")])
# Do the resampling - we specify that we want resampling from two
# populations by using strata:
boot_res <- boot(boot_data,
                 mean_diff_msleep,
                 999,
                 strata = factor(boot_data$vore))


# Compute confidence intervals:
boot.ci(boot_res, type = c("perc", "bca"))
```

~

Exercise 7.10. Let's continue the example with a confidence interval for the difference in how long carnivores and herbivores sleep. How can you create a confidence interval under the assumption that the two groups have equal variances?

7.4.3   Bootstrap hypothesis tests

Writing code for bootstrap hypothesis tests can be a little tricky, because the resampling must be done *under the null hypothesis*. The process is greatly simplified by computing p-values using *confidence interval inversion* instead. This approach exploits the equivalence between confidence intervals and hypothesis tests, detailed in Section 14.2. It relies on the fact that:

- The p-value of the test for the parameter θ is the smallest α such that θ is not contained in the corresponding $1 - \alpha$ confidence interval.
- For a test for the parameter θ with significance level α, the set of values of θ that aren't rejected by the test (when used as the null hypothesis) is a $1 - \alpha$ confidence interval for θ.

Here is an example of how we can use a while loop (Section 6.4.5) for confidence interval inversion, in order to test the null hypothesis that the Pearson correlation between sleep time and brain weight is $\rho = -0.2$. It uses the studentised confidence interval that we created in the previous section:

```
# Compute the studentised confidence interval:
cor_boot_student <- function(data, i, method = "pearson")
{
    sample <- data[i,]

    correlation <- cor(sample[, 1], sample[, 2], method = method)

    inner_boot <- boot(sample, cor_boot, 100)
    variance <- var(inner_boot$t)

    return(c(correlation, variance))
}

library(ggplot2)
```

```r
library(boot)

boot_res <- boot(na.omit(msleep[,c("sleep_total", "brainwt")]),
                 cor_boot_student,
                 999)

# Now, a hypothesis test:
# The null hypothesis:
rho_null <- -0.2

# Set initial conditions:
in_interval <- TRUE
alpha <- 0

# Find the lowest alpha for which rho_null is in the interval:
while(in_interval)
{
    # Increase alpha a small step:
    alpha <- alpha + 0.001

    # Compute the 1-alpha confidence interval, and extract
    # its bounds:
    interval <- boot.ci(boot_res,
                        conf = 1 - alpha,
                        type = "stud")$student[4:5]

    # Check if the null value for rho is greater than the lower
    # interval bound and smaller than the upper interval bound,
    # i.e. if it is contained in the interval:
    in_interval <- rho_null > interval[1] & rho_null < interval[2]
}
# The loop will finish as soon as it reaches a value of alpha such
# that rho_null is not contained in the interval.

# Print the p-value:
alpha
```

The `boot.pval` package contains a function computing p-values through inversion of bootstrap confidence intervals. We can use it to obtain a bootstrap p-value without having to write a `while` loop. It works more or less analogously to `boot.ci`. The arguments to the `boot.pval` function is the `boot` object (`boot_res`), the type of interval to use (`"stud"`), and the value of the parameter under the null hypothesis (`-0.2`):

```r
install.packages("boot.pval")
library(boot.pval)
boot.pval(boot_res, type = "stud", theta_null = -0.2)
```

Confidence interval inversion fails in spectacular ways for certain tests for parameters of discrete distributions (Thulin & Zwanzig, 2017), so be careful if you plan on using this approach with count data.

~

Exercise 7.11. With the data from Exercise 7.10, invert a percentile confidence interval to compute the p-value of the corresponding test of the null hypothesis that there is no difference in means. What are the results?

7.4.4 The parametric bootstrap

In some cases, we may be willing to make distributional assumptions about our data. We can then use the *parametric bootstrap*, in which the resampling is done not from the original sample, but the theorised distribution (with parameters estimated from the original sample). Here is an example for the bootstrap correlation test, where we assume a multivariate normal distribution for the data. Note that we no longer include an index as an argument in the function `cor_boot`, because the bootstrap samples won't be drawn directly from the original data:

```
cor_boot <- function(data, method = "pearson")
{
    return(cor(data[, 1], data[, 2], method = method))
}


library(MASS)
generate_data <- function(data, mle)
{
    return(mvrnorm(nrow(data), mle[[1]], mle[[2]]))
}


library(ggplot2)
library(boot)

filtered_data <- na.omit(msleep[,c("sleep_total", "brainwt")])
boot_res <- boot(filtered_data,
                 cor_boot,
                 R = 999,
                 sim = "parametric",
                 ran.gen = generate_data,
                 mle = list(colMeans(filtered_data),
                            cov(filtered_data)))

# Show bootstrap distribution:
plot(boot_res)

# Compute bootstrap percentile confidence interval:
boot.ci(boot_res, type = "perc")
```

The BCa interval implemented in `boot.ci` is not valid for parametric bootstrap samples, so running `boot.ci(boot_res)` without specifying the interval `type` will render an error[4]. Percentile intervals work just fine, though.

---

[4]If you really need a BCa interval for the parametric bootstrap, you can find the formulas for it in Davison & Hinkley (1997).

8

Regression models

Regression models, in which explanatory variables are used to model the behaviour of a response variable, are without a doubt the most commonly used class of models in the statistical toolbox. In this chapter, we will have a look at different types of regression models tailored to many different sorts of data and applications.

After reading this chapter, you will be able to use R to:

- Fit and evaluate linear models, including linear regression and ANOVA,
- Fit and evaluate generalised linear models, including logistic regression and Poisson regression,
- Use multiple imputation to handle missing data,
- Fit and evaluate mixed models, and
- Create matched samples.

8.1 Linear models

Being flexible enough to handle different types of data, yet simple enough to be useful and interpretable, linear models are among the most important tools in the statistics toolbox. In this section, we'll discuss how to fit and evaluate linear models in R.

Linear regression is used to model the relationship between two or more variables, where one variable is called the *response variable* (or outcome, or dependent variable), and the other variables are called the *explanatory variables* (or predictors, independent variables, features, or covariates). The goal of linear regression is to find the best linear relationship between the response variable and the explanatory variables, which can then be used to make predictions or infer the strength and direction of the relationship between the variables.

In a *simple linear regression*, where there is a single explanatory variable, the relationship between the response variable and the explanatory variable is modelled as a straight line. The equation for this line is represented as

$$y = \beta_0 + \beta_1 x,$$

where

- y is the response variable,
- x is the explanatory variable,
- β_1 is the slope of the line, describing how much y changes when x changes one unit, and
- β_0 is the intercept; the value y attains when $x = 0$.

To find the best fit line, the *least squares method* is used, which involves finding the line that minimizes the sum of squared differences between the observed values and the fitted line. This yields fitted, or estimated, values of β_0 and β_1.

Multiple linear regression is an extension of simple linear regression, where there are multiple explanatory variables. The equation for the multiple linear regression model is represented as

$$y = \beta_0 + \beta_1 x_1 + \beta_2 x_2 + \ldots + \beta_p x_p,$$

where

- y is the response variable,
- x_i are the p explanatory variables,
- β_i are the coefficients or the slopes of the lines, and
- β_0 is the intercept; the value y attains when all $x_i = 0$.

Even if the relationship described by the linear regression formula is true, observations of y will deviate from the line due to random variation. To formally state the model, we need to include these random errors. Given n observations of p explanatory variables and the response variable y, the linear model is:

$$y_i = \beta_0 + \beta_1 x_{i1} + \beta_2 x_{i2} + \cdots + \beta_p x_{ip} + \epsilon_i, \qquad i = 1, \ldots, n$$

where the ϵ_i are independent random errors with mean 0, meaning that the model also can be written as:

$$E(y_i) = \beta_0 + \beta_1 x_{i1} + \beta_2 x_{i2} + \cdots + \beta_p x_{ip}, \qquad i = 1, \ldots, n$$

8.1.1   Fitting linear models

The `mtcars` data from Henderson and Velleman (1981) has become one of the classic datasets in R, and a part of the initiation rite for new R users is to use the `mtcars` data to fit a linear regression model. The data describes fuel consumption, number of cylinders, and other information about cars from the 1970s:

```
?mtcars
View(mtcars)
```

Let's have a look at the relationship between gross horsepower (`hp`) and fuel consumption (`mpg`) in the `mtcars` data:

```
library(ggplot2)
ggplot(mtcars, aes(hp, mpg)) +
      geom_point()
```

The relationship doesn't appear to be perfectly linear; nevertheless, we can try fitting a linear regression model to the data. This can be done using `lm`. We fit a model with `mpg` as the response variable and `hp` as the explanatory variable as follows:

```
m <- lm(mpg ~ hp, data = mtcars)
```

The first argument is a formula, saying that mpg is a function of hp, i.e.,

$$mpg = \beta_0 + \beta_1 \cdot hp.$$

A summary of the model is obtained using summary:

```
summary(m)
```

```
##
## Call:
## lm(formula = mpg ~ hp, data = mtcars)
##
## Residuals:
##     Min      1Q  Median      3Q     Max
## -5.7121 -2.1122 -0.8854  1.5819  8.2360
##
## Coefficients:
##               Estimate Std. Error t value           Pr(>|t|)
## (Intercept) 30.09886    1.63392   18.421 < 0.0000000000000002 ***
## hp          -0.06823    0.01012   -6.742          0.000000179 ***
## ---
## Signif. codes:  0 '***' 0.001 '**' 0.01 '*' 0.05 '.' 0.1 ' ' 1
##
## Residual standard error: 3.863 on 30 degrees of freedom
## Multiple R-squared:  0.6024, Adjusted R-squared:  0.5892
## F-statistic: 45.46 on 1 and 30 DF,  p-value: 0.0000001788
```

The summary contains a lot of information about the model:

- Call: information about the formula and dataset used to fit the model.
- Residuals: information about the model residuals (which we'll talk more about in Section 8.1.4).
- Coefficients: a table showing the fitted model coefficients (here, the fitted intercept $\beta_0 \approx 30.1$ and the fitted slope $\beta_1 \approx -0.07$), their standard errors (more on these below), t-statistics (used for computing p-values, and of no interest in their own right), and p-values. The p-values are used to test the null hypothesis that the β_i are 0 against the two-sided alternative that they are non-zero. In this case, both p-values are small, so we reject the null hypotheses and conclude that we have statistical evidence that the coefficients are non-zero.
- Information about the model fit, including the the Multiple R-squared, or coefficient of determination, R^2, which describes how much of the variance of y is described by x (in this case, 60.24%).

The standard errors reported in the table are often thought of as quantifying the uncertainty in the coefficient estimates. This is correct, but the standard errors are often misinterpreted. It is much better to report confidence intervals for the model coefficients instead. These can be obtained using confint:

```
confint(m)
```

We can add the fitted line to the scatterplot by using `geom_abline`, which lets us add a straight line with a given intercept and slope – we take these to be the coefficients from the fitted model, given by `coef`:

```
# Check model coefficients:
coef(m)

# Add regression line to plot:
ggplot(mtcars, aes(hp, mpg)) +
      geom_point() +
      geom_abline(aes(intercept = coef(m)[1], slope = coef(m)[2]),
                  colour = "red")
```

The resulting plot can be seen in Figure 8.1.

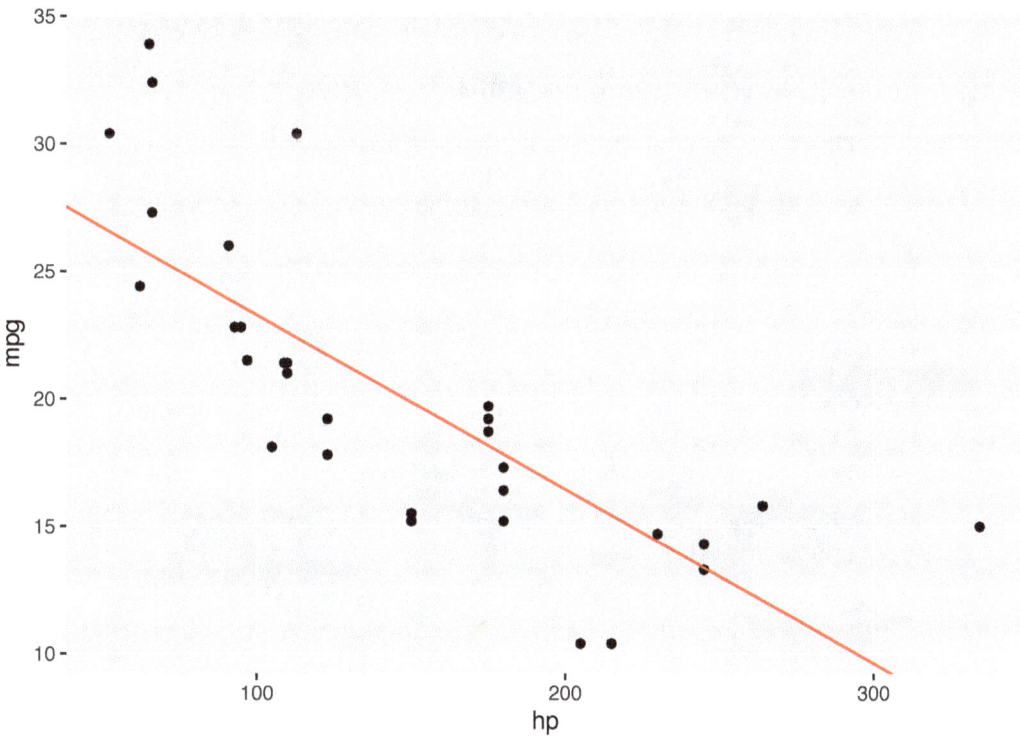

FIGURE 8.1 Linear regression model for the mtcars data.

If we wish to add more variables to the model, creating a multiple linear regressions model, we simply add them to the right-hand side of the formula in the function call:

```
m2 <- lm(mpg ~ hp + wt, data = mtcars)
summary(m2)
```

In this case, the model becomes

$$mpg = \beta_0 + \beta_1 \cdot hp + \beta_2 \cdot wt.$$

Next, we'll look at what more R has to offer when it comes to regression. Before that though, it's a good idea to do a quick exercise to make sure that you now know how to fit linear models.

~

Exercise 8.1. The `sales-weather.csv` data from Section 5.12 describes the weather in a region during the first quarter of 2020. Download the file from the book's web page[1]. Fit a linear regression model with `TEMPERATURE` as the response variable and `SUN_HOURS` as an explanatory variable. Plot the results. Is there a connection?

You'll return to and expand this model in the next few exercises, so make sure to save your code.

Exercise 8.2. Fit a linear model to the `mtcars` data using the formula `mpg ~ .`. What happens? What is `~ .` a shorthand for?

8.1.2   Publication-ready summary tables

Just as for frequency tables and contingency tables, we can use the `gtsummary` package to obtain publication-ready summary tables for regression models. The `tbl_regression` function gives a nicely formatted summary table, showing the fitted β_i along with their confidence intervals and p-values:

```
m2 <- lm(mpg ~ hp + wt, data = mtcars)

library(gtsummary)
tbl_regression(m2)
```

For more on how to export `gtsummary` tables, see Section 3.2.2.

`tbl_regression` has options for including the intercept, changing the confidence level, and replacing the variable names by labels:

```
tbl_regression(m2,
               intercept = TRUE,
               conf.level = 0.99,
               label = list(hp ~ "Gross horsepower", wt ~ "Weight (1000 lbs)"))
```

See `?tbl_regression` for even more options.

8.1.3   Dummy variables and interactions

Categorical variables can be included in regression models by using *dummy variables*. A dummy variable takes the values 0 and 1, indicating that an observation either belongs to a category (1) or not (0). For instance, consider a categorical variable z with two levels, A and B. We introduce the dummy variable x_2, defined as follows:

- $x_2 = 0$ if $z = A$,

---

[1]http://www.modernstatisticswithr.com/data.zip

- $x_2 = 1$ if $z = B$.

Now, consider a regression model with two explanatory variables: a numeric variable x_1 and the dummy variable x_2 described above. The model equation is

$$y = \beta_0 + \beta_1 x_1 + \beta_2 x_2.$$

If $z = A$, then $x_2 = 0$ and the model reduces to

$$y = \beta_0 + \beta_1 x_1 + \beta_2 \cdot 0 = \beta_0 + \beta_1 x_1.$$

This describes a line where the intercept is β_0 and the slope is β_1. However, if $z = B$ we have $x_2 = 1$ and the model becomes

$$y = \beta_0 + \beta_1 x_1 + \beta_2 \cdot 1 = (\beta_0 + \beta_2) + \beta_1 x_1.$$

This describes a line where the intercept is $\beta_0 + \beta_2$ and the slope is β_1. That is, the intercept has changed. β_2 describes how the intercept changes when $z = B$, compared to when $z = A$. Because the effect is relative to A, A is said to be the *reference category* for this dummy variable.

If the original categorical variable has more than two categories, c categories, say, the number of dummy variables included in the regression model should be $c - 1$ (with the last category corresponding to all dummy variables being 0). If we have a categorical variable z with three levels, A, B, and C, we would need to define two dummy variables:

- $x_2 = 1$ if $z = B$ and $x_2 = 0$ otherwise.
- $x_3 = 1$ if $z = C$ and $x_3 = 0$ otherwise.

Note that x_2 and x_3 both are 0 when $z = A$. Again, this makes A the reference category for this dummy variable, and the coefficients corresponding to x_2 and x_3 will describe the effect of B and C relative to A.

You can probably imagine that creating dummy variables can be a lot of work if we have categorical variables with many levels. Fortunately, R does this automatically for us if we include a `factor` variable in a regression model:

```
# Make cyl a categorical variable:
mtcars$cyl <- factor(mtcars$cyl)

m <- lm(mpg ~ hp + wt + cyl, data = mtcars)
summary(m)
```

Note how only two categories, 6 cylinders and 8 cylinders, are shown in the summary table. The third category, 4 cylinders, corresponds to both of those dummy variables being 0. Therefore, the coefficient estimates for `cyl6` and `cyl8` are relative to the remaining reference category `cyl4`. For instance, compared to `cyl4` cars, `cyl6` cars have a higher fuel consumption, with their `mpg` being 3.36 lower.

We can control which category is used as the reference category by setting the order of the `factor` variable, as in Section 5.4. The first `factor` level is always used as the reference, so if for instance we want to use `cyl6` as our reference category, we'd do the following:

```
# Make cyl a categorical variable with cyl6 as
# reference variable:
mtcars$cyl <- factor(mtcars$cyl, levels =
                        c(6, 4, 8))

m <- lm(mpg ~ hp + wt + cyl, data = mtcars)
summary(m)
```

Dummy variables are frequently used for modelling differences between different groups. Including only the dummy variable corresponds to using different intercepts for different groups. If we also include an *interaction* with the dummy variable, we can get different slopes for different groups. As the name implies, *interaction terms* in regression models describe how different explanatory variables interact. To include an interaction between x_1 and x_2 in our model, we add $x_1 \cdot x_2$ as an explanatory variable.

This yields the model

$$y = \beta_0 + \beta_1 x_1 + \beta_2 x_2 + \beta_{12} x_1 x_2,$$

where x_1 is numeric and x_2 is a dummy variable. Then the intercept and slope change depending on the value of x_2 as follows:

$$y = \beta_0 + \beta_1 x_1, \qquad \text{if } x_2 = 0,$$
$$y = (\beta_0 + \beta_2) + (\beta_1 + \beta_{12}) x_1, \qquad \text{if } x_2 = 1.$$

This yields a model where both the intercept and the slope differ between the two groups that x_2 represents. β_2 describes how the slope for B differs from that for A, and β_{12} describes how the slope for B differs from that for A.

To include an interaction term in our regression model, we can use either of the following equivalent formulas:

```
m <- lm(mpg ~ hp*cyl, data = mtcars)

m <- lm(mpg ~ hp + cyl + hp:cyl, data = mtcars)

summary(m)
```

In this model, cyl6 is the reference category. The variables cyl4 and cyl8 in the summary table describe how the number of cylinders affect the intercept. For 6 cylinders, the intercept is 20.7 mpg. For 4 cylinders, the intercept is 15.3 units higher than for 6 cylinders (remember that all effects are relative to the reference category!). For 8 cylinders, it is 2.6 units lower than for 6 cylinders.

The hp:cyl4 and hp:cyl8 variables describe how the slope for hp is affected by the number of cylinders. For 6 cylinders, the slope for hp is -0.007613. For 4 cylinders, it is $-0.007613 - 0.105163 = -0.112776$, and for 8 cylinders it is $-0.007613 - 0.006631 = -0.014244$. Note that in this case we have a non-monotone effect of the number of cylinders – the slope is the lowest for 6 cylinders, and higher for 4 and 8 cylinders. This would not have been possible if we had included the number of cylinders as a numeric variable instead.

$\sim$

Exercise 8.3. R automatically creates dummy variables for us when we include categorical variables as explanatory variables in regression models. Sometimes, we may wish to create them manually, e.g., when creating a categorical variable from a numerical one. Return to the weather model from Exercise 8.1. Create a dummy variable for precipitation (zero precipitation (0) or non-zero precipitation (1)) and add it to your model. Also include an interaction term between the precipitation dummy and the number of sun hours. Are any of the coefficients significantly non-zero?

8.1.4 Model diagnostics

There are a few different ways in which we can plot the fitted model. First, we can of course make a scatterplot of the data and add a curve showing the fitted values corresponding to the different points. These can be obtained by running `predict(m)` with our fitted model `m`.

```
# Fit two models:
mtcars$cyl <- factor(mtcars$cyl)
m1 <- lm(mpg ~ hp + wt, data = mtcars) # Simple model
m2 <- lm(mpg ~ hp*wt + cyl, data = mtcars) # Complex model

# Create data frames with fitted values:
m1_pred <- data.frame(hp = mtcars$hp, mpg_pred = predict(m1))
m2_pred <- data.frame(hp = mtcars$hp, mpg_pred = predict(m2))

# Plot fitted values:
library(ggplot2)
ggplot(mtcars, aes(hp, mpg)) +
    geom_point() +
    geom_line(data = m1_pred, aes(x = hp, y = mpg_pred),
            colour = "red") +
    geom_line(data = m2_pred, aes(x = hp, y = mpg_pred),
            colour = "blue")
```

We could also plot the observed values against the fitted values:

```
n <- nrow(mtcars)
models <- data.frame(Observed = rep(mtcars$mpg, 2),
                  Fitted = c(predict(m1), predict(m2)),
                  Model = rep(c("Model 1", "Model 2"), c(n, n)))

ggplot(models, aes(Fitted, Observed)) +
    geom_point(colour = "blue") +
    facet_wrap(~ Model, nrow = 3) +
    geom_abline(intercept = 0, slope = 1) +
    xlab("Fitted values") + ylab("Observed values")
```

Linear models are fitted and analysed using a number of assumptions, most of which are assessed by looking at plots of the model residuals, $y_i - \hat{y}_i$, where $\hat{y}_i$ is the fitted value for observation i. Some important assumptions are:

- *The model is linear in the parameters*: we check this by looking for non-linear patterns in the residuals, or in the plot of observed values against fitted values.

- *The observations are independent*: which can be difficult to assess visually. We'll look at models that are designed to handle correlated observations in Sections 8.8 and 11.6.
- *Homoscedasticity*: which means that the random errors all have the same variance. We check this by looking for non-constant variance in the residuals. The opposite of homoscedasticity is heteroscedasticity.
- *Normally distributed random errors*: this assumption is important if we want to use the traditional parametric p-values, confidence intervals, and prediction intervals. If we use permutation p-values or bootstrap confidence intervals (as we will later in this chapter), we no longer need this assumption.

Additionally, residual plots can be used to find influential points that (possibly) have a large impact on the model coefficients (influence is measured using *Cook's distance* and potential influence using *leverage*). We've already seen that we can use plot(m) to create some diagnostic plots. To get more and better-looking plots, we can use the autoplot function for lm objects from the ggfortify package:

```
library(ggfortify)
autoplot(m1, which = 1:6, ncol = 2, label.size = 3)
```

In each of the plots in Figure 8.2, we look for the following:

- Residuals vs. fitted: look for patterns that can indicate non-linearity, e.g., that the residuals all are high in some areas and low in others. The blue line is there to aid the eye – it should ideally be relatively close to a straight line (in this case, it isn't perfectly straight, which could indicate a mild non-linearity).
- Normal Q-Q: see if the points follow the line, which would indicate that the residuals (which we for this purpose can think of as estimates of the random errors) follow a normal distribution.
- Scale-Location: similar to the residuals vs. fitted plot, this plot shows whether the residuals are evenly spread for different values of the fitted values. Look for patterns in how much the residuals vary – if, e.g., they vary more for large fitted values, then that is a sign of heteroscedasticity. A horizontal blue line is a sign of homoscedasticity.
- Cook's distance: look for points with high values. A commonly cited rule-of-thumb (Cook & Weisberg, 1982) says that values above 1 indicate points with a high influence.
- Residuals vs. leverage: look for points with a high residual and high leverage. Observations with a high residual but low leverage deviate from the fitted model but don't affect it much. Observations with a high residual and a high leverage likely have a strong influence on the model fit, meaning that the fitted model could be quite different if these points were removed from the dataset.
- Cook's distance vs. leverage: look for observations with a high Cook's distance and a high leverage, which are likely to have a strong influence on the model fit.

A formal test for heteroscedasticity, the Breusch-Pagan test, is available in the car package as a complement to graphical inspection. A low p-value indicates statistical evidence for heteroscedasticity. To run the test, we use ncvTest (where "ncv" stands for non-constant variance):

```
install.packages("car")
library(car)
ncvTest(m1)
```

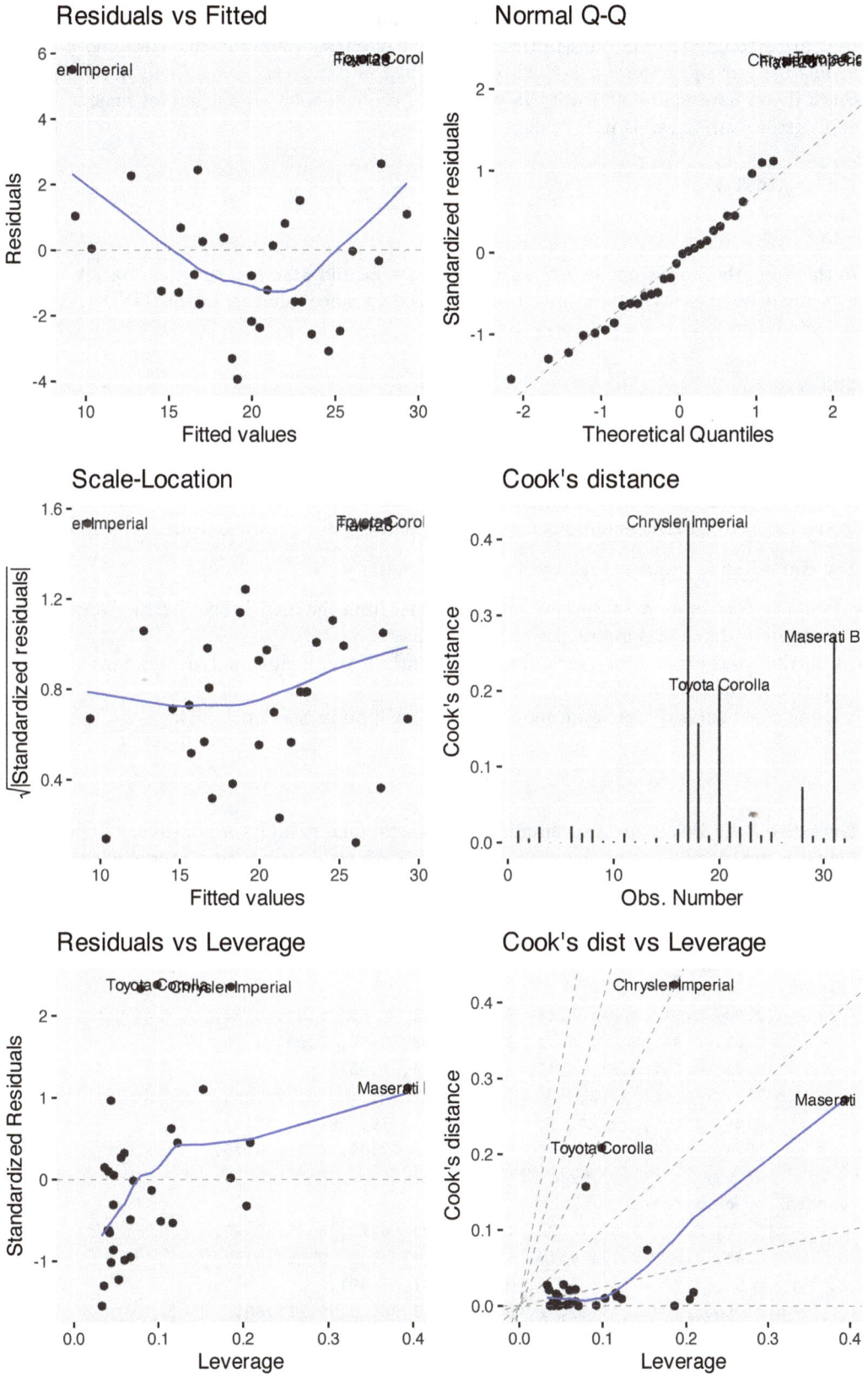

FIGURE 8.2 Diagnostic plots for linear regression.

A common problem in linear regression models is multicollinearity, i.e., explanatory variables that are strongly correlated. Multicollinearity can cause your β coefficients and p-values to change greatly if there are small changes in the data, rendering them unreliable. To check if you have multicollinearity in your data, you can create a scatterplot matrix of your explanatory variables, as in Section 4.9.1:

```
library(GGally)
ggpairs(mtcars[, -1])
```

In this case, there are some highly correlated pairs, hp and disp among them. As a numerical measure of collinearity, we can use the generalised variance inflation factor (GVIF), given by the vif function in the car package:

```
library(car)
m <- lm(mpg ~ ., data = mtcars)
vif(m)
```

A high GVIF indicates that a variable is highly correlated with other explanatory variables in the dataset. Recommendations for what a "high GVIF" is varies, from 2.5 to 10 or more.

You can mitigate problems related to multicollinearity by:

- Removing one or more of the correlated variables from the model (because they are strongly correlated, they measure almost the same thing anyway!),
- Centring your explanatory variables (particularly if you include polynomial terms) – more on this in the next section,
- Using a regularised regression model (which we'll do in Section 11.4).

~

Exercise 8.4. Below are two simulated datasets. One exhibits a non-linear dependence between the variables, and the other exhibits heteroscedasticity. Fit a model with y as the response variable and x as the explanatory variable for each dataset, and make some residual plots. Which dataset suffers from which problem?

```
exdata1 <- data.frame(
    x = c(2.99, 5.01, 8.84, 6.18, 8.57, 8.23, 8.48, 0.04, 6.80,
          7.62, 7.94, 6.30, 4.21, 3.61, 7.08, 3.50, 9.05, 1.06,
          0.65, 8.66, 0.08, 1.48, 2.96, 2.54, 4.45),
    y = c(5.25, -0.80, 4.38, -0.75, 9.93, 13.79, 19.75, 24.65,
          6.84, 11.95, 12.24, 7.97, -1.20, -1.76, 10.36, 1.17,
          15.41, 15.83, 18.78, 12.75, 24.17, 12.49, 4.58, 6.76,
          -2.92))
exdata2 <- data.frame(
    x = c(5.70, 8.03, 8.86, 0.82, 1.23, 2.96, 0.13, 8.53, 8.18,
          6.88, 4.02, 9.11, 0.19, 6.91, 0.34, 4.19, 0.25, 9.72,
          9.83, 6.77, 4.40, 4.70, 6.03, 5.87, 7.49),
    y = c(21.66, 26.23, 19.82, 2.46, 2.83, 8.86, 0.25, 16.08,
          17.67, 24.86, 8.19, 28.45, 0.52, 19.88, 0.71, 12.19,
```

```
        0.64, 25.29, 26.72, 18.06, 10.70, 8.27, 15.49, 15.58,
        19.17))
```

Exercise 8.5. We continue our investigation of the weather models from Exercises 8.1 and 8.3.

> 1. Plot the observed values against the fitted values for the two models that you've fitted. Does either model seem to have a better fit?

> 2. Create residual plots for the second model from Exercise 8.3. Are there any influential points? Any patterns? Any signs of heteroscedasticity?

8.1.5 Bootstrap and permutation tests

Non-normal regression errors can sometimes be an indication that you need to transform your data, that your model is missing an important explanatory variable, that there are interaction effects that aren't accounted for, or that the relationship between the variables is non-linear. But sometimes, you get non-normal errors simply because the errors are non-normal.

The p-values reported by `summary` are computed under the assumption of normally distributed regression errors and can be sensitive to deviations from normality. I usually prefer to use bootstrap p-values (and confidence intervals) instead, as these don't rely on the assumption of normality. In Section 8.2.3, we'll go into detail about how these are computed. A convenient approach is to use `boot_summary` from the `boot.pval` package. It provides a data frame with estimates, bootstrap confidence intervals, and bootstrap p-values (computed using interval inversion; see Section 7.4.3) for the model coefficients. The arguments specify the number of bootstrap samples drawn, the interval type, and what resampling strategy to use (more on the latter). We set the number of bootstrap samples to 9,999, but otherwise use the default settings:

```
library(boot.pval)
boot_summary(m, R = 9999)
```

To convert this to a publication ready-table, use `summary_to_gt` or `summary_to_flextable`:

```
# gt table:
boot_summary(m, R = 9999) |>
  summary_to_gt()

# To export to Word:
library(flextable)
boot_summary(m, R = 9999) |>
  summary_to_flextable() |>
  save_as_docx(path = "my_table.docx")
```

When bootstrapping regression models, we usually rely on *residual resampling*, where the model residuals are resampled. This is explained in detail in Section 8.2.3. There is also an alternative bootstrap scheme for regression models, often referred to as *case resampling*, in which the observations (or cases) $(y_i, x_{i1}, \ldots, x_{ip})$ are resampled instead of the residuals. This approach can be applied when the explanatory variables can be treated as being random (but measured without error) rather than fixed. It can also be useful for models with heteroscedasticity, as it doesn't rely on assumptions about constant variance (which, on the other hand, makes it less efficient if the errors actually are homoscedastic). In other words, it allows us to drop both the normality assumption and the homoscedasticity assumption from regression models. To use case resampling, we simply add `method = "case"` to `boot_summary`:

```
boot_summary(m, R = 9999, method = "case")
```

In this case, the resulting confidence intervals are similar to what we obtained with residual resampling.

An alternative to using the bootstrap is to compute p-values using permutation tests. This can be done using the `lmp` function from the `lmPerm` package. Note that, like the bootstrap, this doesn't affect the model fitting in any way – the only difference is how the p-values are computed. Moreover, the syntax for `lmp` is identical to that of `lm`. The `Ca` and `maxIter` arguments are added to ensure that we compute sufficiently many permutations to get a reliable estimate of the permutation p-value:

```
# First, install lmPerm:
install.packages("lmPerm")

# Get summary table with permutation p-values:
library(lmPerm)
m <- lmp(mpg ~ hp + wt,
         data = mtcars,
         Ca = 1e-3,
         maxIter = 1e6)
summary(m)
```

∼

Exercise 8.6. Refit your model from Exercise 8.3 using `lmp`. Are the two main effects still significant?

8.1.6   Transformations

If your data displays signs of non-linearity, heteroscedasticity, or non-normal residuals, you can sometimes use transformations of the response variable to mitigate those problems. It is for instance common to apply a log-transformation and fit the model

$$\ln(y_i) = \beta_0 + \beta_1 x_{i1} + \beta_2 x_{i2} + \cdots + \beta_p x_{ip} + \epsilon_i, \qquad i = 1, \ldots, n.$$

Likewise, a square-root transformation is sometimes used:

$$\sqrt{y_i} = \beta_0 + \beta_1 x_{i1} + \beta_2 x_{i2} + \cdots + \beta_p x_{ip} + \epsilon_i, \qquad i = 1, \ldots, n.$$

Both of these are non-linear models, because y is no longer a linear function of x_i. We can still use `lm` to fit them though, with $\ln(y)$ or $\sqrt{y}$ as the response variable.

How then do we know which transformation to apply? A useful method for finding a reasonable transformation is the Box-Cox method (Box & Cox, 1964). This defines a transformation using a parameter λ. The transformation is defined as $\frac{y_i^\lambda - 1}{\lambda}$ if $\lambda \neq 0$ and $\ln(y_i)$ if $\lambda = 0$. $\lambda = 1$ corresponds to no transformation at all, $\lambda = 1/2$ corresponds to a square-root transformation, $\lambda = 1/3$ corresponds to a cubic-root transformation, $\lambda = 2$ corresponds to a square transformation, and so on.

The `boxcox` function in `MASS` lets us find a good value for λ. It is recommended to choose a "nice" λ (e.g., $0, 1, 1/2$ or a similar value), that is close to the peak (inside the interval indicated by the outer dotted lines) of the curve plotted by `boxcox`:

```
m <- lm(mpg ~ hp + wt, data = mtcars)

library(MASS)
boxcox(m)
```

In this case, because 0 is close to the peak, the curve indicates that $\lambda = 0$, which corresponds to a log-transformation, could be a good choice. Let's give it a go:

```
mtcars$logmpg <- log(mtcars$mpg)
m_bc <- lm(logmpg ~ hp + wt, data = mtcars)
summary(m_bc)

library(ggfortify)
autoplot(m_bc, which = 1:6, ncol = 2, label.size = 3)
```

The model fit seems to have improved after the transformation. The downside is that we now are modelling the log-mpg rather than mpg, which make the model coefficients a little difficult to interpret: the beta coefficients now describe how the log-mpg changes when the explanatory variables are changed one unit. A positive β still indicates a positive effect, and a negative β a negative effect, but the size of the coefficients is trickier to interpret.

It is also possible to transform the explanatory variables. We'll see some examples of that in Section 8.2.2.

$\sim$

Exercise 8.7. Run `boxcox` with your model from Exercise 8.3. Why do you get an error? How can you fix it? Does the Box-Cox method indicate that a transformation can be useful for your model?

8.1.7 Prediction

An important use of linear models is prediction. In R, this is done using `predict`. By providing a fitted model and a new dataset, we can get predictions.

Let's use one of the models that we fitted to the `mtcars` data to make predictions for two cars that aren't from the 1970s. Below, we create a data frame with data for a 2009 Volvo XC90 D3 AWD (with a fuel consumption of 29 mpg) and a 2019 Ferrari Roma (15.4 mpg):

```
new_cars <- data.frame(hp = c(161, 612), wt = c(4.473, 3.462),
                       row.names = c("Volvo XC90", "Ferrari Roma"))
```

To get the model predictions for these new cars, we run the following:

```
predict(m, new_cars)
```

predict also lets us obtain prediction intervals for our prediction, under the assumption of normality. Prediction intervals provide interval estimates for the new observations. They incorporate both the uncertainty associated with our model estimates, and the fact that the new observation is likely to deviate slightly from its expected value. To get 90% prediction intervals, we add interval = "prediction" and level = 0.9:

```
m <- lm(mpg ~ hp + wt, data = mtcars)
predict(m, new_cars,
        interval = "prediction",
        level = 0.9)
```

If we were using a transformed *y*-variable, we'd probably have to transform the predictions back to the original scale for them to be useful:

```
mtcars$logmpg <- log(mtcars$mpg)
m_bc <- lm(logmpg ~ hp + wt, data = mtcars)

preds <- predict(m_bc, new_cars,
                 interval = "prediction",
                 level = 0.9)

# Predictions for log-mpg:
preds
# Transform back to original scale:
exp(preds)
```

8.1.8   ANOVA

Linear models are also used for analysis of variance (*ANOVA*) models to test whether there are differences among the means of different groups. We'll use the mtcars data to give some examples of this. Let's say that we want to investigate whether the mean fuel consumption (mpg) of cars differs depending on the number of cylinders (cyl), and that we want to include the type of transmission (am) as a blocking variable.

To get an ANOVA table for this problem, we must first convert the explanatory variables to factor variables, as the variables in mtcars all numeric (despite some of them being categorical). We can then use aov to fit the model, and then summary:

```
# Convert variables to factors:
mtcars$cyl <- factor(mtcars$cyl)
mtcars$am <- factor(mtcars$am)
```

```
# Fit model and print ANOVA table:
m <- aov(mpg ~ cyl + am, data = mtcars)
summary(m)
```

(aov actually uses lm to fit the model, but by using aov we specify that we want an ANOVA table to be printed by summary.)

When there are different numbers of observations in the groups in an ANOVA, so that we have an unbalanced design, the sums of squares used to compute the test statistics can be computed in at least three different ways, commonly called types I, II, and III. See Herr (1986) for an overview and discussion of this.

summary prints a type I ANOVA table, which isn't the best choice for unbalanced designs. We can, however, get type II or III tables by instead using Anova from the car package to print the table:

```
library(car)
Anova(m, type = "II")
Anova(m, type = "III") # Default in SAS and SPSS.
```

As a guideline, for unbalanced designs, you should use type II tables if there are no interactions, and type III tables if there are interactions. To look for interactions, we can use interaction.plot to create a two-way interaction plot:

```
interaction.plot(mtcars$am, mtcars$cyl, response = mtcars$mpg)
```

In this case, as seen in Figure 8.3, there is no clear sign of an interaction between the two variables, as the lines are more or less parallel. A type II table is therefore probably the best choice here.

We can obtain diagnostic plots the same way we did for other linear models:

```
library(ggfortify)
autoplot(m, which = 1:6, ncol = 2, label.size = 3)
```

To find which groups have significantly different means, we can use a post hoc test like Tukey's HSD, available through the TukeyHSD function:

```
TukeyHSD(m)
```

We can visualise the results of Tukey's HSD with plot, which shows 95% confidence intervals for the mean differences:

```
# When the difference isn't significant, the dashed line indicating
# "no differences" falls within the confidence interval for
# the difference:
plot(TukeyHSD(m, "am"))
```

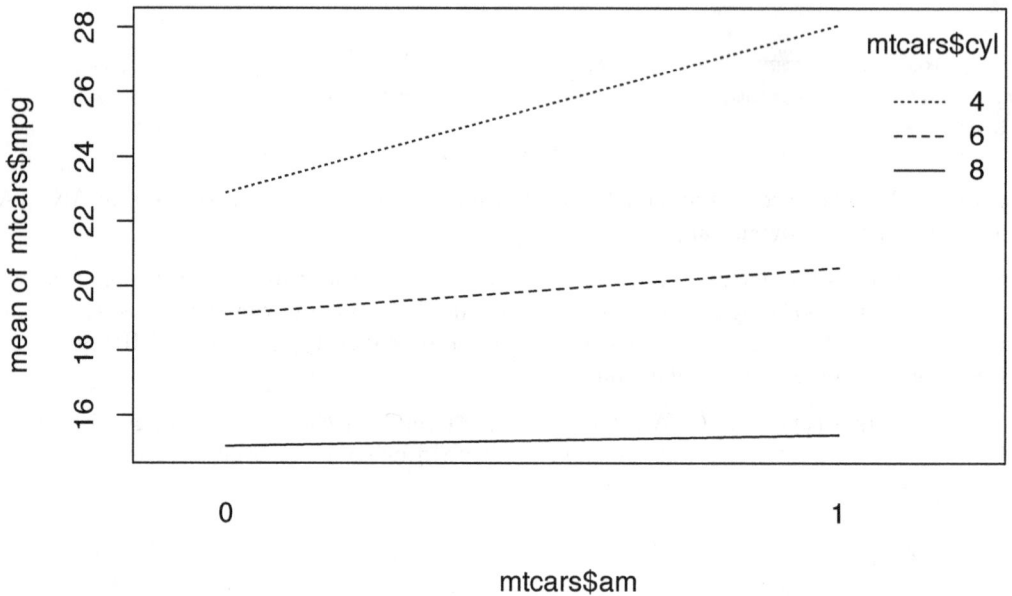

FIGURE 8.3 Two-way interaction plot.

```
# When the difference is significant, the dashed line does not
# fall within the confidence interval:
plot(TukeyHSD(m, "cyl"))
```

~

Exercise 8.8. Return to the residual plots that you created with `autoplot`. Figure out how you can plot points belonging to different `cyl` groups in different colours.

Exercise 8.9. The `aovp` function in the `lmPerm` package can be utilised to perform permutation tests instead of the classical parametric ANOVA tests. Rerun the analysis in the example above, using `aovp` instead. Do the conclusions change? What happens if you run your code multiple times? Does using `summary` on a model fitted using `aovp` generate a type I, II, or III table by default? Can you change what type of table it produces?

Exercise 8.10. In the case of a one-way ANOVA (i.e., ANOVA with a single explanatory variable), the Kruskal-Wallis test can be used as a nonparametric option. It is available in `kruskal.test`. Use the Kruskal-Wallis test to run a one-way ANOVA for the `mtcars` data, with `mpg` as the response variable and `cyl` as an explanatory variable.

8.2 Linear models: advanced topics

8.2.1 Robust estimation

The least squares method used by `lm` is not the only way to fit a regression model. Another option is to use a *robust* regression model based on M-estimators. Such models tend to be less sensitive to outliers and can be useful if you are concerned about the influence of deviating points. The `rlm` function in MASS is used for this. It is identical to that of `lm`:

```
library(MASS)
m <- rlm(mpg ~ hp + wt, data = mtcars)
summary(m)
```

8.2.2 Interactions between numerical variables

In the fuel consumption model, it seems plausible that there could be an *interaction* between gross horsepower and weight, both of which are numeric. Just as when we had a categorical variable, we can include an interaction term by adding `hp:wt` to the formula:

```
m <- lm(mpg ~ hp + wt + hp:wt, data = mtcars)
summary(m)
```

Alternatively, to include the main effects of `hp` and `wt` along with the interaction effect, we can use `hp*wt` as a shorthand for `hp + wt + hp:wt` to write the model formula more concisely:

```
m <- lm(mpg ~ hp*wt, data = mtcars)
summary(m)
```

When our regression model includes interactions between numeric variables, it is often recommended to centre the explanatory variables, i.e., to shift them so that they all have mean 0.

There are a number of benefits to this: for instance that the intercept then can be interpreted as the expected value of the response variable when all explanatory variables are equal to their means, i.e., in an average case[2]. It can also reduce any multicollinearity – i.e., dependence between the explanatory variables – in the data, particularly when including interactions or polynomial terms in the model. Finally, it can reduce problems with numerical instability that may arise due to floating point arithmetics. Note, however, that there is no need to centre the response variable. On the contrary, doing so will usually only serve to make interpretation more difficult.

Centring the explanatory variables can be done using `scale`.

Without pipes:

```
# Create a new data frame, leaving the response variable mpg
# unchanged, while centring the explanatory variables:
mtcars_scaled <- data.frame(mpg = mtcars[,1],
                            cyl = mtcars[,2],
                            scale(mtcars[,-1], scale = FALSE))
```

With pipes:

---

[2]If the variables aren't centred, the intercept is the expected value of the response variable when all explanatory variables are 0. This isn't always realistic or meaningful.

```
# Create a new data frame, leaving the response variable mpg
# unchanged, while centring the explanatory variables:
library(dplyr)
mtcars |> mutate(across(c(where(is.numeric), -mpg),
                        ~ as.numeric(scale(., scale = FALSE)))) -> mtcars_scaled
```

We then refit the model using the centred data:

```
m <- lm(mpg ~ hp*wt, data = mtcars_scaled)
summary(m)
```

If we wish to add a polynomial term to the model, we can do so by wrapping the polynomial in `I()`. For instance, to add a quadratic effect in the form of the square weight of a vehicle to the model, we'd use:

```
m <- lm(mpg ~ hp*wt + I(wt^2), data = mtcars_scaled)
summary(m)
```

8.2.3   Bootstrapping regression coefficients

In this section, we'll have a closer look at how bootstrapping works in linear regression models. In most cases, we'll simply use `boot_summary` for this, but if you want to know more about what that function does behind the scenes, then read on.

First, note that the only random part in the linear model

$$y_i = \beta_0 + \beta_1 x_{i1} + \beta_2 x_{i2} + \cdots + \beta_p x_{ip} + \epsilon_i, \qquad i = 1, \ldots, n$$

is the error term ϵ_i. In most cases, it is therefore this term (and this term only) that we wish to resample. The explanatory variables should remain constant throughout the resampling process; the inference is conditioned on the values of the explanatory variables.

To achieve this, we'll resample from the model residuals and add those to the values predicted by the fitted function, which creates new bootstrap values of the response variable. We'll then fit a linear model to these values, from which we obtain observations from the bootstrap distribution of the model coefficients.

It turns out that the bootstrap performs better if we resample not from the original residuals $e_1, \ldots, e_n$, but from scaled and centred residuals $r_i - \bar{r}$, where each r_i is a scaled version of residual e_i, scaled by the leverage h_i:

$$r_i = \frac{e_i}{\sqrt{1 - h_i}},$$

see Chapter 6 of Davison & Hinkley (1997) for details. The leverages can be computed using `lm.influence`.

We implement this procedure in the code below (and will then have a look at convenience functions that help us achieve the same thing more easily). It makes use of `formula`, which can be used to extract the model formula from regression models:

```r
library(boot)

coefficients <- function(formula, data, i, predictions, residuals) {
  # Create the bootstrap value of response variable by
  # adding a randomly drawn scaled residual to the value of
  # the fitted function for each observation:
  data[,all.vars(formula)[1]] <- predictions + residuals[i]

  # Fit a new model with the bootstrap value of the response
  # variable and the original explanatory variables:
  m <- lm(formula, data = data)
  return(coef(m))
}

# Fit the linear model:
m <- lm(mpg ~ hp + wt, data = mtcars)

# Compute scaled and centred residuals:
res <- residuals(m)/sqrt(1 - lm.influence(m)$hat)
res <- res - mean(res)

# Run the bootstrap, extracting the model formula and the
# fitted function from the model m:
boot_res <- boot(data = mtcars, statistic = coefficients,
                 R = 999, formula = formula(m),
                 predictions = predict(m),
                 residuals = res)

# Compute 95% confidence intervals:
boot.ci(boot_res, type = "perc", index = 1) # Intercept
boot.ci(boot_res, type = "perc", index = 2) # hp
boot.ci(boot_res, type = "perc", index = 3) # wt
```

The argument `index` in `boot.ci` should be the row number of the parameter in the table given by `summary`. The intercept is on the first row, and so its `index` is 1, `hp` is on the second row and its `index` is 2, and so on.

Clearly, the above code is a little unwieldy. Fortunately, the `car` package contains a function called `Boot` that can be used to bootstrap regression models in the exact same way:

```r
library(car)

boot_res <- Boot(m, method = "residual", R = 9999)

# Compute 95% confidence intervals:
confint(boot_res, type = "perc")
```

Finally, the most convenient approach is to use `boot_summary` from the `boot.pval` package, as mentioned earlier:

```
library(boot.pval)
boot_summary(m, type = "perc", method = "residual", R = 9999)
```

~

Exercise 8.11. Refit your model from Exercise 8.3 using a robust regression estimator with `rlm`. Compute bootstrap confidence intervals for the coefficients of the robust regression model.

8.2.4   Prediction intervals using the bootstrap

It is often desirable to have prediction intervals, a type of confidence for predictions. Prediction intervals try to capture two sources of uncertainty:

- *Model uncertainty*, which we will capture by resampling the data and make predictions for the expected value of the observation,
- *Random noise*, i.e., that almost all observations deviate from their expected value. We will capture this by resampling residuals from the fitted bootstrap models.

Next, we will compute bootstrap prediction intervals. Based on the above, the value that we generate in each bootstrap replication will be the sum of a prediction and a resampled residual (see Davison & Hinkley (1997), section 6.3, for further details):

```
boot_pred <- function(data, new_data, model, i,
                      formula, predictions, residuals){
    # Resample residuals and fit new model:
    data[,all.vars(formula)[1]] <- predictions + residuals[i]
    m_boot <- lm(formula, data = data)

    # We use predict to get an estimate of the
    # expectation of new observations, and then
    # add resampled residuals to also include the
    # natural variation around the expectation:
    predict(m_boot, newdata = new_data) +
      sample(residuals, nrow(new_data))
}

library(boot)

m <- lm(mpg ~ hp + wt, data = mtcars)

# Compute scaled and centred residuals:
res <- residuals(m)/sqrt(1 - lm.influence(m)$hat)
res <- res - mean(res)

boot_res <- boot(data = m$model,
                 statistic = boot_pred,
                 R = 999,
                 model = m,
```

```
                    new_data = new_cars,
                    formula = formula(m),
                    predictions = predict(m),
                    residuals = res)

# 90% bootstrap prediction intervals:
boot.ci(boot_res, type = "perc", index = 1, conf = 0.9) # Volvo
boot.ci(boot_res, type = "perc", index = 2, conf = 0.9) # Ferrari
```

~

Exercise 8.12. Use your model from Exercise 8.3 to compute a bootstrap prediction interval for the temperature on a day with precipitation but no sun hours.

8.2.5   Prediction for multiple datasets

In certain cases, we wish to fit different models to different subsets of the data. Functionals like `apply` and `map` (Section 6.5) are handy when you want to fit several models at once. Below is an example of how we can use `split` (Section 5.2.1) and tools from the `purrr` package (Section 6.5.3) to fit the models simultaneously, as well as for computing the fitted values in a single line of code. These work best with the `magrittr` pipe `%>%`:

```
# Split the dataset into three groups depending on the
# number of cylinders:
library(magrittr)
mtcars_by_cyl <- mtcars %>% split(.$cyl)

# Fit a linear model to each subgroup:
library(purrr)
models <- mtcars_by_cyl |> map(~ lm(mpg ~ hp + wt, data = .))

# Compute the fitted values for each model:
map2(models, mtcars_by_cyl, predict)
```

We'll make use of this approach when we study linear mixed models in Section 8.8.

8.2.6   Alternative summaries with `broom`

The `broom` package contains some useful functions when working with linear models (and many other common models), which allow us to get various summaries of the model fit in useful formats. Let's install it:

```
install.packages("broom")
```

A model fitted with `m` is stored as a `list` with lots of elements:

```
m <- lm(mpg ~ hp + wt, data = mtcars)
str(m)
```

How can we access the information about the model? For instance, we may want to get the summary table from summary, but as a data frame rather than as printed text. Here are two ways of doing this, using summary and the tidy function from broom:

```
# Using base R:
summary(m)$coefficients
```

```
# Using broom:
library(broom)
tidy(m)
```

tidy is the better option if you want to retrieve the table as part of a pipeline. For instance, if you want to adjust the p-values for multiplicity using Bonferroni correction (Section 3.7), you could do as follows:

```
lm(mpg ~ hp + wt, data = mtcars) |>
    tidy() |>
    p.adjust(p.value, method = "bonferroni")
```

If you prefer bootstrap p-values, you can use boot_summary from boot.pval similarly. That function also includes an argument for adjusting the p-values for multiplicity:

```
library(boot.pval)
lm(mpg ~ hp + wt, data = mtcars) |>
    boot_summary(adjust.method = "bonferroni")
```

Another useful function in broom is glance, which lets us get some summary statistics about the model:

```
glance(m)
```

Finally, augment can be used to add predicted values, residuals, and Cook's distances to the dataset used for fitting the model, which of course can be very useful for model diagnostics:

```
# To get the data frame with predictions and residuals added:
augment(m)
```

```
# To plot the observed values against the fitted values:
library(ggplot2)
mtcars |>
    lm(mpg ~ hp + wt, data = _) |>
    augment() |>
    ggplot(aes(.fitted, mpg)) +
      geom_point() +
      xlab("Fitted values") + ylab("Observed values")
```

8.2.7 Variable selection

A common question when working with linear models is what variables to include in your model. Common practices for variable selection include stepwise regression methods, where variables are added to or removed from the model depending on p-values, R^2 values, or information criteria like the Akaike information criterion (AIC) or the Bayesian information criterion (BIC).

Don't ever do this if your main interest is p-values. Stepwise regression increases the risk of type I errors, renders the p-values of your final model invalid, and can lead to over-fitting; see, e.g., Smith (2018). Instead, you should let your research hypothesis guide your choice of variables or base your choice on a pilot study.

If your main interest is prediction, then that is a completely different story. For predictive models, it is usually recommended that variable selection and model fitting be done simultaneously. This can be done using regularised regression models, to which Section 11.4 is devoted.

8.2.8 Bayesian estimation of linear models

We can fit Bayesian linear models using the `rstanarm` package. To fit a model to the `mtcars` data using all explanatory variables, we can use `stan_glm` in place of `lm` as follows:

```r
library(rstanarm)
m <- stan_glm(mpg ~ ., data = mtcars)

# Print the estimates:
coef(m)
```

Next, we can plot the posterior distributions of the effects:

```r
plot(m, "dens", pars = names(coef(m)))
```

To get 95% credible intervals for the effects, we can use `posterior_interval`:

```r
posterior_interval(m,
                   pars = names(coef(m)),
                   prob = 0.95)
```

We can also plot them using `plot`:

```r
plot(m, "intervals",
     pars = names(coef(m)),
     prob = 0.95)
```

Finally, we can use $\hat{R}$ to check model convergence. It should be less than 1.1 if the fitting has converged:

```r
plot(m, "rhat")
```

Like for `lm`, `residuals(m)` provides the model residuals, which can be used for diagnostics. For instance, we can plot the residuals against the fitted values to look for signs of non-linearity, adding a curve to aid the eye:

```
model_diag <- data.frame(Fitted = predict(m),
                         Residual = residuals(m))

library(ggplot2)
ggplot(model_diag, aes(Fitted, Residual)) +
   geom_point() +
   geom_smooth(se = FALSE)
```

For fitting ANOVA models, we can instead use `stan_aov` with the argument `prior = R2(location = 0.5)` to fit the model.

8.3 Modelling proportions: logistic regression

8.3.1 Generalised linear models

Generalised linear models (GLMs) are (yes) a generalisation of the linear model, that can be used when your response variable has a non-normal error distribution. Typical examples are when your response variable is binary (only takes two values, e.g., 0 or 1), or a count of something. Fitting GLMs is more or less entirely analogous to fitting linear models in R, but model diagnostics are very different. In this and the following section, we will look at some examples of how it can be done. We'll start with logistic regression, which is used to model proportions.

8.3.2 Fitting logistic regression models

As the first example of binary data, we will consider the wine quality dataset `wine` from Cortez et al. (2009), which is available in the UCI Machine Learning Repository at http://archive.ics.uci.edu/ml/datasets/Wine+Quality. It contains measurements on white and red vinho verde wine samples from northern Portugal.

We start by loading the data. It is divided into two separate `.csv` files, one for white wines and one for red, which we have to merge.

Without pipes:

```
# Import data about white and red wines:
white <- read.csv("https://tinyurl.com/winedata1",
                  sep = ";")
red <- read.csv("https://tinyurl.com/winedata2",
                sep = ";")

# Add a type variable:
white$type <- "white"
red$type <- "red"
```

```
# Merge the datasets:
wine <- rbind(white, red)
# Convert the type variable to a factor (this is
# needed when we fit the logistic regression model later):
wine$type <- factor(wine$type)

# Check the result:
summary(wine)
```

With pipes:

```
library(dplyr)

# Import data about white and red wines.
# Add a type variable describing the colour.
white <- read.csv("https://tinyurl.com/winedata1",
             sep = ";") |> mutate(type = "white")
red <- read.csv("https://tinyurl.com/winedata2",
             sep = ";") |> mutate(type = "red")

# Merge the datasets and convert the type variable to
# a factor (this is needed when we fit the logistic
# regression model later):
white |> bind_rows(red) |>
    mutate(type = factor(type)) -> wine

# Check the result:
summary(wine)
```

We are interested in seeing if measurements like pH (pH) and alcohol content (alcohol) can be used to determine the colours of the wine. The colour is represented by the type variable, which is binary. In our model, it will be included as a dummy variable, where 0 corresponds to "red" and 1 corresponds to "white". By default, R uses the class that is last in alphabetical order as $y = 1$. Here, that class is "white", as can be seen by running levels(wine$type).

Our model is that the type of a randomly selected wine is binomial $Bin(1, \pi_i)$-distributed (Bernoulli distributed), where π_i depends on explanatory variables like pH and alcohol content. A common model for this situation is a *logistic regression model*. Given n observations of p explanatory variables, the model is:

$$\log\left(\frac{\pi_i}{1 - \pi_i}\right) = \beta_0 + \beta_1 x_{i1} + \beta_2 x_{i2} + \cdots + \beta_p x_{ip}, \qquad i = 1, \ldots, n$$

Where in linear regression models we model the expected value of the response variable as a linear function of the explanatory variables, we now model the expected value of a *function of the expected value of the response variable* (that is, a function of π_i). In GLM terminology, this function is known as a *link function*.

Logistic regression models can be fitted using the glm function. To specify what our model is, we use the argument family = binomial:

```
m <- glm(type ~ pH + alcohol, data = wine, family = binomial)
summary(m)
```

The p-values presented in the summary table are based on a Wald test known to have poor performance unless the sample size is very large (Agresti, 2013). In this case, with a sample size of 6,497, it is probably safe to use; but, for smaller sample sizes, it is preferable to use a bootstrap test instead, which you will do in Exercise 8.15.

The coefficients of a logistic regression model aren't as straightforward to interpret as those in a linear model. If we let β denote a coefficient corresponding to an explanatory variable x, then:

- If β is positive, then π_i increases when x_i increases.
- If β is negative, then π_i decreases when x_i increases.
- e^β is the *odds ratio*, which shows how much the odds $\frac{\pi_i}{1-\pi_1}$ change when x_i is increased 1 step.

We can extract the coefficients and odds ratios using `coef`:

```
coef(m)       # Coefficients, beta
exp(coef(m))  # Odds ratios
```

Just as for linear models, publication-ready tables can be obtained using `tbl_regression`:

```
tbl_regression(m)
```

To get odds ratios instead of the β_i coefficients (called $\log(OR)$ in the table), we use the argument `exponentiate`:

```
tbl_regression(m, exponentiate = TRUE)
```

To find the fitted probability that an observation belongs to the second class ("white") we can use `predict(m, type = "response")`.

Without pipes:

```
# Get fitted probabilities:
probs <- predict(m, type = "response")

# Check what the average prediction is for
# the two groups:
mean(probs[wine$type == "red"])
mean(probs[wine$type == "white"])
```

With pipes:

```
# Check what the average prediction is for each group:
# We use the augment function from the broom package to add
# the predictions to the original data frame.
```

```
library(broom)
m |> augment(type.predict = "response") |>
    group_by(type) |>
    summarise(AvgPrediction = mean(.fitted))
```

It turns out that the model predicts that most wines are white – even the red ones! The reason may be that we have more white wines (4,898) than red wines (1,599) in the dataset. Adding more explanatory variables could perhaps solve this problem. We'll give that a try in the next section.

~

Exercise 8.13. Download `sharks.csv` file from the book's web page[3]. It contains information about shark attacks in South Africa. Using data on attacks that occurred in 2000 or later, fit a logistic regression model to investigate whether the age and sex of the individual that was attacked affect the probability of the attack being fatal.

Note: save the code for your model, as you will return to it in the subsequent exercises.

Exercise 8.14. In Section 8.2.6 we saw how some functions from the `broom` package could be used to get summaries of linear models. Try using them with the `wine` data model that we created above. Do the `broom` functions work for generalised linear models as well?

8.3.3 Bootstrap confidence intervals

In a logistic regression, the response variable y_i is a binomial (or Bernoulli) random variable with success probability π_i. In this case, we don't want to resample residuals to create confidence intervals, as it turns out that this can lead to predicted probabilities outside the range $(0, 1)$. Instead, we can either use the case resampling strategy described in Section 8.1.5 or use a parametric bootstrap approach where we generate new binomial variables (Section 7.1.2) to construct bootstrap confidence intervals.

To use case resampling, we can use `boot_summary` from `boot.pval`:

```
library(boot.pval)

m <- glm(type ~ pH + alcohol, data = wine, family = binomial)

boot_summary(m, type = "perc", method = "case")
```

In the parametric approach, for each observation, the fitted success probability from the logistic model will be used to sample new observations of the response variable. This method can work well if the model is well specified but tends to perform poorly for misspecified models; so, make sure to carefully perform model diagnostics (as described in the next section) before applying it. To use the parametric approach, we can do as follows:

---

[3]http://www.modernstatisticswithr.com/data.zip

```
library(boot)

coefficients <- function(formula, data, predictions, ...) {
    # Check whether the response variable is a factor or
    # numeric, and then resample:
    if(is.factor(data[,all.vars(formula)[1]])) {
      # If the response variable is a factor:
      data[,all.vars(formula)[1]] <-
        factor(levels(data[,all.vars(formula)[1]])[1 + rbinom(nrow(data),
                1, predictions)]) } else {
                # If the response variable is numeric:
                data[,all.vars(formula)[1]] <-
                  unique(data[,all.vars(formula)[1]])[1 + rbinom(nrow(data),
                                                        1, predictions)]}

    m <- glm(formula, data = data, family = binomial)
    return(coef(m))
}

m <- glm(type ~ pH + alcohol, data = wine, family = binomial)

boot_res <- boot(data = wine, statistic = coefficients,
                 R = 999,
                 formula = formula(m),
                 predictions = predict(m, type = "response"))

# Compute confidence intervals:
boot.ci(boot_res, type = "perc", index = 1) # Intercept
boot.ci(boot_res, type = "perc", index = 2) # pH
boot.ci(boot_res, type = "perc", index = 3) # Alcohol
```

~

Exercise 8.15. Use the model that you fitted to the sharks.csv data in Exercise 8.13 for the following:

1. When the MASS package is loaded, you can use confint to obtain (asymptotic) confidence intervals for the parameters of a GLM. Use it to compute confidence intervals for the parameters of your model for the sharks.csv data.

2. Compute parametric bootstrap confidence intervals and p-values for the parameters of your logistic regression model for the sharks.csv data. Do they differ from the intervals obtained using confint? Note that there are a lot of missing values for the response variable. Think about how that will affect your bootstrap confidence intervals and adjust your code accordingly.

3. Use the confidence interval inversion method of Section 7.4.3 to compute bootstrap p-values for the effect of age.

8.3.4 Model diagnostics

It is notoriously difficult to assess model fit for GLMs, because the behaviour of the residuals is very different from residuals in ordinary linear models. In the case of logistic regression, the response variable is always 0 or 1, meaning that there will be two bands of residuals:

```
# Store deviance residuals:
m <- glm(type ~ pH + alcohol, data = wine, family = binomial)
res <- data.frame(Predicted = predict(m),
                  Residuals = residuals(m, type ="deviance"),
                  Index = 1:nrow(m$data),
                  CooksDistance = cooks.distance(m))

# Plot fitted values against the deviance residuals:
library(ggplot2)
ggplot(res, aes(Predicted, Residuals)) +
    geom_point()

# Plot index against the deviance residuals:
ggplot(res, aes(Index, Residuals)) +
    geom_point()
```

Plots of raw residuals are of little use in logistic regression models. A better option is to use a binned residual plot, in which the observations are grouped into bins based on their fitted value. The average residual in each bin can then be computed, which will tell us which parts of the model have a poor fit. A function for this is available in the `arm` package:

```
install.packages("arm")

library(arm)
binnedplot(predict(m, type = "response"),
           residuals(m, type = "response"))
```

The grey lines show confidence bounds which are supposed to contain about 95% of the bins. If too many points fall outside these bounds, it's a sign that we have a poor model fit. In this case, there are a few points outside the bounds. Most notably, the average residuals are fairly large for the observations with the lowest fitted values, i.e., among the observations with the lowest predicted probability of being white wines.

Let's compare the above plot to that for a model with more explanatory variables:

```
m2 <- glm(type ~ pH + alcohol + fixed.acidity + residual.sugar,
          data = wine, family = binomial)

binnedplot(predict(m2, type = "response"),
           residuals(m2, type = "response"))
```

The results are shown in Figure 8.4. They look better – adding more explanatory variable appears to have improved the model fit.

Model: m1 (two explanatory variables)

Model: m2 (four explanatory variables)

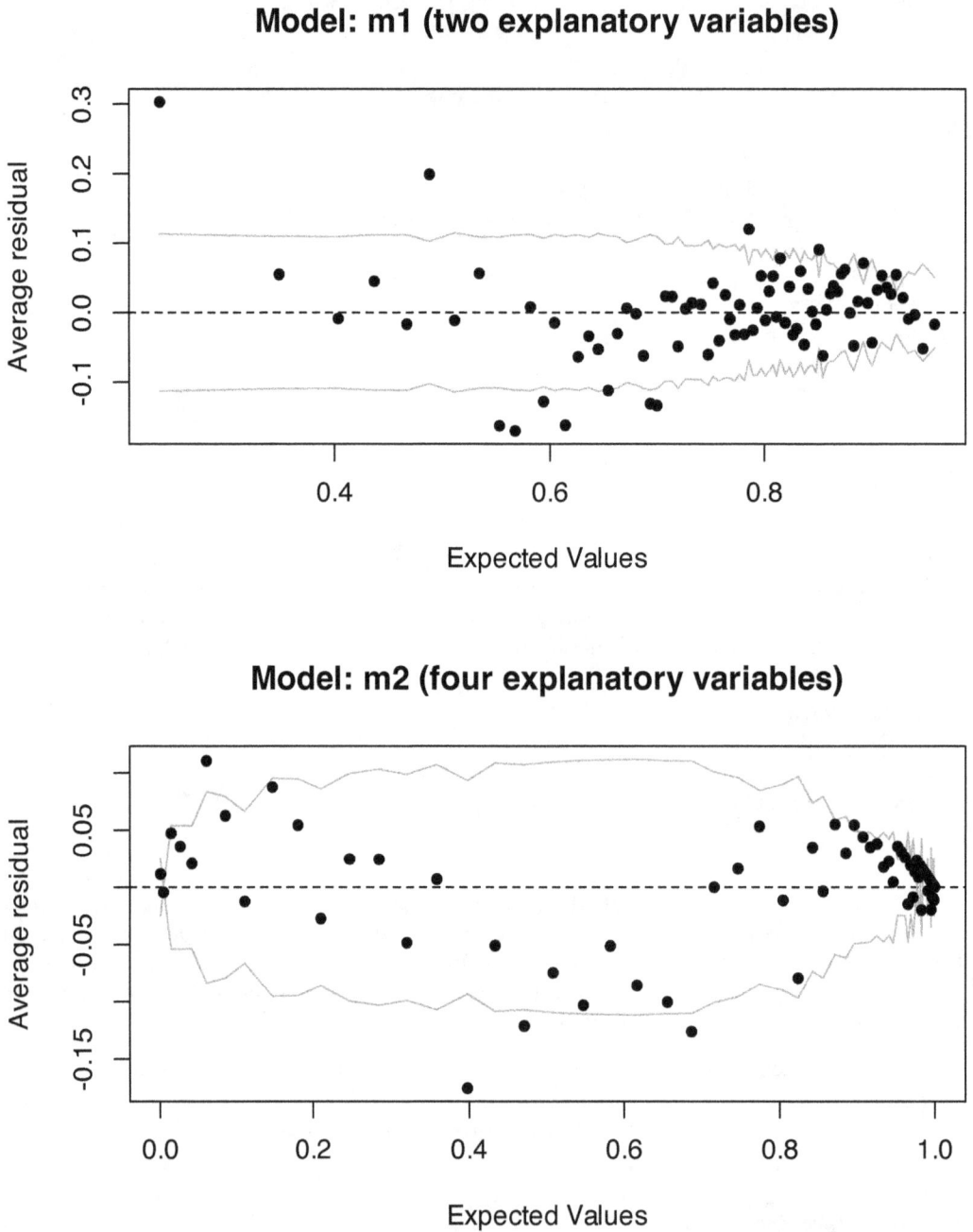

FIGURE 8.4 Binned residual plots for logistic regression models.

It's worth repeating that if your main interest is hypothesis testing, you shouldn't fit multiple models and then pick the one that gives the best results. However, if you're doing an exploratory analysis or are interested in predictive modelling, you can and should try different models. It can then be useful to do a formal hypothesis test of the null hypothesis that m and m2 fit the data equally well, against the alternative that m2 has a better fit. If both fit the data equally well, we'd prefer m, since it is a simpler model. We can use anova to perform a likelihood ratio *deviance test* (see Section 14.4 for details), which tests this:

```
anova(m, m2, test = "LRT")
```

The p-value is very low, and we conclude that m2 has a better model fit.

Another useful function is cooks.distance, which can be used to compute the Cook's distance for each observation, which is useful for finding influential observations. In this case, I've chosen to print the row numbers for the observations with a Cook's distance greater than 0.004 – this number has been arbitrarily chosen in order only to highlight the observations with the highest Cook's distance.

```
res <- data.frame(Index = 1:length(cooks.distance(m)),
                  CooksDistance = cooks.distance(m))

# Plot index against the Cook's distance to find
# influential points:
ggplot(res, aes(Index, CooksDistance)) +
  geom_point() +
  geom_text(aes(label = ifelse(CooksDistance > 0.004,
                               rownames(res), "")),
            hjust = 1.1)
```

~

Exercise 8.16. Investigate the residuals for your sharks.csv model. Are there any problems with the model fit? Any influential points?

8.3.5 Prediction

Just as for linear models, we can use predict to make predictions for new observations using a GLM. To begin with, let's randomly sample 10 rows from the wine data and fit a model using all data except those 10 observations:

```
# Randomly select 10 rows from the wine data:
rows <- sample(1:nrow(wine), 10)

m <- glm(type ~ pH + alcohol, data = wine[-rows,], family = binomial)
```

We can now use predict to make predictions for the 10 observations:

```
preds <- predict(m, wine[rows,])
preds
```

Those predictions look a bit strange though – what are they? By default, `predict` returns predictions on the scale of the link function. That's not really what we want in most cases; instead, we are interested in the predicted probabilities. To get those, we have to add the argument `type = "response"` to the call:

```
preds <- predict(m, wine[rows,], type = "response")
preds
```

Logistic regression models are often used for prediction, in what is known as classification. Section 11.1.7 is concerned with how to evaluate the predictive performance of logistic regression and other classification models.

8.4 Regression models for multicategory response variables

Logistic regression can be extended to situations where our response variable is categorical with three or more levels. Such variables can be of two types:

- **Nominal**: the categories do not have a natural ordering (e.g., sex, regions).
- **Ordinal**: the categories have a natural ordering (e.g., education levels, risk levels, answers on a scale that ranges from strongly disagree to strongly agree).

We'll consider logistic regression models for both types of variables, starting with ordinal ones.

8.4.1 Ordinal response variables

The method that we'll use for ordinal response variables is called *ordinal logistic regression*.

Let's say that our response variable Y has k levels: $1, 2, 3, \ldots, k$. Let's also say that we have p explanatory variables $x_1, \ldots, x_p$ that we believe are connected to Y.

A dataset with such data is found in the `ordinalex.csv` data file, which can be downloaded from the book's web page[4].

```
# Load the data:
ordinalex <- read.csv2("ordinalex.csv")

# Have a look at the data:
View(ordinalex)
```

The variable that we'll use as the response in our model is `Risk`. For these data, we have $k = 4$ levels for Y: $1 =$ no risk, $2 =$ low risk, $3 =$ moderate risk, $4 =$ high risk.

The ordinal logistic regression model is that:

$$P(Y \leq j) = f\Big(\alpha_j - (\beta_1 x_1 + \beta_2 x_2 + \cdots + \beta_p x_p)\Big), \qquad j = 1, 2, \ldots, k$$

[4]http://www.modernstatisticswithr.com/data.zip

This can then be used to compute $P(Y = j)$ for any j. α_j is connected to the base probability that $Y \leq j$. The β_i coefficients describe how the probability is affected by the explanatory variables.

To make this look more like a logistic regression model, we can rewrite the model as:

$$\ln\left(\frac{P(Y \leq j)}{1 - P(Y \leq j)}\right) = \alpha_j - (\beta_1 x_1 + \beta_2 x_2 + \cdots + \beta_p x_p)$$

The β_i are interpreted as follows:

- If $\beta_i = 0$, then x_i does not affect the probability,
- If $\beta_i > 0$, then the probability of Y belonging to a high j increases when x_i increases,
- If $\beta_i < 0$, then the probability of Y belonging to a high j decreases when x_i increases,
- e^{β_i} gives the odds ratio for obtaining a higher j for x_i.

In this model, the effect of x_i is the same for all j, an assumption known as *proportional odds*. The explanatory variables affect the general direction of Y rather than the individual categories. The assumption of proportional odds can be tested as part of the model diagnostics. Non-proportional odds can arise, e.g., when an x_i causes Y to more often end up in extreme j (i.e., very low or very high j) instead of pushing Y to only higher values or only lower values.

The `polr` function from MASS can be used for fitting ordinal logistic regression models. We'll use `Age` and `Sex` as explanatory variables. First, we set the order of the categories.

Without pipes:

```
ordinalex$Risk <- factor(ordinalex$Risk, c("No risk", "Low risk",
                                "Moderate risk", "High risk"),
                    ordered = TRUE)
```

With pipes:

```
library(dplyr)
ordinalex |>
  mutate(Risk = factor(Risk, c("No risk", "Low risk",
                      "Moderate risk","High risk"),
                ordered = TRUE)) -> ordinalex
```

We can now fit an ordinal logistic regression model:

```
library(MASS)
m <- polr(Risk ~ Age + Sex, data = ordinalex)

# Model summary:
summary(m)
```

Note that no p-values are presented. The reason is that the traditional Wald p-values for this model are unreliable unless we have very large sample sizes. It is still possible to compute them using the `coeftest` function from the AER function, but I recommend using the (albeit slower) bootstrap approach for computing p-values instead:

```
# Wald p-values (not recommended):
library(AER)
coeftest(m)
```

```
# Bootstrap p-values (takes a while to run):
library(boot.pval)
boot_summary(m, method = "case") # Raw coefficients
boot_summary(m, method = "case", coef = "exp") # Odds ratios
```

The assumption of proportional odds can be tested using Brant's test. The null hypothesis is that the odds are proportional, so a low p-value would indicate that we don't have proportional odds:

```
library(car)
poTest(m)
```

In this case, the p-value for Age is low, which would indicate that we don't have proportional odds for this variable. We might therefore consider other models, such as the nominal regression model described in the next section.

Finally, we can visualise the results from the model as follows (and as shown in Figure 8.5):

```
library(tidyr)
library(ggplot2)
```

```
# Make predictions for males and females aged 15-50:
plotdata <- expand.grid(Age = 15:50,
                        Sex = c("Female", "Male"))
plotdata |>
  bind_cols(predict(m, plotdata, type = "probs")) -> plotdata
```

```
# Reformat the predictions to be suitable for creating a ggplot:
plotdata |> pivot_longer(c("No risk", "Low risk", "Moderate risk", "High risk"),
                         names_to = "Risk") |>
  rename(Probability = value) |>
  mutate(Risk = factor(Risk, levels = levels(ordinalex$Risk))) -> plotdata
```

```
# Plot the predicted risk for women:
plotdata |> filter(Sex == "Female") |>
  ggplot(aes(Age, Probability, colour = Risk)) +
  geom_line(linewidth = 2)
```

```
# Plot the predicted risk for both sexes:
plotdata |>
  ggplot(aes(Age, Probability, colour = Risk, linetype = Sex)) +
  geom_line(linewidth = 2)
```

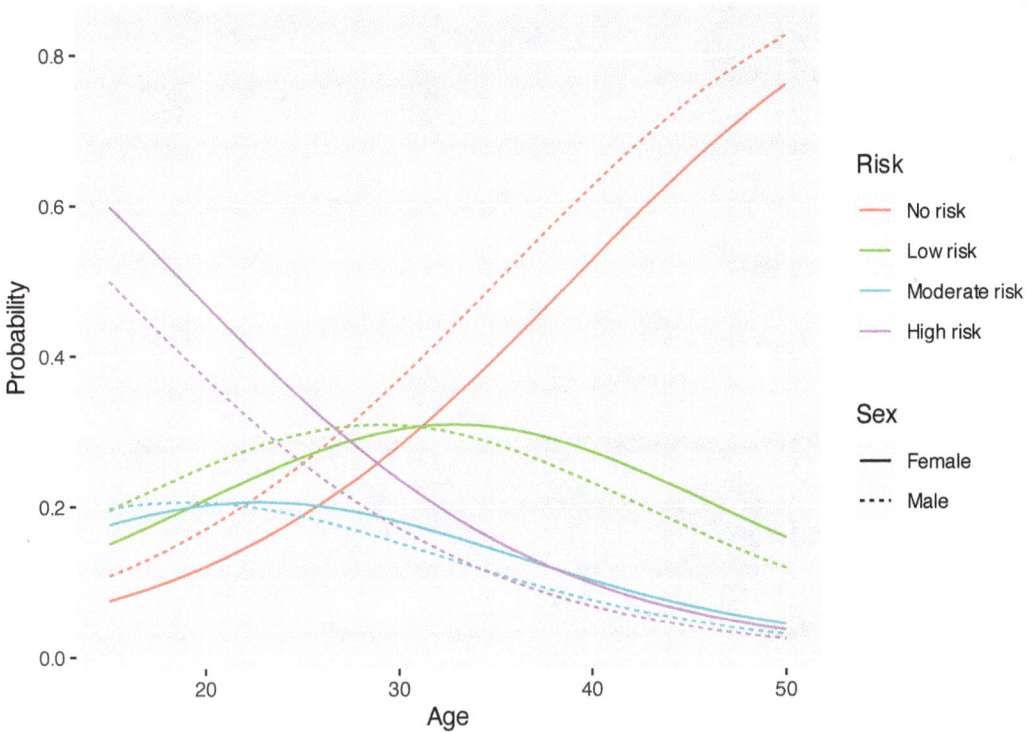

FIGURE 8.5 Model predictions for an ordinal logistic regression model.

8.4.2 Nominal response variables

Multinomial logistic regression, also known as softmax regression or MaxEnt regression, can be used when our response variable is a nominal categorical variable. It can also be used for ordinal response variables when the proportional odds assumption isn't met. A multinomial logistic regression model for a response variable with k categories is essentially a collection of $k - 1$ ordinary logistic regressions, one for each category except the reference category, which are combined into a single model. The multinom function in nnet can be used to fit such a model. We'll have a look at an example using the same dataset as in the previous section:

```
library(nnet)
m <- multinom(Risk ~ Age + Sex, data = ordinalex)
```

The summary table shows the β coefficients for the $k - 1$ non-reference categories. In this case, No risk is the reference category, so all coefficients are relative to No risk and can be interpreted as in an ordinary logistic regression:

```
summary(m)
```

If we make predictions using this model, we get the probabilities of belonging to the different categories:

```
new_data <- data.frame(Age = 18, Sex = "Female")
predict(m, new_data, type = "probs")
```

Finally, we can visualise the fitted model as follows:

```
library(tidyr)
library(ggplot2)

# Make predictions for males and females aged 15-50:
plotdata <- expand.grid(Age = 15:50,
                        Sex = c("Female", "Male"))
plotdata |>
  bind_cols(predict(m, plotdata, type = "probs")) -> plotdata

# Reformat the predictions to be suitable for creating a ggplot:
plotdata |> pivot_longer(c("No risk", "Low risk", "Moderate risk", "High risk"),
                         names_to = "Risk") |>
  rename(Probability = value) |>
  mutate(Risk = factor(Risk, levels = levels(ordinalex$Risk))) -> plotdata

plotdata |>
  ggplot(aes(Age, Probability, colour = Risk, linetype = Sex)) +
  geom_line(linewidth = 2)
```

The results are similar to what we got with an ordinal logistic regression, but there are some differences, for instance for the probability of moderate or high risk for 50 year olds.

8.5   Modelling count data

Logistic regression is but one of many types of GLMs used in practice. One important example is Cox regression, which is used for survival data. We'll return to that model in Section 9.1. For now, we'll consider count data instead.

8.5.1   Poisson and negative binomial regression

Let's have a look at the shark attack data in sharks.csv, available on the book's website. It contains data about shark attacks in South Africa, downloaded from The Global Shark Attack File (http://www.sharkattackfile.net/incidentlog.htm). To load it, we download the file and set file_path to the path of sharks.csv.

```
sharks <- read.csv(file_path, sep =";")
```

We then compute number of attacks per year, and filter the data to only keep observations from 1960 to 2019:

Without pipes:

```
# Compute number of attacks per year:
attacks <- aggregate(Type ~ Year, data = sharks, FUN = length)
names(attacks)[2] <- "n"

# Keep data for 1960-2019:
attacks <- subset(attacks, Year >= 1960)
```

With pipes:

```
library(dplyr)
sharks |>
    mutate(Year = as.numeric(Year)) |>
    group_by(Year) |>
    count() |>
    filter(Year >= 1960) -> attacks
```

The number of attacks in a year is not binary but a count that, in principle, can take any non-negative integer as its value. Are there any trends over time for the number of reported attacks?

```
# Plot data from 1960-2019:
library(ggplot2)
ggplot(attacks, aes(Year, n)) +
    geom_point() +
    ylab("Number of attacks")
```

No trend is evident. To confirm this, let's fit a regression model with n (number of attacks) as the response variable and Year as an explanatory variable. For count data like this, a good first model to use is *Poisson regression*. Let μ_i denote the expected value of the response variable given the explanatory variables. Given n observations of p explanatory variables, the Poisson regression model is:

$$\log(\mu_i) = \beta_0 + \beta_1 x_{i1} + \beta_2 x_{i2} + \cdots + \beta_p x_{ip}, \qquad i = 1, \ldots, n$$

To fit it, we use glm as before, but this time with family = poisson:

```
m <- glm(n ~ Year, data = attacks, family = poisson)
summary(m)
```

We can add the curve corresponding to the fitted model to our scatterplot as follows (and as shown in Figure 8.6):

```
attacks_pred <- data.frame(Year = attacks$Year, at_pred =
                            predict(m, type = "response"))

ggplot(attacks, aes(Year, n)) +
  geom_point() +
  ylab("Number of attacks") +
```

```
geom_line(data = attacks_pred, aes(x = Year, y = at_pred),
         colour = "red")
```

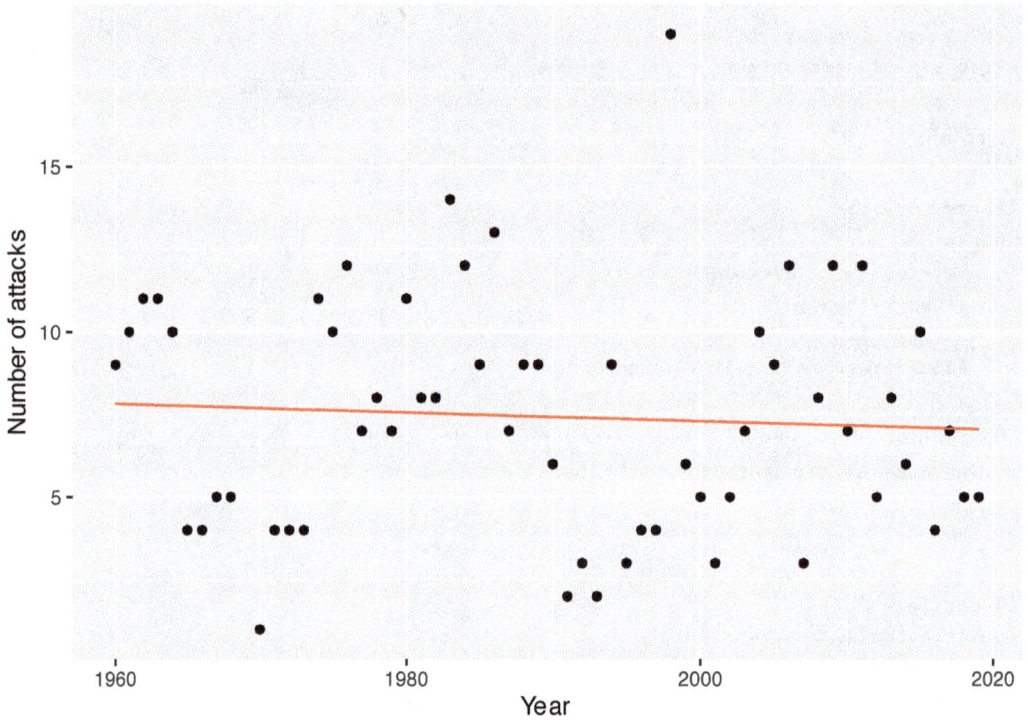

FIGURE 8.6 Poisson regression model for the sharks data.

The fitted model seems to confirm our view that there is no trend over time in the number of attacks.

For model diagnostics, we can use a binned residual plot and a plot of Cook's distance to find influential points:

```
# Binned residual plot:
library(arm)
binnedplot(predict(m, type = "response"),
           residuals(m, type = "response"))



# Plot index against the Cook's distance to find
# influential points:
res <- data.frame(Index = 1:nrow(m$data),
                  CooksDistance = cooks.distance(m))
ggplot(res, aes(Index, CooksDistance)) +
  geom_point() +
  geom_text(aes(label = ifelse(CooksDistance > 0.1,
                               rownames(res), "")),
            hjust = 1.1)
```

A common problem in Poisson regression models is excess zeroes, i.e., more observations with value 0 than what is predicted by the model. To check the distribution of counts in the data, we can draw a histogram:

```
ggplot(attacks, aes(n)) +
  geom_histogram(binwidth = 1, colour = "black")
```

If there are a lot of zeroes in the data, we should consider using another model, such as a hurdle model or a zero-inflated Poisson regression. Both of these are available in the `pscl` package.

Another common problem is *overdispersion*, which occurs when there is more variability in the data than what is predicted by the GLM. A formal test of overdispersion (Cameron & Trivedi, 1990) is provided by `dispersiontest` in the `AER` package. The null hypothesis is that there is no overdispersion, and the alternative is that there is overdispersion:

```
install.packages("AER")

library(AER)
dispersiontest(m, trafo = 1)
```

There are several alternative models that can be considered in the case of overdispersion. One of them is *negative binomial regression*, which uses the same link function as Poisson regression. We can fit it using the `glm.nb` function from `MASS`:

```
library(MASS)
m_nb <- glm.nb(n ~ Year, data = attacks)

summary(m_nb)
```

For the shark attack data, the predictions from the two models are virtually identical, meaning that both are equally applicable in this case:

```
attacks_pred <- data.frame(Year = attacks$Year, at_pred =
                              predict(m, type = "response"))
attacks_pred_nb <- data.frame(Year = attacks$Year, at_pred =
                              predict(m_nb, type = "response"))

ggplot(attacks, aes(Year, n)) +
  geom_point() +
  ylab("Number of attacks") +
  geom_line(data = attacks_pred, aes(x = Year, y = at_pred),
            colour = "red") +
  geom_line(data = attacks_pred_nb, aes(x = Year, y = at_pred),
            colour = "blue", linetype = "dashed")
```

Finally, we can obtain bootstrap confidence intervals, e.g., using case resampling, using `boot_summary`:

```
library(boot.pval)
boot_summary(m_nb, type = "perc", method = "case")
```

$\sim$

Exercise 8.17. The quakes dataset, available in base R, contains information about seismic events off Fiji. Fit a Poisson regression model with stations as the response variable and mag as an explanatory variable. Are there signs of overdispersion? Does using a negative binomial model improve the model fit?

8.5.2 Modelling rates

Poisson regression models, and related models like negative binomial regression, can not only be used to model count data. They can also be used to model *rate data*, such as the number of cases per capita or the number of cases per unit area. In that case, we need to include an *exposure* variable N that describes, e.g., the population size or area corresponding to each observation. The model will be that:

$$\log(\mu_i/N_i) = \beta_0 + \beta_1 x_{i1} + \beta_2 x_{i2} + \cdots + \beta_p x_{ip}, \qquad i = 1, \ldots, n.$$

Because $\log(\mu_i/N_i) = \log(\mu_i) - \log(N_i)$, this can be rewritten as:

$$\log(\mu_i) = \beta_0 + \beta_1 x_{i1} + \beta_2 x_{i2} + \cdots + \beta_p x_{ip} + \log(N_i), \qquad i = 1, \ldots, n.$$

In other words, we should include $\log(N_i)$ on the right-hand side of our model, *with a known coefficient* equal to 1. In regression, such a term is known as an *offset*. We can add it to our model using the offset function.

As an example, we'll consider the ships data from the MASS package. It describes the number of damage incidents for different ship types operating in the 1960s and 1970s, and it includes information about how many months each ship type was in service (i.e., each ship type's *exposure*):

```
library(MASS)
?ships
View(ships)
```

For our example, we'll use ship type as the explanatory variable, incidents as the response variable, and service as the exposure variable. First, we remove observations with 0 exposure (by definition, these can't be involved in incidents, and so there is no point in including them in the analysis).

Without pipes:

```
ships <- ships[ships$service != 0,]
```

With pipes:

```
library(dplyr)
ships |> filter(service != 0) -> ships
```

Then, we fit the model using `glm` and `offset`:

```
m <- glm(incidents ~ type + offset(log(service)),
         data = ships,
         family = poisson)
```

```
summary(m)
```

Model diagnostics can be performed as in the previous sections.

Rate models are usually interpreted in terms of the rate ratios e^{β_j}, which describe the multiplicative increases of the intensity of rates when x_j is increased by one unit. To compute the rate ratios for our model, we use `exp`:

```
exp(coef(m))
```

$\sim$

Exercise 8.18. Compute bootstrap confidence intervals for the rate ratios in the model for the `ships` data.

8.6   Bayesian estimation of generalised linear models

We can fit a Bayesian GLM with the `rstanarm` package, using `stan_glm` in the same way we did for linear models. Let's look at an example with the `wine` data. First, we load and prepare the data:

```
# Import data about white and red wines:
white <- read.csv("https://tinyurl.com/winedata1",
                  sep = ";")
red <- read.csv("https://tinyurl.com/winedata2",
                sep = ";")
white$type <- "white"
red$type <- "red"
wine <- rbind(white, red)
wine$type <- factor(wine$type)
```

Now, we fit a Bayesian logistic regression model:

```
library(rstanarm)
m <- stan_glm(type ~ pH + alcohol, data = wine, family = binomial)

# Print the estimates:
coef(m)
```

Next, we can plot the posterior distributions of the effects:

```
plot(m, "dens", pars = names(coef(m)))
```

To get 95% credible intervals for the effects, we can use `posterior_interval`. We can also use `plot` to visualise them:

```
posterior_interval(m,
        pars = names(coef(m)),
        prob = 0.95)

plot(m, "intervals",
        pars = names(coef(m)),
        prob = 0.95)
```

Finally, we can use $\hat{R}$ to check model convergence. It should be less than 1.1 if the fitting has converged:

```
plot(m, "rhat")
```

8.7 Missing data and multiple imputation

A common problem is *missing data*, where some values in the dataset are missing. This can be handled by *multiple imputation*, where the missing values are artificially replaced with simulated values. These simulated values are usually predicted using other (non-missing) variables as explanatory variables. Because the impact of the simulated values on the model fit can be large, the process is repeated several times, with different imputed values in each iteration, and a model is fitted to each imputed dataset. These can then be combined to create a single model.

The `mice` package allows us to perform multiple imputation. Let's install it:

```
install.packages("mice")
```

8.7.1 Multiple imputation

For our examples, we'll use the `estates` dataset. Download the `estates.xlsx` data from the book's web page[5]. It describes the selling prices (in thousands of Swedish kronor (SEK)) of houses in and near Uppsala, Sweden, along with a number of variables describing the location, size, and standard of the house. We set `file_path` to the path to `estates.xlsx` and then load the data:

```
library(openxlsx)
estates <- read.xlsx(file_path)
```

---

[5]http://www.modernstatisticswithr.com/data.zip

```
View(estates)
```

As you can see, several values are missing. We are interested in creating a regression model where `selling_price` is the response variable, and `living_area` (size of house) and `location` (location of house) as explanatory variables, and would like to use multiple imputation.

Regression modelling using multiple imputation is carried out in five steps:

1. Select the variables you want to use for imputation.
2. Perform imputation m times.
3. Fit models to each imputed sample.
4. Pool the results: combine the estimates from the m fitted models to obtain a single set of estimates, standard errors, and p-values; see Rubin (1987) and Barnard & Rubin (1999) for details.
5. Create a summary.

This is well suited to a pipeline, where each line corresponds to one of the steps above:

```
library(dplyr)
library(mice)

estates |>
    select(selling_price, living_area, location, plot_area,
            supplemental_area, tax_value) |>
    mice(m = 5, print = FALSE) |>
    with(lm(selling_price ~ living_area + location)) |>
    pool() |>
    summary()
```

By default, `mice` uses a method called predictive mean matching when imputing numerical variables, logistic regression when imputing binary variables, polytomous regression when imputing nominal categorical variables, and ordered logistic regression when imputing ordinal categorical variables. Many other methods are available; see `?mice` for a list. We can for instance use random forest imputation instead (we'll discuss random forest methods in Section 11.5.2), by using the `method` argument in `mice`:

```
estates |>
    select(selling_price, living_area, location, plot_area,
            supplemental_area, tax_value) |>
    mice(m = 5, print = FALSE, method = "rf") |>
    with(lm(selling_price ~ living_area + location)) |>
    pool() |>
    summary()
```

If we like, we can do imputation only for some of the variables. We then specify a formula for each such variable, showing what variables to use for imputation of that variable. The formula follows the syntax for regression formulas and can include interactions.

For example, if we only want to do imputation for `living_area` and `plot_area`:

```
# Create a list of formulas for the variables we want to do imputation for:
imp_formulas <- list(
    living_area = formula(living_area ~ .), # Use all other variables
    plot_area = formula(plot_area ~ location * living_area)) # Use location and
                                                             # living area, with
                                                             # an interaction


# Run the multiple imputation:
estates |>
    select(selling_price, living_area, location, plot_area,
           supplemental_area, tax_value) |>
    mice(m = 5, print = FALSE, formulas = imp_formulas) |>
    with(lm(selling_price ~ living_area + location + plot_area)) |>
    pool() |>
    summary()
```

8.7.2   The effect of missing data

First, let's discuss what effect missing data has on a regression analysis. We'll illustrate this using the penguins dataset from the palmerpenguins package. We fit a linear regression model with two explanatory variables:

```
library(palmerpenguins)
library(dplyr)

# Data set without missing values:
penguins |> select(flipper_length_mm, body_mass_g, species) |>
            na.omit() -> penguins

m <- lm(flipper_length_mm ~ body_mass_g + species, data = penguins)
summary(m)
```

To see how missing data affects the results, we run a simulation. We use the ampute functions from mice to randomly remove some observations from the dataset. We set the proportion of missing data to 25%. ampute converts the categorical species variable to numeric, so we also convert that back:

```
library(mice)
penguins_missing <- ampute(penguins, prop = 0.25)$amp
penguins_missing$species <- factor(penguins_missing$species)
```

We can now check the results to see if the fitted model differs from that using the complete dataset:

```
m2 <- lm(flipper_length_mm ~ body_mass_g + species,
         data = penguins_missing)
summary(m2)
```

```
# How large is the relative change of the coefficients?
coef(m2)/coef(m)
```

```
# How large is the absolute change of the p-values?
abs(summary(m2)$coef[,4]-summary(m)$coef[,4])
```

To see how large these effects can be, we can run this simulation 1,000 times, similar to the simulations we ran in Section 7.2:

```
B <- 1000
res <- data.frame(coef1 = vector("numeric", B),
                  coef2 = vector("numeric", B),
                  coef3 = vector("numeric", B),
                  coef4 = vector("numeric", B))

for(i in 1:B)
{
  penguins_missing <- ampute(penguins, prop = 0.25)$amp
  penguins_missing$species <- factor(penguins_missing$species)

  m2 <- lm(flipper_length_mm ~ body_mass_g + species,
           data = penguins_missing)

  # How large is the relative change of the coefficients?
  res[i,] <- coef(m2)/coef(m)
}
```

```
# How many % was the largest relative decrease?
(1-apply(res, 2, min))*100
```

```
# How many % was the largest relative increase?
(apply(res, 2, max)-1)*100
```

In my run, the estimated values for the third coefficient (the dummy variable for the Chinstrap species) ranged from 21% smaller to 19% larger than the estimate for the complete data. That's substantial, so, it is clear that missing data can skew the results.

8.7.3 The effect of multiple imputation

So what happens if we use multiple imputation to estimate the coefficients instead of just using complete observations, as we did in our previous simulation?

We run the same simulation again, but this time using multiple imputation. It therefore takes a little longer to run.

```
B <- 1000
res <- data.frame(coef1 = vector("numeric", B),
                  coef2 = vector("numeric", B),
                  coef3 = vector("numeric", B),
```

```
                      coef4 = vector("numeric", B))

# Start progress bar:
pbar <- txtProgressBar(min = 0, max = B, style = 3)

for(i in 1:B)
{
  penguins_missing <- ampute(penguins, prop = 0.25)$amp
  penguins_missing$species <- factor(penguins_missing$species)

  penguins_missing |>
    mice(m = 5, print = FALSE) |>
    with(lm(flipper_length_mm ~ body_mass_g + species)) |>
    pool() |>
    summary() |>
    select(estimate) -> m2

  # How large is the relative change of the coefficients?
  res[i,] <- m2/coef(m)

  # Update progress bar
  setTxtProgressBar(pbar, i)
}

close(pbar)

# How many % was the largest relative decrease?
(1-apply(res, 2, min))*100

# How many % was the largest relative increase?
(apply(res, 2, max)-1)*100
```

The results are much better than before. In my run, the estimated values for the third coefficient (the dummy variable for the Chinstrap `species`) ranged from 2.2% smaller to 2.6% larger than the estimate for the complete data (which should be compared to the 21% and 19% I got when we didn't use multiple imputation). The lesson here is that the multiple imputation lets us get estimates that are closer to what we would get if we had had complete data.

8.8 Mixed models

Mixed models are used in regression problems where measurements have been made on clusters of related units. As the first example of this, we'll use a dataset from the `lme4` package, which also happens to contain useful methods for mixed models. Let's install it:

```
install.packages("lme4")
```

The `sleepstudy` dataset from `lme4` contains data from a study on reaction times in a sleep deprivation study. The participants were restricted to 3 hours of sleep per night, and their average reaction time on a series of tests were measured each day during the 9 days that the study lasted:

```
library(lme4)
?sleepstudy
str(sleepstudy)
```

Let's start our analysis by making boxplots showing reaction times for each subject. We'll also superimpose the observations for each participant on top of their boxplots:

```
library(ggplot2)
ggplot(sleepstudy, aes(Subject, Reaction)) +
      geom_boxplot() +
      geom_jitter(aes(colour = Subject),
                      position = position_jitter(0.1))
```

We are interested in finding out if the reaction times increase when the participants have been starved for sleep for a longer period. Let's try plotting reaction times against days, adding a regression line:

```
ggplot(sleepstudy, aes(Days, Reaction, colour = Subject)) +
  geom_point() +
  geom_smooth(method = "lm", colour = "black", se = FALSE)
```

As we saw in the boxplots and can see in this plot too, some participants always have comparatively high reaction times, whereas others always have low values. There are clear differences between individuals, and the measurements for each individual will be correlated. This violates a fundamental assumption of the traditional linear model, namely that all observations are independent.

In addition to this, it also seems that the reaction times change in different ways for different participants, as can be seen if we facet the plot by test subject:

```
ggplot(sleepstudy, aes(Days, Reaction, colour = Subject)) +
  geom_point() +
  theme(legend.position = "none") +
  facet_wrap(~ Subject, nrow = 3) +
  geom_smooth(method = "lm", colour = "black", se = FALSE)
```

The results are shown in Figure 8.7. Both the intercept and the slope of the average reaction time differs between individuals. Because of this, the fit given by the single model can be misleading. Moreover, the fact that the observations are correlated will cause problems for the traditional intervals and tests. We need to take this into account when we estimate the overall intercept and slope.

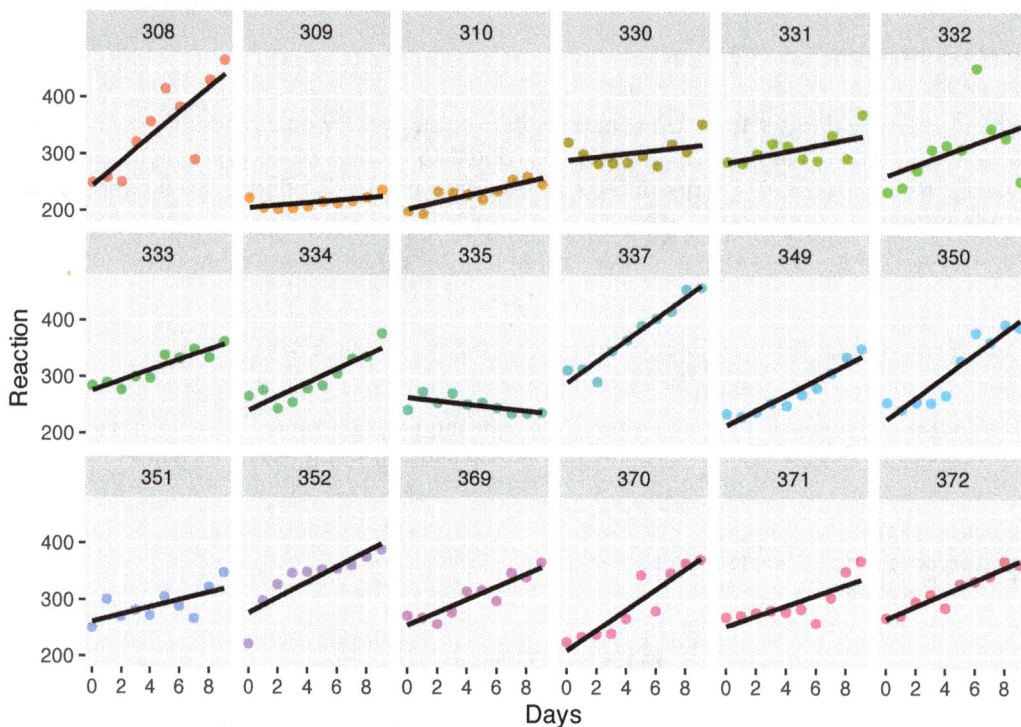

FIGURE 8.7 Comparison of the trends for the 18 sleep study test subjects.

One approach could be to fit a single model for each subject. That doesn't seem very useful though. We're not really interested in these particular test subjects but in how sleep deprivation affects reaction times in an average person. It would be much better to have a single model that somehow incorporates the correlation between measurements made on the same individual. That is precisely what a linear mixed regression model does.

8.8.1 Fitting a linear mixed model

A linear mixed model (LMM) has two types of effects (explanatory variables):

- *Fixed effects*, which are non-random. These are usually the variables of primary interest in the data. In the `sleepstudy` example, `Days` is a fixed effect.
- *Random effects*, which represent nuisance variables that cause measurements to be correlated. These are usually not of interest in and of themselves, but are something that we need to include in the model to account for correlations between measurements. In the `sleepstudy` example, `Subject` is a random effect.

Linear mixed models can be fitted using `lmer` from the `lme4` package. The syntax is the same as for `lm`, with the addition of random effects. These can be included in different ways. Let's have a look at them.

First, we can include a *random intercept*, which gives us a model where the intercept (but not the slope) varies between test subjects. In our example, the formula for this is:

```
library(lme4)
m1 <- lmer(Reaction ~ Days + (1|Subject), data = sleepstudy)
```

Alternatively, we could include a *random slope* in the model, in which case the slope (but not the intercept) varies between test subjects. The formula would be:

```
m2 <- lmer(Reaction ~ Days + (0 + Days|Subject), data = sleepstudy)
```

Finally, we can include both a random intercept and random slope in the model. This can be done in two different ways, as we can model the intercept and slope as being correlated or uncorrelated:

```
# Correlated random intercept and slope:
m3 <- lmer(Reaction ~ Days + (1 + Days|Subject), data = sleepstudy)

# Uncorrelated random intercept and slope:
m4 <- lmer(Reaction ~ Days + (1|Subject) + (0 + Days|Subject),
           data = sleepstudy)
```

Which model should we choose? Are the intercepts and slopes correlated? It could of course be the case that individuals with a high intercept have a smaller slope – or a greater slope! To find out, we can fit different linear models to each subject and then make a scatterplot of their intercepts and slopes. To fit a model to each subject, we use split and map as in Section 8.2.5:

```
# Collect the coefficients from each linear model:
library(purrr)
sleepstudy |> split(sleepstudy$Subject) |>
             map(~ lm(Reaction ~ Days, data = .)) |>
             map(coef) -> coefficients

# Convert to a data frame:
coefficients <- data.frame(matrix(unlist(coefficients),
                     nrow = length(coefficients),
                     byrow = TRUE),
                     row.names = names(coefficients))
names(coefficients) <- c("Intercept", "Days")

# Plot the coefficients:
ggplot(coefficients, aes(Intercept, Days,
                         colour = row.names(coefficients))) +
     geom_point() +
     geom_smooth(method = "lm", colour = "black", se = FALSE) +
     labs(fill = "Subject")

# Test the correlation:
cor.test(coefficients$Intercept, coefficients$Days)
```

The correlation test is not significant, and judging from the plot, there is little indication that the intercept and slope are correlated. We saw earlier that both the intercept and the slope seem to differ between subjects, and so `m4` seems like the best choice here. Let's stick with that and look at a summary table for the model.

```
summary(m4, correlation = FALSE)
```

I like to add `correlation = FALSE` here, which suppresses some superfluous output from `summary`.

You'll notice that unlike the `summary` table for linear models, there are no p-values! This is a deliberate design choice from the `lme4` developers, who argue that the approximate tests available aren't good enough for small sample sizes (Bates et al., 2015).

Summary tables, including bootstrap p-values, for the fixed effects are available through `boot_summary`:

```
library(boot.pval)
boot_summary(m4, type = "perc")
```

Using the bootstrap is usually the best approach for mixed models. If you *really* want some quick p-values, you can load the `lmerTest` package, which adds p-values computed using the Satterthwaite approximation (Kuznetsova et al., 2017). This is better than the usual approximate test but still not as good as bootstrap p-values.

```
install.packages("lmerTest")
```

```
library(lmerTest)
m4 <- lmer(Reaction ~ Days + (1|Subject) + (0 + Days|Subject),
           data = sleepstudy)
summary(m4, correlation = FALSE)
```

If we need to extract the model coefficients, we can do so using `fixef` (for fixed effects) and `ranef` (for random effects):

```
fixef(m4)
ranef(m4)
```

If we want to extract the variance components from the model, we can use `VarCorr`:

```
VarCorr(m4)
```

Let's add the lines from the fitted model to our facetted plot, to compare the results of our mixed model to the lines that were fitted separately for each individual:

```
mixed_mod <- coef(m4)$Subject
mixed_mod$Subject <- row.names(mixed_mod)
```

```
ggplot(sleepstudy, aes(Days, Reaction)) +
```

```
    geom_point() +
    theme(legend.position = "none") +
    facet_wrap(~ Subject, nrow = 3) +
    geom_smooth(method = "lm", colour = "cyan", se = FALSE,
                size = 0.8) +
    geom_abline(aes(intercept = `(Intercept)`, slope = Days,
                    color = "magenta"),
                data = mixed_mod, size = 0.8)
```

Notice that the lines differ. The intercept and slopes have been *shrunk* toward the global effects, i.e., toward the average of all lines.

~

Exercise 8.19. Consider the Oxboys data from the nlme package. Does a mixed model seem appropriate here? If so, are the intercept and slope correlated for different subjects? Fit a suitable model, with height as the response variable.

Save the code for your model, as you will return to it in the next few exercises.

Exercise 8.20. The broom.mixed package allows you to get summaries of mixed models as data frames, just as broom does for linear and generalised linear models. Install it and use it to get the summary table for the model for the Oxboys data that you created in the previous exercise. How are fixed and random effects included in the table?

8.8.2 Model diagnostics

As for any linear model, residual plots are useful for diagnostics for linear mixed models. Of particular interest are signs of heteroscedasticity, as homoscedasticity is assumed in the mixed model. We'll use fortify.merMod to turn the model into an object that can be used with ggplot2, and then create some residual plots:

```
library(ggplot2)
fm4 <- fortify.merMod(m4)

# Plot residuals:
ggplot(fm4, aes(.fitted, .resid)) +
  geom_point() +
  geom_hline(yintercept = 0) +
  xlab("Fitted values") + ylab("Residuals")

# Compare the residuals of different subjects:
ggplot(fm4, aes(Subject, .resid)) +
  geom_boxplot() +
  coord_flip() +
  ylab("Residuals")

# Observed values versus fitted values:
ggplot(fm4, aes(.fitted, Reaction)) +
```

```
    geom_point(colour = "blue") +
    facet_wrap(~ Subject, nrow = 3) +
    geom_abline(intercept = 0, slope = 1) +
    xlab("Fitted values") + ylab("Observed values")

## Q-Q plot of residuals:
ggplot(fm4, aes(sample = .resid)) +
  geom_qq() + geom_qq_line()

## Q-Q plot of random effects:
ggplot(ranef(m4)$Subject, aes(sample = `(Intercept)`)) +
  geom_qq() + geom_qq_line()
ggplot(ranef(m4)$Subject, aes(sample = `Days`)) +
  geom_qq() + geom_qq_line()
```

The normality assumption appears to be satisfied, but there are some signs of heteroscedasticity in the boxplots of the residuals for the different subjects.

~

Exercise 8.21. Return to your mixed model for the Oxboys data from Exercise 8.19. Make diagnostic plots for the model. Are there any signs of heteroscedasticity or non-normality?

8.8.3   Nested random effects and multilevel/hierarchical models

In many cases, a random factor is *nested* within another. To see an example of this, consider the Pastes data from lme4:

```
library(lme4)
?Pastes
str(Pastes)
```

We are interested in the strength of a chemical product. There are 10 delivery batches (batch), and three casks within each delivery (cask). Because of variations in manufacturing, transportation, storage, and so on, it makes sense to include random effects for both batch and cask in a linear mixed model. However, each cask only appears within a single batch, which makes the cask effect *nested* within batch.

Models that use nested random factors are commonly known as *multilevel models* (because the random factors in the model exist at different "levels"), or *hierarchical models* (because there is a hierarchy between the random factors in the model). These aren't really any different from other mixed models, but depending on how the data is structured, we may have to be a bit careful to get the nesting right when we fit the model with lmer.

If the two effects weren't nested, we could fit a model using:

```
# Incorrect model:
m1 <- lmer(strength ~ (1|batch) + (1|cask),
           data = Pastes)
summary(m1, correlation = FALSE)
```

However, because the casks are labelled a, b, and c within each batch, we've now fitted a model where casks from different batches are treated as being equal! To clarify that the labels a, b, and c belong to different casks in different batches, we need to include the nesting in our formula. This is done as follows:

```
# Cask is nested within batch:
m2 <- lmer(strength ~ (1|batch/cask),
          data = Pastes)
summary(m2, correlation = FALSE)
```

Equivalently, we can also use:

```
m3 <- lmer(strength ~ (1|batch) + (1|batch:cask),
          data = Pastes)
summary(m3, correlation = FALSE)
```

8.8.4   ANOVA with random effects

The lmerTest package provides ANOVA tables that allow us to use random effects in ANOVA models. To use it, simply load lmerTest before fitting a model with lmer, and then run anova(m, type = "III") (or replace III with II or I if you want a type II or type I ANOVA table instead).

As an example, consider the TVbo data from lmerTest. Three types of TV sets were compared by eight assessors for four different pictures. To see if there is a difference in the mean score for the colour balance of the TV sets, we can fit a mixed model. We'll include a random intercept for the assessor. This is a balanced design (in which case the results from all three types of tables coincide):

```
library(lmerTest)

# TV data:
?TVbo

# Fit model with both fixed and random effects:
m <- lmer(Colourbalance ~ TVset*Picture + (1|Assessor),
          data = TVbo)

# View fitted model:
m

# All three types of ANOVA table give the same results here:
anova(m, type = "III")
anova(m, type = "II")
anova(m, type = "I")
```

The interaction effect is significant at the 5% level. As for other ANOVA models, we can visualise this with an interaction plot:

```
interaction.plot(TVbo$TVset, TVbo$Picture,
                 response = TVbo$Colourbalance)
```

~

Exercise 8.22. Fit a mixed effects ANOVA to the TVbo data, using Coloursaturation as the response variable, TVset and Picture as fixed effects, and Assessor as a random effect. Does there appear to be a need to include the interaction between Assessor and TVset as a random effect? If so, do it.

8.8.5   Generalised linear mixed models

Everything that we have just done for the linear mixed models carries over to *generalised linear mixed models* (GLMMs), which are GLMs with both fixed and random effects.

A common example is the *item response model*, which plays an important role in psychometrics. This model is frequently used in psychological tests containing multiple questions or sets of questions ("items"), where both the subject and the item are considered random effects. As an example, consider the VerbAgg data from lme4:

```
library(lme4)
?VerbAgg
View(VerbAgg)
```

We'll use the binary version of the response, r2, and fit a logistic mixed regression model to the data, to see if it can be used to explain the subjects' responses. The formula syntax is the same as for linear mixed models, but now we'll use glmer to fit a GLMM. We'll include Anger and Gender as fixed effects (we are interested in seeing how these affect the response) and item and id as random effects with random slopes (we believe that answers to the same item and answers from the same individual may be correlated):

```
m <- glmer(r2 ~ Anger + Gender + (1|item) + (1|id),
           data = VerbAgg, family = binomial)
summary(m, correlation = FALSE)
```

We can plot the fitted random effects for item to verify that there appear to be differences between the different items:

```
mixed_mod <- coef(m)$item
mixed_mod$item <- row.names(mixed_mod)

ggplot(mixed_mod, aes(`(Intercept)`, item)) +
    geom_point() +
    xlab("Random intercept")
```

The situ variable, describing situation type, also appears interesting. Let's include it as a fixed effect. Let's also allow different situational (random) effects for different respondents. It seems reasonable that such responses are random rather than fixed (as in the solution to

Exercise 8.22), and we do have repeated measurements of these responses. We'll therefore also include situ as a random effect nested within id:

```
m <- glmer(r2 ~ Anger + Gender + situ + (1|item) + (1|id/situ),
            data = VerbAgg, family = binomial)
summary(m, correlation = FALSE)
```

Finally, we'd like to obtain bootstrap confidence intervals for fixed effects. Because this is a fairly large dataset ($n = 7,584$), this can take a very long time to run; so, stretch your legs and grab a cup of coffee or two while you wait:

```
library(boot.pval)
boot_summary(m, type = "perc", R = 100)
# Ideally, R should be greater, but for the sake of
# this example, we'll use a low number.
```

$\sim$

Exercise 8.23. Consider the grouseticks data from the lme4 package (Elston et al., 2001). Fit a mixed Poisson regression model to the data, with TICKS as the response variable and YEAR and HEIGHT as fixed effects. What variables are suitable to use for random effects? Compute a bootstrap confidence interval for the effect of HEIGHT.

8.8.6 Bayesian estimation of mixed models

From a numerical point of view, using Bayesian modelling with rstanarm is preferable to frequentist modelling with lme4 if you have complex models with many random effects. Indeed, for some models, lme4 will return a warning message about a singular fit, basically meaning that the model is too complex, whereas rstanarm, powered by the use of a prior distribution, always will return a fitted model regardless of complexity.

After loading rstanarm, fitting a Bayesian linear mixed model with a weakly informative prior is as simple as substituting lmer with stan_lmer:

```
library(lme4)
library(rstanarm)
m4 <- stan_lmer(Reaction ~ Days + (1|Subject) + (0 + Days|Subject),
                data = sleepstudy)

# Print the results:
m4
```

To plot the posterior distributions for the coefficients of the fixed effects, we can use plot, specifying which effects we are interested in using pars:

```
plot(m4, "dens", pars = c("(Intercept)", "Days"))
```

To get 95% credible intervals for the fixed effects, we can use posterior_interval as follows:

```
posterior_interval(m4,
                    pars = c("(Intercept)", "Days"),
                    prob = 0.95)
```

We can also plot them using `plot`:

```
plot(m4, "intervals",
     pars = c("(Intercept)", "Days"),
     prob = 0.95)
```

Finally, we'll check that the model fitting has converged:

```
plot(m4, "rhat")
```

8.9 Creating matched samples

Matching is used to balance the distribution of explanatory variables in the groups that are being compared. This is often required in observational studies, where the treatment variable is not randomly assigned but determined by some external factor(s) that may be related to the treatment. For instance, if you wish to study the effect of smoking on mortality, you can recruit a group of smokers and non-smokers and follow them for a few years. But both mortality and smoking are related to *confounding* variables such as age and gender, meaning that imbalances in the age and gender distributions of smokers and non-smokers can bias the results. There are several methods for creating balanced or *matched samples* that seek to mitigate this bias, including *propensity score matching*, which we'll use here. The `MatchIt` and `optmatch` packages contain the functions that we need for this.

To begin with, let's install the two packages:

```
install.packages(c("MatchIt", "optmatch"))
```

We will illustrate the use of the packages using the `lalonde` dataset that is shipped with the `MatchIt` package:

```
library(MatchIt)
data(lalonde)
?lalonde
View(lalonde)
```

Note that the data has row names, which are useful, e.g., for identifying which individuals have been paired. We can access them using `rownames(lalonde)`.

8.9.1 Propensity score matching

To perform automated propensity score matching, we will use the `matchit` function, which computes propensity scores and then matches participants from the treatment and control groups using these. Matches can be found in several ways. We'll consider two of them here. As input, the `matchit` function takes a formula describing the treatment variable and potential confounders, what datasets to use, which method to use, and what ratio of control to treatment participants to use.

A common method is *nearest neighbour matching*, where each participant is matched to the participant in the other group with the most similar propensity score. By default, it starts by finding a match for the participant in the treatment group that has the largest propensity score, then it finds a match for the participant in the treatment groups with the second largest score, and so on. Two participants cannot be matched with the same participant in the control group. The nearest neighbour match is *locally optimal* in the sense that it finds the best (still) available match for each participant in the treatment group, ignoring if that match in fact would be even better for another participant in the treatment group.

To perform propensity score matching using nearest neighbour matching with one match each, evaluate the results, and then extract the matched samples, we can use `matchit` as follows:

```
matches <- matchit(treat ~ re74 + re75 + age + educ + married,
                   data = lalonde, method = "nearest", ratio = 1)

summary(matches)
plot(matches)
plot(matches, type = "hist")

matched_data <- match.data(matches)
summary(matched_data)
```

To view the matched pairs, you can use:

```
matches$match.matrix
```

To view the values of the `re78` variable of the matched pairs, use:

```
varName <- "re78"
resMatrix <- lalonde[row.names(matches$match.matrix), varName]
for(i in 1:ncol(matches$match.matrix))
{
  resMatrix <- cbind(resMatrix, lalonde[matches$match.matrix[,i],
                                    varName])
}
rownames(resMatrix) <- row.names(matches$match.matrix)
View(resMatrix)
```

As an alternative to nearest neighbour matching, *optimal matching* can be used. This is similar to nearest neighbour matching, but strives to obtain *globally optimal* matches rather than locally optimal. This means that each participant in the treatment group is paired with a participant in the control group, while also taking into account how similar the latter participant is to other participants in the treatment group.

To perform propensity score matching using optimal matching with two matches each:

```
matches <- matchit(treat ~ re74 + re75 + age + educ + married,
                   data = lalonde, method = "optimal", ratio = 2)

summary(matches)
plot(matches)
plot(matches, type = "hist")

matched_data <- match.data(matches)
summary(matched_data)
```

You may also want to find all controls that match participants in the treatment group exactly. This is called exact matching:

```
matches <- matchit(treat ~ re74 + re75 + age + educ + married,
                   data = lalonde, method = "exact")

summary(matches)
plot(matches)
plot(matches, type = "hist")

matched_data <- match.data(matches)
summary(matched_data)
```

Participants with no exact matches won't be included in matched_data.

8.9.2   Stepwise matching

At times you will want to combine the above approaches. For instance, you may want to have an exact match for age, and then an approximate match using the propensity scores for other variables. This is also achievable but requires the matching to be done in several steps. To first match the participant exactly on age and then 1-to-2 via nearest neighbour propensity score matching on re74 and re75, we can use a loop:

```
# Match exactly on age:
matches <- matchit(treat ~ age, data = lalonde, method = "exact")
matched_data <- match.data(matches)

# Match the first subclass 1-to-2 via nearest neighbour propensity
# score matching:
matches2 <- matchit(treat ~ re74 + re75,
                    data = matched_data[matched_data$subclass == 1,],
                    method = "nearest", ratio = 2)
matched_data2 <- match.data(matches2, weights = "weights2",
                            subclass = "subclass2")
matchlist <- matches2$match.matrix

# Match the remaining subclasses in the same way:
for(i in 2:max(matched_data$subclass))
{
```

```
    matches2 <- matchit(treat ~ re74 + re75,
              data = matched_data[matched_data$subclass == i,],
              method = "nearest", ratio = 2)
    matched_data2 <- rbind(matched_data2, match.data(matches2,
                                      weights = "weights2",
                                      subclass = "subclass2"))
    matchlist <- rbind(matchlist, matches2$match.matrix)
}

# Check results:
View(matchlist)
View(matched_data2)
```

8.10   Ethical issues in regression modelling

The p-hacking problem, discussed in Section 3.11.2, is perhaps particularly prevalent in regression modelling. Regression analysis often involves a large number of explanatory variables, and practitioners often try out several different models (e.g., by performing stepwise variable selection; see Section 8.2.7). Because so many hypotheses are tested, often in many different but similar models, there is a large risk of false discoveries.

In any regression analysis, there is a risk of finding *spurious relationships*. These are dependencies between the response variable and an explanatory variable that either are non-causal or are purely coincidental. As an example of the former, consider the number of deaths by drowning, which is strongly correlated with ice cream sales. Not because ice cream causes people to drown, but because both are affected by the weather: we are more likely to go swimming or buy ice cream on hot days. Lurking variables, like the temperature in the ice cream drowning example, are commonly referred to as *confounding factors*. An effect may be statistically significant, but that does not necessarily mean that it is meaningful.

∼

Exercise 8.24. *Discuss the following.* You are tasked with analysing a study on whether Vitamin D protects against the flu. One group of patients is given Vitamin D supplements, and one group is given a placebo. You plan on fitting a regression model to estimate the effect of the vitamin supplements, but note that some confounding factors that you have reason to believe are of importance, such as age and ethnicity, are missing from the data. You can therefore not include them as explanatory variables in the model. Should you still fit the model?

Exercise 8.25. *Discuss the following.* You are fitting a linear regression model to a dataset from a medical study on a new drug which potentially can have serious side effects. The test subjects take a risk by participating in the study. Each observation in the dataset corresponds to a test subject. Like all ordinary linear regression models, your model gives more weight to observations that deviate from the average (and have a high leverage or Cook's distance). Given the risks involved for the test subjects, is it fair to give different weight to data from different individuals? Is it OK to remove outliers because they influence the results too much, meaning that the risk that the subject took was for nought?

9

Survival analysis and censored data

Survival analysis, or time-to-event analysis, often involves censored data. Censoring also occurs in measurements with detection limits, often found in biomarker data and environmental data. This chapter is concerned with methods for analysing such data.

After reading this chapter, you will be able to use R to:

- Visualise survival data,
- Fit survival analysis models, and
- Analyse data with left-censored observations.

9.1 The basics of survival analysis

Many studies are concerned with the time until an event happens: time until a machine fails, time until a patient diagnosed with a disease dies, and so on. In this section we will consider some methods for *survival analysis* (also known as reliability analysis in engineering and duration analysis in economics), which is used for analysing such data. The main difficulty here is that studies often end before all participants have had events, meaning that some observations are *right-censored* – for these observations, we don't know when the event happened, but only that it happened after the end of the study.

The `survival` package contains a number of useful methods for survival analysis. Let's install it:

```
install.packages("survival")
```

We will study the lung cancer data in `lung`:

```
library(survival)
?lung
View(lung)
```

The survival times of the patients consist of two parts: `time` (time from diagnosis until either death or the end of the study) and `status` (1 if the observation is censored, 2 if the patient died before the end of the study). To combine these so that they can be used in a survival analysis, we must create a `Surv` object:

```
Surv(lung$time, lung$status)
```

Here, a + sign after a value indicates right-censoring.

9.1.1   Visualising survival

Survival times are best visualised using Kaplan-Meier curves that show the proportion
of surviving patients. Let's compare the survival times of women and men. We first fit a
survival model using `survfit`, and then draw the Kaplan-Meier curve using `ggsurvplot` from
`survminer`:

```
install.packages("survminer")
library(survival)
library(survminer)
# Fit a survival model:
m <- survfit(Surv(time, status) ~ sex, data = lung)

# Draw the Kaplan-Meier curves:
ggsurvplot(m)
```

The plot shows the proportion of patients alive at different time points, starting at 100% at
time 0. Tick marks along the lines show when censoring has occurred.

The average survival time is typically quantified using the median. Simply computing the
median of the survival times is not a good idea, as this doesn't take censoring into account.
We can, however, get a reliable estimate of the median survival time for each group in our
dataset from the Kaplan-Meier curves. The estimates are computed as the time at which
the curves show that 50% of the patients have died.

```
# Print the estimated median survival in each group:
m
```

To print the values for the survival curves at different time points, we can use `summary`. This
way, we can get measures such as the one-year survival:

```
# Survival at all time points:
summary(m)

# One- and two-year survival (365.25 days and 730.5 days, respectively):
summary(m, times = c(365.25, 730.5))
```

We can customise the Kaplan-Meier plot, e.g., by adding confidence intervals to the estimates,
adding a table showing the number at risk (number of patients at risk at different timepoints),
and highlighting the median survival in each group:

```
ggsurvplot(m,
           conf.int = TRUE, # Add confidence intervals,
           risk.table = TRUE, # Show number at risk
           surv.median.line = "hv" # Show median survival
           )
```

The resulting plot is shown in Figure 9.1. See `?ggsurvplot` for more options and examples.

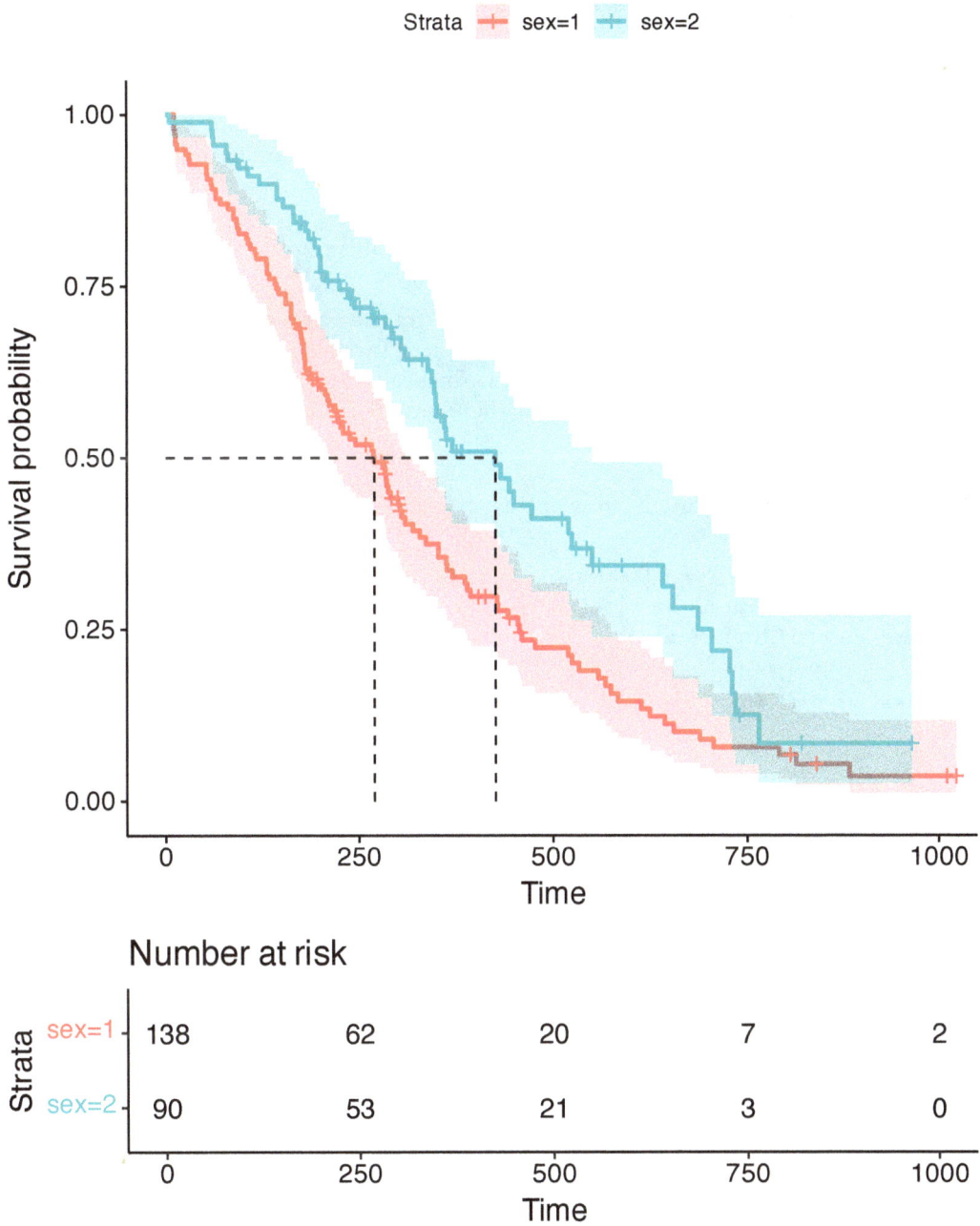

FIGURE 9.1 Kaplan-Meier plot for the lung data.

9.1.2   Testing for group differences

To test whether the survival curves of two groups differ, we can use the logrank test (also known as the Mantel-Cox or Mantel-Haenszel test), given by `survdiff`:

```
survdiff(Surv(time, status) ~ sex, data = lung)
```

Another option is the Peto-Peto test, which puts more weight on early events (deaths, in the case of the `lung` data), and therefore is suitable when such events are of greater interest. In contrast, the logrank test puts equal weights on all events regardless of when they occur. The Peto-Peto test is obtained by adding the argument `rho = 1`:

```
survdiff(Surv(time, status) ~ sex, rho = 1, data = lung)
```

If we wish to compare more than two groups, we can use a two-step procedure similar to ANOVA (Section 8.1.8). In the `lung` data, the `ph.ecog` variable is a measure of the patients' well-being (where lower values are better). It divides the patients into four groups: 0, 1, 2, and 3. To test whether at least one of these groups differs from the others, we do a four-sample version of the logrank test:

```
survdiff(Surv(time, status) ~ ph.ecog, data = lung)
```

The p-value is low ($7^{.-5} = 0.00007$) and we reject the null hypothesis that the survival curves are the same in all four groups. Next, we can test the pairwise differences using `pairwise_survdiff` from `survminer`:

```
pairwise_survdiff(Surv(time, status) ~ ph.ecog, data = lung)
```

There is no difference between groups 0 and 1 or between groups 2 and 3, but the differences between all other pairs of groups are significant. By default, the p-values presented in the table are adjusted for multiplicity using the Benjamini-Hochberg method (Section 3.7). You can control what method is used for the adjustment using the argument `p.adjust.method`; see `?pairwise_survdiff` for details.

In addition to tests, we're also interested in confidence intervals. The `Hmisc` package contains a function for obtaining confidence intervals based on the Kaplan-Meier estimator, called `bootkm`. This allows us to get confidence intervals for the quantiles (including the median) of the survival distribution for different groups, as well as for differences between the quantiles of different groups. First, let's install it:

```
install.packages("Hmisc")
```

We can now use `bootkm` to compute bootstrap confidence intervals for survival times based on the `lung` data. We'll compute an interval for the median survival time for females, and one for the difference in median survival time between females and males:

```
library(Hmisc)

# Create a survival object:
```

```
survobj <- Surv(lung$time, lung$status)

# Get bootstrap replicates of the median survival time for
# the two groups:
median_surv_time_female <- bootkm(survobj[lung$sex == 2],
                                  q = 0.5, B = 999)
median_surv_time_male <- bootkm(survobj[lung$sex == 1],
                                q = 0.5, B = 999)

# 95% bootstrap confidence interval for the median survival time
# for females:
quantile(median_surv_time_female,
         c(.025,.975), na.rm=TRUE)

# 95% bootstrap confidence interval for the difference in median
# survival time:
quantile(median_surv_time_female - median_surv_time_male,
         c(.025,.975), na.rm=TRUE)
```

To obtain confidence intervals for other quantiles, we simply change the argument q in bootkm.

~

Exercise 9.1. Consider the ovarian data from the survival package.

1. Plot Kaplan-Meier curves comparing the two treatment groups.
2. What are the median survival times in the two groups? Why do we get a strange estimate for the group where rx is 2?
3. Compute a bootstrap confidence interval for the difference in the 90% quantile for the survival time for the two groups.

9.1.3 Hazard functions

The hazard function describes the rate of events at time t if a subject has survived until time t. The higher the hazard, the greater the probability of an event. Hazard rates play an integral part in survival analysis, particularly in regression models.

Using ggsurvplot, we can plot the cumulative hazard function, estimated from the Kaplan-Meier survival curve, which shows the total hazard, or total amount of risk that has been accumulated, up until each timepoint:

```
m <- survfit(Surv(time, status) ~ sex, data = lung)
ggsurvplot(m, fun = "cumhaz")
```

The resulting plot is shown in Figure 9.2.

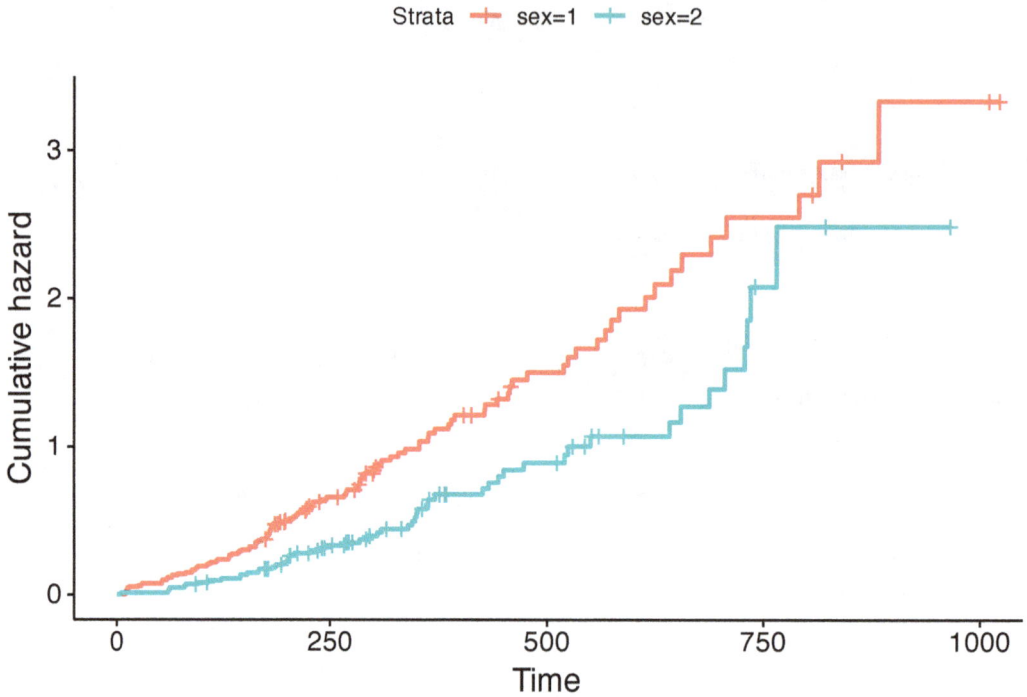

FIGURE 9.2 Cumulative hazards for the lung data.

9.2   Regression models

In regression models for survival data, the response variable is a survival time. Using these models, we can study the effect of different variables on survival and predict survival.

9.2.1   The Cox proportional hazards model

The most commonly used regression model for survival data is the *Cox proportional hazards model* (Cox, 1972), fitted using `coxph`:

```
m <- coxph(Surv(time, status) ~ age + sex, data = lung)
summary(m)
```

The exponentiated coefficients show the hazard ratios, i.e., the relative increases (values greater than 1) or decreases (values below 1) of the hazard rate when a covariate is increased one step, while all others are kept fixed:

```
exp(coef(m))
```

In this case, the hazard increases with age (multiply the hazard by 1.017 for each additional year that the person has lived), and is lower for women (`sex=2`) than for men (`sex=1`).

As before, we can get a publication-ready table using `tbl_regression`:

```
library(gtsummary)
tbl_regression(m, exponentiate = TRUE)
```

The `censboot_summary` function from `boot.pval` provides a table of estimates, bootstrap confidence intervals, and bootstrap p-values for the model coefficients. The `coef` argument can be used to specify whether to print confidence intervals for the coefficients or for the exponentiated coefficients (i.e., hazard ratios):

```
# censboot_summary requires us to use model = TRUE
# when fitting our regression model:
m <- coxph(Surv(time, status) ~ age + sex,
          data = lung, model = TRUE)

library(boot.pval)
# Original coefficients:
censboot_summary(m, coef = "raw")


# Exponentiated coefficients:
censboot_summary(m, coef = "exp")


# Export to Word:
library(flextable)
censboot_summary(m, coef = "exp") |>
  summary_to_flextable() |>
  save_as_docx(path = "my_table.docx")
```

As the name implies, the Cox proportional hazards model relies on the assumption of *proportional hazards*, which essentially means that the effect of the explanatory variables is constant over time. In the case of `sex` in our `lung` example, it states that the hazard ratio between the sexes is constant over time, i.e., that there is no interaction between `sex` and time.

The proportional hazards assumption can be assessed visually by plotting the model residuals, using `cox.zph` and the `ggcoxzph` function from the `survminer` package. Specifically, we will plot the scaled Schoenfeld (1982) residuals, which measure the difference between the observed covariates and the expected covariates given the risk at the time of an event. If the proportional hazards assumption holds, then there should be no trend over time for these residuals. Use the trend line to aid the eye:

```
library(survminer)

ggcoxzph(cox.zph(m), var = 1) # age
ggcoxzph(cox.zph(m), var = 2) # sex

# Formal p-values for a test of proportional
# hazards, for each variable:
cox.zph(m)
```

In this case, there are no apparent trends over time (which is in line with the corresponding formal hypothesis tests), indicating that the proportional hazards model could be applicable here.

$$\sim$$

Exercise 9.2. Consider the `ovarian` data from the `survival` package.

> 1. Use a Cox proportional hazards regression to test whether there is a difference between the two treatment groups, adjusted for age.
> 2. Compute bootstrap confidence interval for the hazard ratio of age.

9.2.2 Repeated observation and frailty models

In some cases, we have repeated observations on the same individuals. An example can be found in the `retinopathy` data from the `survival` package:

```
library(survival)
?retinopathy
```

In this dataset, there are two observations for each patient, because each patient has two eyes. If we ignore this information, we ignore the correlations between observations from the same patient, which can cause standard errors and p-values to become misleading.

There are two main approaches to dealing with repeated measurements in Cox models. The first approach is to use robust standard errors that take the correlations into account. For the `retinopathy` data, this can be achieved by adding `cluster = id` to the call to `coxph`. The second approach is to use a simple mixed model, called a *frailty model*, in which each individual has a unique excess risk (or "frailty"). To run such a model, we add `+ frailty(id)` to the model formula. You'll get to try both approaches in the following two exercises.

$$\sim$$

Exercise 9.3. Consider the `retinopathy` data from the `survival` package. In this dataset, there are multiple observations for each patient, and like in a mixed model, we need to take this into account when computing standard errors and p-values. `id` is used to identify patients and `type`, `trt`, and `age` are the explanatory variables we are interested in. Fit a Cox proportional hazards regression model, adding `cluster = id` to the call to `coxph` to include the information that observations from the same individuals are correlated. Is the assumption of proportional hazards fulfilled?

Exercise 9.4. Repeat the analysis in the previous exercise, but use a frailty model instead. Does the conclusion change?

9.2.3 Accelerated failure time models

In many cases, the proportional hazards assumption does not hold. In such cases we can turn to *accelerated failure time models* (Wei, 1992), or AFT models for short. These are regression models that don't rely on the assumption of proportional hazards. In AFT models, the effect of covariates is to accelerate or decelerate the life course of a subject.

While the proportional hazards model is semiparametric, accelerated failure time models are typically fully parametric, and thus involve stronger assumptions about an underlying distribution. In this case, we have to make assumptions about the distribution of the survival times.

Two common choices are the Weibull distribution and the log-logistic distribution. The Weibull distribution is commonly used in engineering, e.g., in reliability studies. The hazard function of a Weibull model is always monotonic, i.e., either always increasing or always decreasing. In contrast, the log-logistic distribution allows the hazard function to be non-monotonic, making it more flexible, and often more appropriate for medical studies. The distribution curves of these models are shown in Figure 9.3.

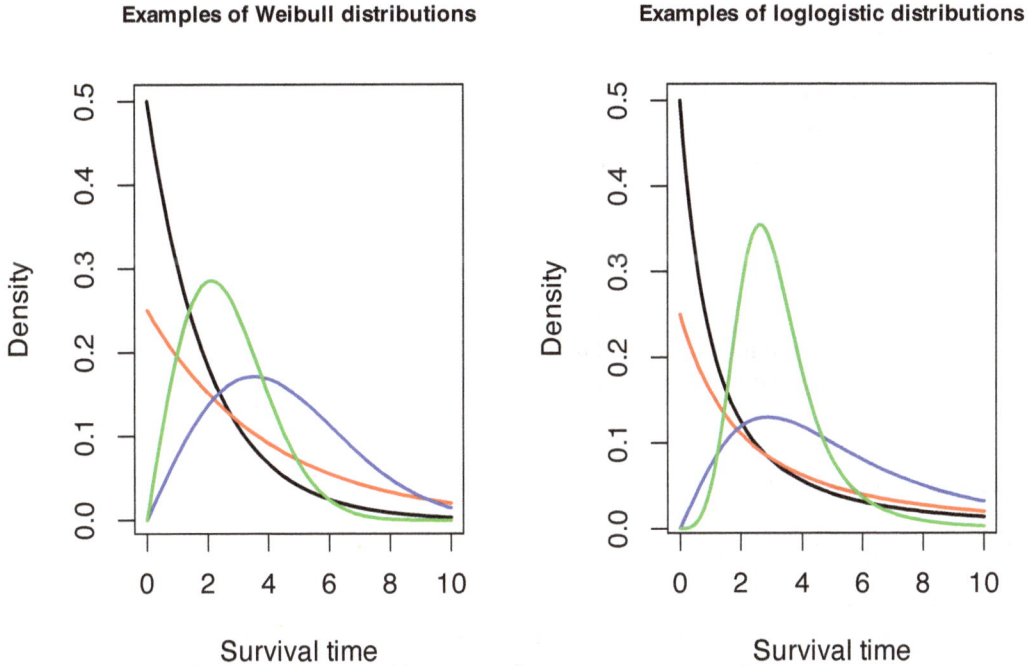

FIGURE 9.3 Distributions used for accelerated failure time models.

Let's fit both types of models to the lung data and have a look at the results:

```
# Fit Weibull model:
m_w <- survreg(Surv(time, status) ~ age + sex, data = lung,
               dist = "weibull", model = TRUE)

m_w

# Fit log-logistic model:
m_ll <- survreg(Surv(time, status) ~ age + sex, data = lung,
                dist = "loglogistic", model = TRUE)

m_ll
```

Interpreting the coefficients of accelerated failure time models is easier than interpreting coefficients from proportional hazards models. The exponentiated coefficients show the relative increase or decrease in the expected survival times when a covariate is increased one step, while all others are kept fixed:

```
exp(coef(m_ll))
```

In this case, according to the log-logistic model, the expected survival time decreases by 1.4% (i.e., multiply by 0.986) for each additional year that the patient has lived. The expected survival time for females (sex=2) is 61.2% higher than for males (multiply by 1.612). This also means that the median survival time is 61.2% higher.

To obtain bootstrap confidence intervals and p-values for the effects, we follow the same procedure as for the Cox model, using censboot_summary. Here is an example for the log-logistic accelerated failure time model:

```
library(boot.pval)
# Original coefficients:
censboot_summary(m_ll, coef = "raw")
# Exponentiated coefficients:
censboot_summary(m_ll, coef = "exp")
```

∼

Exercise 9.5. Consider the ovarian data from the survival package. Fit a log-logistic accelerated failure time model to the data, using all available explanatory variables. What is the estimated difference in survival times between the two treatment groups?

9.3 Competing risks

Survival studies often involve competing and mutually exclusive events that can occur. An example is found in the Melanoma data from the MASS package, where patients can die either from melanoma (one type of event) or die from other causes (another type of event):

```
library(MASS)
?Melanoma
data(Melanoma)
```

This situation is commonly referred to as *competing risks*. In this section, we'll see how we can visualise and analyse competing risks. First, we need to recode the status variable in the Melanoma data, which uses a non-standard coding that won't work with the functions we will use. We also add some labels while we're at it.

The coding we want to use for competing risks data is that 0 represents censoring (patient is alive), 1 represents the event of main interest (patient died of melanoma) and 2 represents the competing event (patient died from other causes).

Without pipes:

```
library(dplyr)
Melanoma$status <- factor(recode(Melanoma$status, "2" = 0, "1" = 1, "3" = 2),
```

```
                            labels = c("Alive",
                                    "Died from melanoma",
                                    "Died from other causes"))
Melanoma$sex <- factor(Melanoma$sex, levels = c(0, 1), labels = c("Female", "Male"))
```

With pipes:

```
library(dplyr)
Melanoma |>
  mutate(status = factor(recode(status, "2" = 0, "1" = 1, "3" = 2),
         labels = c("Alive", "Died from melanoma", "Died from other causes")),
         sex = factor(sex, levels = c(0, 1), labels = c("Female", "Male"))
         ) -> Melanoma
```

Next, let's install two packages that we'll need for the analysis:

```
install.packages(c("tidycmprsk", "ggsurvfit"))
```

Data with competing risks is sometimes analysed using a cause-specific approach, where each event type is analysed separately. Survival times for patients that have had other types of events are then censored by the other event. This allows us to study each type of event, but a major caveat is that the censoring due to other events needs to be independent from the rate of the type of event that we are studying in order for the analysis to be valid. That is, if we study death from melanoma, patients that are censored due to the end of the study need to have the same death rate from melanoma as patients who die from other causes. A different way of saying this is to say that the competing events need to be independent, which is a fairly strong assumption. If we are willing to make this assumption, we can draw Kaplan-Meier curves and fit regression models separately for each type of event, analysing them one by one.

A different approach uses the *cumulative incidence function* (CIF) which estimates the marginal probability for each type of event without assuming independence. In R, we can compute the CIF using cuminc, and then plot it using ggcuminc:

```
# Compute the CIF:
library(tidycmprsk)
m <- cuminc(Surv(time, status) ~ 1, data = Melanoma)

# Plot the CIF:
library(ggsurvfit)
ggcuminc(m, outcome = c("Died from melanoma", "Died from other causes"))
```

We can also compute the CIF stratified by some factor. For instance, we can compare the cumulative incidences for males and females (Figure 9.4):

```
cuminc(Surv(time, status) ~ sex, data = Melanoma) |>
  ggcuminc(outcome = c("Died from melanoma", "Died from other causes"))
```

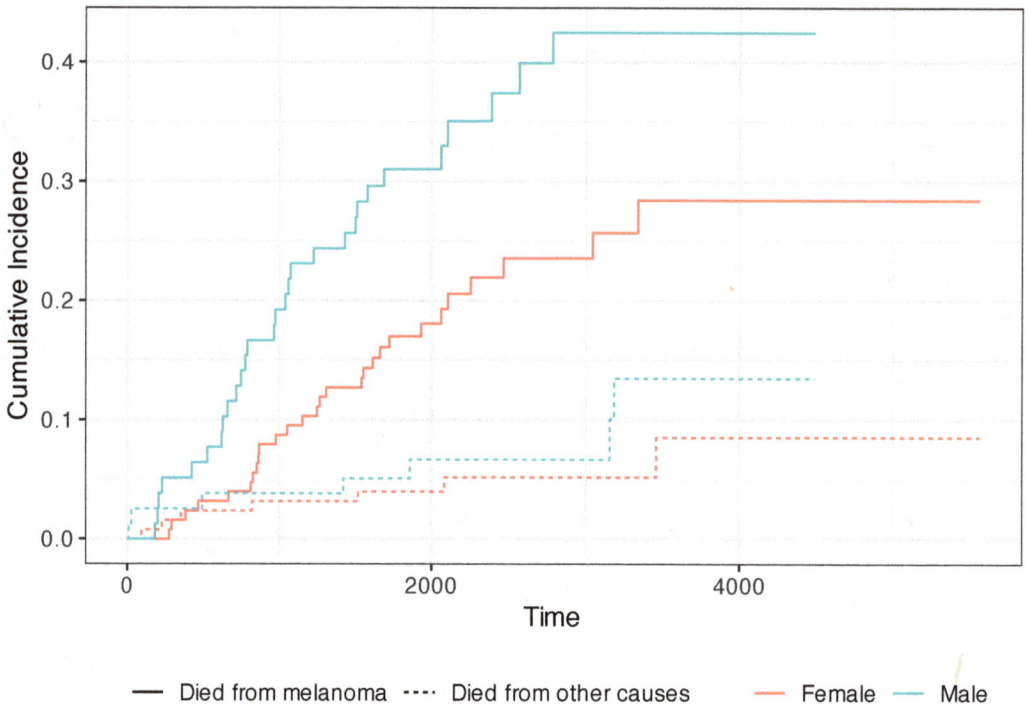

FIGURE 9.4 Cumulative incidence functions for the Melanoma data.

If we wish to model how the CIF is affected by explanatory variables, we can use the Fine-Gray regression model (Fine & Gray, 1999), which is an analogue to the Cox proportional hazards model based on the CIF. It is available through the `crr` function from the `tidycmprsk` package. The `failcode` argument specifies which event we are interested in modelling the CIF for:

```
# CIF for death from melanoma (status = 1):
m1 <- crr(Surv(time, status) ~ sex + age + ulcer,
          failcode = "Died from melanoma",
          data = Melanoma)
m1
# The presence of an ulcer increases the incidence.

# CIF for death from other causes (status = 2):
m2 <- crr(Surv(time, status) ~ sex + age + ulcer,
          failcode = "Died from other causes",
          data = Melanoma)
m2
# Age increases the incidence.
```

The coefficients are interpreted as in a Cox proportional hazards model, meaning it is useful to look at the exponentiated coefficients to get the hazard ratios:

```
exp(coef(m1))
```

As per usual, a publication-ready table can be obtained using `tbl_regression`:

```
library(gtsummary)
tbl_regression(m1, exponentiate = TRUE)
```

9.4 Recurrent events

Another common type of survival data concerns *recurrent events*. A classic example of such data is the `bladder2` dataset in the `survival` package, which describes recurrences of bladder cancer in 178 patients. Let's have a look at it:

```
library(survival)
?bladder2
View(bladder2)
```

Note that there are two times associated with each event: a start time (`start`) and a stop time (`stop`). For patients with multiple events, e.g., the patient with `id` 5 that is listed on rows 5 and 6 in the data frame, the starting time for the second event is the stop time for the first event, and so on. We are interested in the difference between the two treatments shown in the `rx` variable.

For this kind of data, it is common that the time until a recurrence differs depending on how many previous recurrences an individual has had. In our example, the number of recurrences is listed in the `enum` variable. We can draw Kaplan-Meier curves grouped by the number of recurrences using the code below. Note that when we create the survival model, we need to specify both starting times and stop times in our `Surv` object.

```
m <- survfit(Surv(start, stop, event) ~ rx, data = bladder2)

# Draw the Kaplan-Meier curves, grouped by the enum variable:
ggsurvplot_group_by(m, bladder2, "enum") |>
  arrange_ggsurvplots(ncol = 2, nrow = 2)
```

In this case, as seen in Figure 9.5, there are only a small number of patients with `rx=2` that had three or four recurrences, making it difficult to compare the Kaplan-Meier curves to the right in the figure. We can, however, note that there appears to be a trend where the time to an event is shorter the more recurrences a patient has had. This could also be due to the fact that patients who for some reason have a shorter expected time between events tend to get more recurrences.

The Anderson-Gill model (Anderson & Gill, 1982) is an extension of the Cox proportional hazards model that handles recurrent events data. It is straightforward to fit this model using `coxph`:

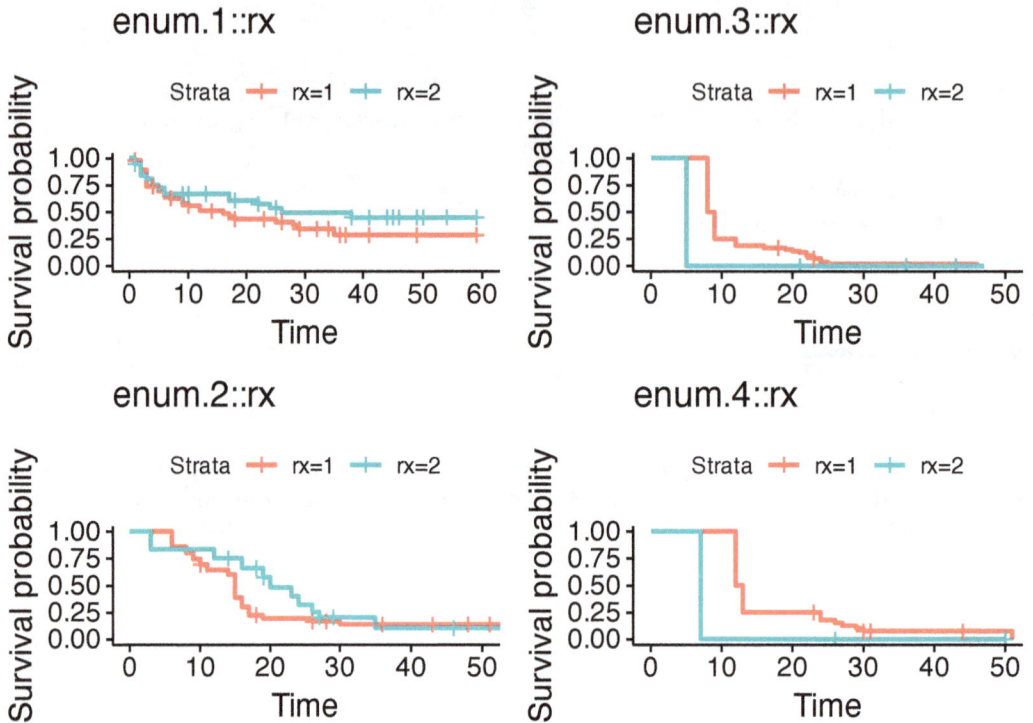

FIGURE 9.5 Kaplan-Meier curves grouped by the number of recurrences.

```
m <- coxph(Surv(start, stop, event) ~ rx, data = bladder2)
summary(m)
```

The approach above is a little naïve, as it ignores the fact that we, like in Exercise 9.3, have repeated observations of the same patients. A better option would be to add `cluster = id` to indicate which observations belong to the same patient, or `frailty(id)` to run a frailty model:

```
# Model with robust standard errors:
m <- coxph(Surv(start, stop, event) ~ rx,
           cluster = id,
           data = bladder2)
summary(m)

# Frailty model:
m <- coxph(Surv(start, stop, event) ~ rx + frailty(id),
           data = bladder2)
summary(m)
```

An even better approach might be to stratify the model depending on the number of recurrences, which would allow us to model the fact that the time until an event seems to be lower for patients with multiple recurrences. This is called a Prentice-Williams-Peterson model (Prentice *et al.*, 1981):

```
m <- coxph(Surv(start, stop, event) ~ rx + strata(enum),
           cluster = id,
           data = bladder2)
summary(m)

# A model with interactions between the treatment and the strata:
m <- coxph(Surv(start, stop, event) ~ rx * strata(enum),
           cluster = id,
           data = bladder2)
summary(m)
```

Regardless of which method we use, the treatment effect is not significant in this example. We obtain a publication-ready summary table using `tbl_regression`:

```
library(gtsummary)
tbl_regression(m, exponentiate = TRUE)
```

9.5 Advanced topics

9.5.1 Multivariate survival analysis

Some trials involve multiple time-to-event outcomes that need to be assessed simultaneously in a multivariate analysis. Examples include studies of the time until each of several correlated symptoms or comorbidities occur. This is analogous to the multivariate testing problem of Section 3.7.2, but with right-censored data. To test for group differences for a vector of right-censored outcomes, a multivariate version of the logrank test described in Persson et al. (2019) can be used. It is available through the `MultSurvTests` package:

```
install.packages("MultSurvTests")
```

As an example, we'll use the `diabetes` dataset from `MultSurvTest`. It contains two time-to-event outcomes: time until blindness in a treated eye and in an untreated eye.

```
library(MultSurvTests)
# Diabetes data:
?diabetes
```

We'll compare two groups that received two different treatments. The survival times (time until blindness) and censoring statuses of the two groups are put in a matrices called `z` and `z.delta`, which are used as input for the test function `perm_mvlogrank`:

```
# Survival times for the two groups:
x <- as.matrix(subset(diabetes, LASER==1)[,c(6,8)])
y <- as.matrix(subset(diabetes, LASER==2)[,c(6,8)])
```

```
# Censoring status for the two groups:
delta.x <- as.matrix(subset(diabetes, LASER==1)[,c(7,9)])
delta.y <- as.matrix(subset(diabetes, LASER==2)[,c(7,9)])

# Create the input for the test:
z <- rbind(x, y)
delta.z <- rbind(delta.x, delta.y)

# Run the test with 499 permutations:
perm_mvlogrank(B = 499, z, delta.z, n1 = nrow(x))
```

9.5.2   Bayesian survival analysis

At the time of this writing, the latest release of rstanarm does not contain functions for
fitting survival analysis models. You can check whether this still is the case by running
?rstanarm::stan_surv in the Console. If you don't find the documentation for the stan_surv
function, you will have to install the development version of the package from GitHub (which
contains such functions), using the following code:

```
# Check if the devtools package is installed, and start
# by installing it otherwise:
if (!require(devtools)) {
  install.packages("devtools")
}
library(devtools)
# Download and install the development version of the package:
install_github("stan-dev/rstanarm", build_vignettes = FALSE)
```

Now, let's have a look at how to fit a Bayesian model to the lung data from survival:

```
library(survival)
library(rstanarm)

# Fit proportional hazards model using cubic M-splines (similar
# but not identical to the Cox model!):
m <- stan_surv(Surv(time, status) ~ age + sex, data = lung)
m
```

Fitting a survival model with a random effect works similarly and uses the same syntax as
lme4. Here is an example with the retinopathy data:

```
m <- stan_surv(Surv(futime, status) ~ age + type + trt + (1|id),
               data = retinopathy)
m
```

9.5.3   Power estimates for the logrank test

The `spower` function in `Hmisc` can be used to compute the power of the univariate logrank test in different scenarios using simulation. The helper functions `Weibull2`, `Lognorm2`, and `Gompertz2` can be used to define Weibull, lognormal, and Gomperts distributions to sample from, using survival probabilities at different time points rather than the traditional parameters of those distributions. We'll look at an example involving the Weibull distribution here. Additional examples can be found in the function's documentation (`?spower`).

Let's simulate the power of a three-year follow-up study with two arms (i.e., two groups, control and intervention). First, we define a Weibull distribution for (compliant) control patients. Let's say that their one-year survival is 0.9 and their three-year survival is 0.6. To define a Weibull distribution that corresponds to these numbers, we use `Weibull2` as follows:

```
weib_dist <- Weibull2(c(1, 3), c(.9, .6))
```

We'll assume that the treatment has no effect for the first six months, and that it then has a constant effect, leading to a hazard ratio of 0.75 (so the hazard ratio is 1 if the time in years is less than or equal to 0.5, and 0.75 otherwise). Moreover, we'll assume that there is a constant drop-out rate, such that 20% of the patients can be expected to drop out during the three years. Finally, there is no drop-in. We define a function to simulate survival times under these conditions:

```
# In the functions used to define the hazard ratio, drop-out
# and drop-in, t denotes time in years:
sim_func <- Quantile2(weib_dist,
      hratio = function(t) { ifelse(t <= 0.5, 1, 0.75) },
      dropout = function(t) { 0.2*t/3 },
      dropin = function(t) { 0 })
```

Next, we define a function for the censoring distribution, which is assumed to be the same for both groups. Let's say that each follow-up is done at a random time point between 2 and 3 years. We'll therefore use a uniform distribution on the interval $(2, 3)$ for the censoring distribution:

```
rcens <- function(n)
{
   runif(n, 2, 3)
}
```

Finally, we define two helper functions required by `spower` and then run the simulation study. The output is the simulated power using the settings that we've just created.

```
# Define helper functions:
rcontrol <- function(n) { sim_func(n, "control") }
rinterv  <- function(n) { sim_func(n, "intervention") }

# Simulate power when both groups have sample size 300:
spower(rcontrol, rinterv, rcens, nc = 300, ni = 300,
      test = logrank, nsim = 999)
```

```
# Simulate power when both groups have sample size 450:
spower(rcontrol, rinterv, rcens, nc = 450, ni = 450,
       test = logrank, nsim = 999)

# Simulate power when the control group has size 100
# and the intervention group has size 300:
spower(rcontrol, rinterv, rcens, nc = 100, ni = 300,
       test = logrank, nsim = 999)
```

9.6 Left-censored data and nondetects

Survival data is typically right-censored. Left-censored data, on the other hand, is common in medical research (e.g., in biomarker studies) and environmental chemistry (e.g., measurements of chemicals in water), where some measurements fall below the laboratory's *detection limits* (or limit of detection, LoD). Such data also occur in studies in economics. A measurement below the detection limit, a *nondetect*, is still more informative than having no measurement at all – we may not know the exact value, but we know that the measurement is below a given threshold.

In principle, all methods that are applicable to survival analysis can also be used for left-censored data (although the interpretation of coefficients and parameters may differ), but in practice the distributions of lab measurements and economic variables often differ from those that typically describe survival times. In this section we'll look at methods tailored to the kind of left-censored data that appears in applications in the aforementioned fields.

9.6.1 Estimation

The EnvStats package contains a number of functions that can be used to compute descriptive statistics and estimating parameters of distributions from data with nondetects. Let's install it:

```
install.packages("EnvStats")
```

Estimates of the mean and standard deviation of a normal distribution that take the censoring into account in the right way can be obtained with enormCensored, which allows us to use several different estimators (details surrounding the available estimators can be found using ?enormCensored). Analogous functions are available for other distributions, for instance elnormAltCensored for the lognormal distribution, egammaCensored for the gamma distribution, and epoisCensored for the Poisson distribution.

To illustrate the use of enormCensored, we will generate data from a normal distribution. We know the true mean and standard deviation of the distribution and can compute the estimates for the generated sample. We will then pretend that there is a detection limit for this data, and artificially left-censor about 20% of it. This allows us to compare the estimates for the full sample and the censored sample, to see how the censoring affects the estimates. Try running the code below a few times:

```
# Generate 50 observations from a N(10, 9)-distribution:
x <- rnorm(50, 10, 3)

# Estimate the mean and standard deviation:
mean_full <- mean(x)
sd_full <- sd(x)

# Censor all observations below the "detection limit" 8
# and replace their values by 8:
censored <- x<8
x[censored] <- 8

# The proportion of censored observations is:
mean(censored)

# Estimate the mean and standard deviation in a naïve
# manner, using the ordinary estimators with all
# nondetects replaced by 8:
mean_cens_naive <- mean(x)
sd_cens_naive <- sd(x)

# Estimate the mean and standard deviation using
# different estimators that take the censoring
# into account:

library(EnvStats)
# Maximum likelihood estimate:
estimates_mle <- enormCensored(x, censored,
                                    method = "mle")
# Biased-corrected maximum likelihood estimate:
estimates_bcmle <- enormCensored(x, censored,
                                    method = "bcmle")
# Regression on order statistics, ROS, estimate:
estimates_ros <- enormCensored(x, censored,
                                    method = "ROS")

# Compare the different estimates:
mean_full; sd_full
mean_cens_naive; sd_cens_naive
estimates_mle$parameters
estimates_bcmle$parameters
estimates_ros$parameters
```

The naïve estimators tend to be biased for data with nondetects (sometimes very biased!). Your mileage may vary depending on, e.g., the sample size and the amount of censoring, but in general, the estimators that take censoring into account will fare much better.

After we have obtained estimates for the parameters of the normal distribution, we can plot the data against the fitted distribution to check the assumption of normality:

```r
library(ggplot2)
# Compare to histogram, including a bar for nondetects:
ggplot(data.frame(x), aes(x)) +
    geom_histogram(colour = "black", aes(y = ..density..)) +
    geom_function(fun = dnorm, colour = "red", size = 2,
                args = list(mean = estimates_mle$parameters[1],
                                sd = estimates_mle$parameters[2]))


# Compare to histogram, excluding nondetects:
x_noncens <- x[!censored]
ggplot(data.frame(x_noncens), aes(x_noncens)) +
    geom_histogram(colour = "black", aes(y = ..density..)) +
    geom_function(fun = dnorm, colour = "red", size = 2,
                args = list(mean = estimates_mle$parameters[1],
                                sd = estimates_mle$parameters[2]))
```

To obtain percentile and BCa bootstrap confidence intervals for the mean, we can add the options `ci = TRUE` and `ci.method = "bootstrap"`:

```r
# Using 999 bootstrap replicates:
enormCensored(x, censored, method = "mle",
            ci = TRUE,  ci.method = "bootstrap",
            n.bootstraps = 999)$interval$limits
```

~

Exercise 9.6. Download the `il2rb.csv` data from the book's web page[1]. It contains measurements of the biomarker IL-2RB made in serum samples from two groups of patients. The values that are missing are in fact nondetects, with detection limit 0.25.

Under the assumption that the biomarker levels follow a lognormal distribution, compute bootstrap confidence intervals for the mean of the distribution for the control group. What proportion of the data is left-censored?

9.6.2   Tests of means

When testing the difference between two groups' means, nonparametric tests like the Wilcoxon-Mann-Whitney test often perform very well for data with nondetects, unlike the t-test (Zhang et al., 2009). For data with a high degree of censoring (e.g., more than 50%), most tests perform poorly. For multivariate tests of mean vectors, the situation is the opposite, with Hotelling's T^2 (Section 3.7.2) being a much better option than nonparametric tests (Thulin, 2016).

~

Exercise 9.7. Return to the `il2rb.csv` data from Exercise 9.7. Test the hypothesis that there is no difference in location between the two groups.

---

[1]http://www.modernstatisticswithr.com/data.zip

9.6.3 Censored regression

Censored regression models can be used when the response variable is censored. A common model in economics is the Tobit regression model (Tobin, 1958), which is a linear regression model with normal errors, tailored to left-censored data. It can be fitted using `survreg`.

As an example, consider the `EPA.92c.zinc.df` dataset available in `EnvStats`. It contains measurements of zinc concentrations from five wells, made on eight samples from each well, half of which are nondetects. Let's say that we are interested in comparing these five wells (so that the wells aren't random effects). Let's also assume that the eight samples were collected at different time points, and that we want to investigate whether the concentrations change over time. Such changes could be non-linear, so we'll include the sample number as a factor. To fit a Tobit model to this data, we use `survreg` as follows.

```
library(EnvStats)
?EPA.92c.zinc.df

# Note that in Surv, in the vector describing censoring 0 means
# censoring and 1 no censoring. This is the opposite of the
# definition used in EPA.92c.zinc.df$Censored, so we use the !
# operator to change 0's to 1's and vice versa.
library(survival)
m <- survreg(Surv(Zinc, !Censored, type = "left") ~ Sample + Well,
             data = EPA.92c.zinc.df, dist = "gaussian")
summary(m)
```

Similarly, we can fit a model under the assumption of lognormality:

```
m <- survreg(Surv(Zinc, !Censored, type = "left") ~ Sample + Well,
             data = EPA.92c.zinc.df, dist = "lognormal")
summary(m)
```

Fitting regression models where the explanatory variables are censored is more challenging. For prediction, a good option is models based on decision trees, studied in Section 11.5. For testing whether there is a trend over time, tests based on Kendall's correlation coefficient can be useful. `EnvStats` provides two functions for this: `kendallTrendTest` for testing a monotonic trend, and `kendallSeasonalTrendTest` for testing a monotonic trend within seasons.

10

Structural equation models, factor analysis, and mediation

Factor analysis and structural equation models let us model abstract nonmeasurable concepts, like intelligence and quality of life, called latent variables. These models have a rich theoretical framework which also allows us to extend regression models to include mediators that help explain causal pathways.

After reading this chapter, you will be able to use R to:

- Use exploratory factor analysis to find latent variables in your data,
- Use confirmatory factor analysis to test hypotheses related to latent variables,
- Fit structural analysis to your models, and
- Run mediation analyses to find mediated and moderated effects in regression models.

10.1   Exploratory factor analysis

The purpose of *factor analysis* is to describe and understand the correlation structure for a set of observable variables through a smaller number of unobservable underlying variables, called *factors* or *latent variables*. These are thought to explain the values of the observed variables in a causal manner. Latent variables can represent measurable but unobserved variables, but more commonly represent abstract concepts such as wisdom or quality of life.

Factor analysis is a popular tool in psychometrics. Among other things, it is used to identify latent variables that explain people's results on different tests, e.g., related to personality, intelligence, or attitude.

10.1.1   Running a factor analysis

We'll use the `psych` package, along with the associated package `GPArotation`, for our analyses. Let's install them:

```
install.packages(c("psych", "GPArotation"))
```

For our first example of factor analysis, we'll be using the `attitude` data that comes with R. It describes the outcome of a survey of employees at a financial organisation. Have a look at its documentation to read about the variables in the dataset:

```
?attitude
attitude
```

To fit a factor analysis model to these data, we can use `fa` from `psych`. `fa` requires us to specify the number of factors used in the model. We'll get back to how to choose the number of factors, but for now, let's go with 2:

```
library(psych)
# Fit factor model:
attitude_fa <- fa(attitude, nfactors = 2,
                  rotate = "oblimin", fm = "ml")
```

`fa` does two things for us. First, it fits a factor model to the data, which yields a table of *factor loadings*, i.e., the correlation between the two unobserved factors and the observed variables. However, there is an infinite number of mathematically valid factor models for any given dataset. Therefore, the factors are *rotated* according to some rule to obtain a factor model that hopefully allows for easy and useful interpretation. Several methods can be used to fit the factor model (set using the `fm` argument in `fa`) and for rotation of the solution (set using `rotate`). We'll look at some of the options shortly.

Before that, we'll print the result, showing the factor loadings (after rotation). We'll also plot the resulting model using `fa.diagram`, showing the correlation between the factors and the observed variables:

```
# Print results:
attitude_fa

# Plot results:
fa.diagram(attitude_fa, simple = FALSE)
```

The first factor is correlated to the variables `advance`, `learning`, and `raises`. We can perhaps interpret this factor as measuring the employees' career opportunity at the organisation. The second factor is strongly correlated to `complaints` and (overall) `rating`, but also to a lesser degree correlated to `raises`, `learning`, and `privileges`. This maybe can be interpreted as measuring how the employees feel they are treated at the organisation.

We can also see that the two factors are correlated. In some cases, it makes sense to expect the factors to be uncorrelated. In that case, we can change the rotation method used, from `oblimin` (which yields *oblique rotations*, allowing for correlations - usually a good default) to `varimax`, which yields uncorrelated factors:

```
attitude_fa <- fa(attitude, nfactors = 2,
                  rotate = "varimax", fm = "ml")
fa.diagram(attitude_fa, simple = FALSE)
```

In this case, the results are fairly similar.

The `fm = "ml"` setting means that maximum likelihood estimation of the factor model is performed, under the assumption of a normal distribution for the data. Maximum likelihood estimation is widely recommended for estimation of factor models, and it can often work well even for non-normal data (Costello & Osborne, 2005). However, there are cases where it fails to find useful factors. `fa` offers several different estimation methods. A good alternative is `minres`, which often works well when maximum likelihood fails:

```
attitude_fa <- fa(attitude, nfactors = 2,
                  rotate = "oblimin", fm = "minres")
fa.diagram(attitude_fa, simple = FALSE)
```

Once again, the results are similar to what we saw before. In other examples, the results differ more. When choosing which estimation method and rotation to use, bear in mind that in an exploratory study, there is no harm in playing around with a few different methods. After all, your purpose is to generate hypotheses rather than confirm them, and looking at the data in a few different ways will help you do that (we'll discuss how to test hypotheses related to factors in Section 10.2).

10.1.2   Choosing the number of factors

To determine the number of factors that are appropriate for a particular dataset, we can draw a scree plot with scree. This is interpreted in the same way as for principal components analysis (Section 4.11) and centroid-based clustering (Section 4.12.3) – we look for an "elbow" in the plot, which tells us at which point adding more factors no longer contributes much to the model:

```
scree(attitude, pc = FALSE)
```

A useful alternative version of this is provided by fa.parallel, which adds lines showing what the scree plot would look like for randomly generated uncorrelated data of the same size as the original dataset. As long as the blue line, representing the actual data, is higher than the red line, representing randomly generated data, adding more factors improves the model:

```
fa.parallel(attitude, fm = "ml", fa = "fa")
```

Some older texts recommend that only factors with an eigenvalue (the y-axis in the scree plot) greater than 1 be kept in the model. It is widely agreed that this so-called Kaiser rule is inappropriate (Costello & Osborne, 2005), as it runs the risk of leaving out important factors.

Similarly, some older texts also recommend using principal components analysis to fit factor models. While the two are mathematically similar in that both in some sense reduce the dimensionality of the data, PCA and factor analysis are designed to target different problems. Factor analysis is concerned with an underlying causal structure where the unobserved factors affect the observed variables. In contrast, PCA simply seeks to create a small number of variables that summarise the variation in the data, which can work well even if there are no unobserved factors affecting the variables.

∼

Exercise 10.1. Factor analysis only relies on the covariance or correlation matrix of your data. When using fa and other functions for factor analysis, you can input either a data frame or a covariance/correlation matrix. Read about the ability.cov data that comes shipped with R, and perform a factor analysis of it.

10.1.3 Latent class analysis

When there is a single categorical latent variable, factor analysis overlaps with clustering, which we studied in Section 4.12. Whether we think of the values of the latent variable as clusters, classes, factor levels, or something else is mainly a philosophical question – from a mathematical perspective, it doesn't matter what name we use for them.

When observations from the same cluster are assumed to be uncorrelated, the resulting model is called *latent profile analysis*, which typically is handled using model-based clustering (Section 4.12.5). The special case where the observed variables are categorical is instead known as *latent class analysis*. This is common, e.g., in analyses of survey data, and we'll have a look at such an example in this section. The package that we'll use for our analyses is called poLCA – let's install it:

```
install.packages("poLCA")
```

The National Mental Health Services Survey[1] is an annual survey collecting information about mental health treatment facilities in the US. We'll analyse data from the 2019 survey, courtesy of the Substance Abuse and Mental Health Data Archive, and try to find latent classes. Download nmhss-puf-2019.csv from the book's web page[2], and set file_path to its path. We can then load and look at a summary of the data using:

```
nmhss <- read.csv(file_path)
summary(nmhss)
```

All variables are categorical (except perhaps for the first one, which is an identifier). According to the survey's documentation[3], negative values are used to represent missing values. For binary variables, 0 means no/non-presence, and 1 means yes/presence.

Next, we'll load the poLCA package and read the documentation for the function that we'll use for the analysis.

```
library(poLCA)
?poLCA
```

As you can see in the description of the data argument, the observed variables (called *manifest variables* here) are only allowed to contain consecutive integer values, starting from 1. Moreover, missing values should be represented by NA, and not by negative numbers (just as elsewhere in R!). We therefore need to make two changes to our data:

- Change negative values to NA,
- Change the levels of binary variables so that 1 means no/non-presence, and 2 means yes/presence.

In our example, we'll look at variables describing what treatments are available at the different facilities. Let's create a new data frame for those variables:

---

[1]https://www.datafiles.samhsa.gov/study-dataset/national-mental-health-services-survey-2019-n-mhss-2019-ds0001-nid18959

[2]http://www.modernstatisticswithr.com/data.zip

[3]https://www.datafiles.samhsa.gov/sites/default/files/field-uploads-protected/studies/N-MHSS-2019/N-MHSS-2019-datasets/N-MHSS-2019-DS0001/N-MHSS-2019-DS0001-info/N-MHSS-2019-DS0001-info-codebook.pdf

```
treatments <- nmhss[, names(nmhss)[17:30]]
summary(treatments)
```

To make the changes to the data that we need, we can do the following:

```
# Change negative values to NA:
treatments[treatments < 0] <- NA

# Change binary variables from 0 and 1 to
# 1 and 2:
treatments <- treatments + 1

# Check the results:
summary(treatments)
```

We are now ready to get started with our analysis. To begin with, we will try to find classes based on whether or not the facilities offer the following five treatments:

- TREATPSYCHOTHRPY: The facility offers individual psychotherapy,
- TREATFAMTHRPY: The facility offers couples/family therapy,
- TREATGRPTHRPY: The facility offers group therapy,
- TREATCOGTHRPY: The facility offers cognitive behavioural therapy,
- TREATPSYCHOMED: The facility offers psychotropic medication. The poLCA function needs three inputs: a formula describing what observed variables to use, a data frame containing the observations, and nclass, the number of latent classes to find. To begin with, let's try two classes:

```
m <- poLCA(cbind(TREATPSYCHOTHRPY, TREATFAMTHRPY,
                 TREATGRPTHRPY, TREATCOGTHRPY,
                 TREATPSYCHOMED) ~ 1,
           data = treatments, nclass = 2)
```

The output shows the probabilities of 1s (no/non-presence), and 2s (yes/presence) for the two classes. So, for instance, from the output,

```
$TREATPSYCHOTHRPY
           Pr(1)  Pr(2)
class 1:  0.6628 0.3372
class 2:  0.0073 0.9927
```

we gather that 34% of facilities belonging to the first class offer individual psychotherapy, whereas 99% of facilities from the second class offer individual psychotherapy. Looking at the other variables, we see that the second class always has high probabilities of offering therapies, while the first class doesn't. Interpreting this, we'd say that the second class contains facilities that offer a wide variety of treatments, and the first facilities that only offer some therapies. Finally, we see from the output that 88% of the facilities belong to the second class:

```
Estimated class population shares
 0.1167 0.8833
```

We can visualise the class differences in a plot:

plot(m)

To see which classes different observations belong to, we can use:

m$predclass

Just as in a cluster analysis, it is often a good idea to run the analysis with different numbers of classes. Next, let's try three classes:

```
m <- poLCA(cbind(TREATPSYCHOTHRPY, TREATFAMTHRPY,
                 TREATGRPTHRPY, TREATCOGTHRPY,
                 TREATPSYCHOMED) ~ 1,
           data = treatments, nclass = 3)
```

This time, we run into numerical problems – the model estimation has failed, as indicated by the following warning message:

ALERT: iterations finished, MAXIMUM LIKELIHOOD NOT FOUND

poLCA fits the model using a method known as the *EM algorithm*, which finds maximum likelihood estimates numerically. First, the observations are randomly assigned to the classes. Step by step, the observations are then moved between classes, until the optimal split has been found. It can, however, happen that more steps are needed to find the optimum (by default 1,000 steps are used), or that we end up with unfortunate initial class assignments that prevent the algorithm from finding the optimum. To attenuate this problem, we can increase the number of steps used, or run the algorithm multiple times, each with new initial class assignments. The poLCA arguments for this are maxiter, which controls the number of steps (or iterations) used, and nrep, which controls the number of repetitions with different initial assignments. We'll increase both and see if that helps. Note that this means that the algorithm will take longer to run:

```
m <- poLCA(cbind(TREATPSYCHOTHRPY, TREATFAMTHRPY,
                 TREATGRPTHRPY, TREATCOGTHRPY,
                 TREATPSYCHOMED) ~ 1,
           data = treatments, nclass = 3,
           maxiter = 2500, nrep = 5)
```

These settings should do the trick for this dataset, and you probably won't see a warning message this time. If you do, try increasing either number and run the code again.

The output that you get can differ between runs; in particular, the order of the classes can differ depending on initial assignments. Here is part of the output from my run:

```
$TREATPSYCHOTHRPY
            Pr(1)   Pr(2)
class 1:   0.0076  0.9924
class 2:   0.0068  0.9932
class 3:   0.6450  0.3550
•

$TREATFAMTHRPY
            Pr(1)   Pr(2)
class 1:   0.1990  0.8010
class 2:   0.0223  0.9777
class 3:   0.9435  0.0565

$TREATGRPTHRPY
            Pr(1)   Pr(2)
class 1:   0.0712  0.9288
class 2:   0.3753  0.6247
class 3:   0.4935  0.5065


$TREATCOGTHRPY
            Pr(1)   Pr(2)
class 1:   0.0291  0.9709
class 2:   0.0515  0.9485
class 3:   0.5885  0.4115


$TREATPSYCHOMED
            Pr(1)   Pr(2)
class 1:   0.0825  0.9175
class 2:   1.0000  0.0000
class 3:   0.3406  0.6594


Estimated class population shares
 0.8059 0.0746 0.1196
```

We can interpret this as follows:

- Class 1 (81% of facilities): Offer all treatments, including psychotropic medication.
- Class 2 (7% of facilities): Offer all treatments, except for psychotropic medication.
- Class 3 (12% of facilities): Only offer some treatments, which may include psychotropic medication.

You can either let interpretability guide your choice of how many classes to include in your analysis, or use model fit measures like *AIC* and *BIC*, which are printed in the output and can be obtained from the model using:

```
m$aic
m$bic
```

The lower these are, the better the model fit.

If you like, you can add a covariate to your latent class analysis, which allows you to simultaneously find classes and study their relationship with the covariate. Let's add the variable PAYASST (which says whether a facility offers treatment at no charge or minimal payment to clients who cannot afford to pay) to our data, and then use that as a covariate.

```
# Add PAYASST variable to data, then change negative values
# to NA's:
treatments$PAYASST <- nmhss$PAYASST
treatments$PAYASST[treatments$PAYASST < 0] <- NA

# Run LCA with covariate:
m <- poLCA(cbind(TREATPSYCHOTHRPY, TREATFAMTHRPY,
                 TREATGRPTHRPY, TREATCOGTHRPY,
                 TREATPSYCHOMED) ~ PAYASST,
           data = treatments, nclass = 3,
           maxiter = 2500, nrep = 5)
```

My output from this model includes the following tables:

```
===========================================================
Fit for 3 latent classes:
===========================================================
2 / 1
            Coefficient  Std. error  t value  Pr(>|t|)
(Intercept)     0.10616     0.18197    0.583     0.570
PAYASST         0.43302     0.11864    3.650     0.003
===========================================================
3 / 1
            Coefficient  Std. error  t value  Pr(>|t|)
(Intercept)     1.88482     0.20605    9.147         0
PAYASST         0.59124     0.10925    5.412         0
===========================================================
```

The interpretation is that both class 2 and class 3 differ significantly from class 1 (p-values in the Pr(>|t|) column are low), with the positive coefficients for PAYASST telling us that class 2 and 3 facilities are more likely to offer pay assistance than class 1 facilities.

~

Exercise 10.2. The cheating dataset from poLCA contains students' answers to four questions about cheating, along with their grade point averages (GPA). Perform a latent class analysis using GPA as a covariate. What classes do you find? Does having a high GPA increase the probability of belonging to either class?

10.2 Confirmatory factor analysis

Unlike exploratory factor analysis, which is a technique for investigating relationships between observable variables and latent variables, *confirmatory factor analysis* (CFA), is used to test hypotheses about the structure of such relationships. It is usually used to test a hypothesis arising from theory. The number of latent variables and the observable variables used to measure each latent variable is therefore specified in advance, as are hypotheses about which variables are related. Rotation is not needed.

In the remainder of the chapter, we'll use functions from the lavaan package, so let's begin by installing that:

```
install.packages("lavaan")
```

10.2.1   Running a confirmatory factor analysis

A truly classic dataset for a CFA is the Holzinger and Swineford (1939) data, which contains test scores from nine psychological tests administered to 301 seventh- and eighth-grade children. A subset of this data is available in the lavaan package:

```
library(lavaan)
?HolzingerSwineford1939
View(HolzingerSwineford1939)
```

Let's say that it's been hypothesised that there are three latent variables, each of which is associated with three observed variables, as follows:

- **Visual factor** ξ_1 associated with x_1, x_2, and x_3,
- **Textual factor** ξ_2 associated with x_4, x_5, and x_6,
- **Speed factor** ξ_3 associated with x_7, x_8, and x_9.

Letting δ_i denote error terms, we can write this as a series of regression models, where the terms involving latent variables hypothesised not to be associated with an observed variable are fixed at 0:

$$x_1 = \lambda_{1,1}\xi_1 + 0 \cdot \xi_2 + 0 \cdot \xi_3 + \delta_1$$
$$x_2 = \lambda_{2,1}\xi_1 + 0 \cdot \xi_2 + 0 \cdot \xi_3 + \delta_2$$
$$x_3 = \lambda_{3,1}\xi_1 + 0 \cdot \xi_2 + 0 \cdot \xi_3 + \delta_3$$
$$x_4 = 0 \cdot \xi_1 + \lambda_{4,2}\xi_2 + 0 \cdot \xi_3 + \delta_4$$
$$\vdots$$
$$x_9 = 0 \cdot \xi_1 + 0 \cdot \xi_2 + \lambda_{9,3}\xi_3 + \delta_9$$

Such a series of equations can be difficult to take in, and factor analysis models are therefore often visualised in *path diagrams*, in which latent variables are drawn as circles and observable variables are drawn as rectangles. Single-headed arrows show the direction of an association. Double-headed arrows indicate that two variables are correlated. The path diagram for the Holzinger and Swineford model is shown in Figure 10.1.

In some cases, the error terms δ_1 are included in the path diagram. We omit them here and in other path diagrams, but we always implicitly assume that observable variables have error terms associated with them.

To run a CFA using lavaan, we must first specify the model equations in a format that lavaan recognises. This is similar to, but different from, how regression formulas are specified, e.g., in lm. To say that the variables x1, x2, and x3 are associated with a latent variable called Visual, we write "Visual =~ x1 + x2 + x3". Note that the latent variable is on the

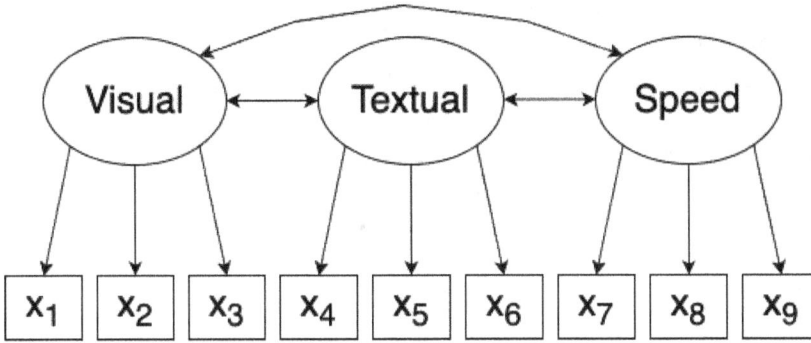

FIGURE 10.1 Path diagram for the Holzinger and Swineford model.

left-hand side, even though it is the explanatory variable in our regression equations. Also note that we use =~ to indicate association between latent and observable variables.

In our example, we write the association between the latent and observable variables involved in a single string, with linebreaks:

```
model <- "Visual   =~ x1 + x2 + x3
          Textual =~ x4 + x5 + x6
          Speed   =~ x7 + x8 + x9"
```

We can then use the cfa function to fit the model and summary to view the results:

```
m <- cfa(model, data = HolzingerSwineford1939)
summary(m)
```

In order to estimate the coefficients $\lambda_{1,1}, \ldots, \lambda_{9,3}$, we need to set the scale of ξ_1, ξ_2, and ξ_3 in some way. The default method in lavaan is to use the so-called marker variable method, in which one $\lambda_{i,j}$ is fixed to 1 for each ξ_j. This is why the estimated $\lambda_{i,j}$ is 1 for the first observed variable associated with each latent variable in the output for the fitted model:

```
Latent Variables:
                   Estimate  Std.Err  z-value  P(>|z|)
  Visual =~
    x1                1.000
    x2                0.554    0.100    5.554    0.000
    x3                0.729    0.109    6.685    0.000
  Textual =~
    x4                1.000
    x5                1.113    0.065   17.014    0.000
    x6                0.926    0.055   16.703    0.000
  Speed =~
    x7                1.000
    x8                1.180    0.165    7.152    0.000
    x9                1.082    0.151    7.155    0.000
```

The p-values in the rightmost column are for tests of the null hypothesis that $\lambda_{i,j} = 0$, i.e., that there is no association between x_i and ξ_j.

The downside to using the marker variable method is that it makes it impossible to test hypotheses about the fixed $\lambda_{i,j}$ (which is why the p-value column in the output is empty for these coefficients). Another common approach is therefore to instead fix the variance of the latent variables to 1, allowing the factor loading of all variables to vary. To use this approach, we add `std.lv = TRUE` to `cfa`:

```
m <- cfa(model, data = HolzingerSwineford1939,
         std.lv = TRUE)
summary(m)
```

10.2.2   Plotting path diagrams

To plot the path diagram of a factor model, we can use the `semPaths` function from the `semPlot` package.

We install the package and plot the path diagram for our model:

```
install.packages("semPlot")
library(semPlot)
semPaths(m)
```

You'll note that there are doubled-arrowed arrows pointing from the latent variables to themselves, and likewise for the observed variables. These are used to denote variances (i.e., the error terms δ_i in the case of observed variables). To remove them, we can add `exoVar = FALSE` (for the latent variables) and `residuals = FALSE` (for the observed variables):

```
semPaths(m, exoVar = FALSE, residuals = FALSE)
```

Fixed parameters are marked using dashed lines. To draw them as filled lines instead, add `fixedStyle = 1`:

```
semPaths(m, fixedStyle = 1)
```

10.2.3   Assessing model fit

An important part of CFA is evaluating how well the model fits the data. We can obtain additional measures of goodness-of-fit by adding `fit.measures = TRUE` when we run `summary` for our model:

```
summary(m, fit.measures = TRUE)
```

Some commonly used measures of goodness-of-fit are:

- **Chi-squared (χ^2) test** (`Model Test User Model`): measures the difference between the covariance structure specified by the model and the empirical covariance structure. A low p-value indicates that these differ, i.e., that the model fails to capture the actual dependence structure. This was traditionally used in CFA but suffers from problems. For small sample sizes, the χ^2 test is sensitive to normality of the δ_i and can have low power. For large sample sizes, it will in contrast virtually always yield a low p-value (because even if your model is close to the truth, it is probably not *exactly* true). For this reason, many practitioners now avoid using the χ^2 test.

- **Comparative Fit Index (CFI):** ranges from 0 to 1, with values close to 1 being better. A value of at least 0.9 is seen as an indicator of a good fit.
- **Tucker-Lewis Index (TLI):** also known as the non-normed fit index (NNFI), (usually) ranges from 0 to 1, with values close to 1 being better. A value of at least 0.95 is seen as an indicator of a good fit.
- **Root Mean Square Error of Approximation (RMSEA):** ranges from 0 to 1, with values close to 0 being better. A value below 0.06 is seen as an indicator of a good fit.
- **Standardized Root Mean Square Residual (SRMR):** ranges from 0 to 1, with values close to 0 being better. A value below 0.08 is seen as an indicator of a good fit.

The rules of thumb for when these measures indicate a good fit are taken from Hu & Bentler (1999).

For our model, the CFI is 0.931 and the SRMR 0.065, indicating a good fit, while a TLI of 0.896 and an RMSEA of 0.092 indicate some problems with the fit. The results are inconclusive. This is fairly common for factor analysis models when the sample size ranges in the hundreds, as it does here. Larger sample sizes tend to lead to more stable results.

~

Exercise 10.3. Consider the `attitude` data that we studied in Section 10.1. Assume that theory stipulates that there are two underlying latent variables associated with the observed variables. The first, "career opportunity", is hypothesised to be associated with `advance`, `learning`, and `raises`. The second, "employee treatment", is associated with `complaints`, `rating`, `raises`, `learning`, and `privileges`. Neither latent variable is associated with `critical`.

Evaluate this hypothesised dependence structure by checking whether a confirmatory factor analysis yields a good fit. Plot the associated path diagram.

10.3   Structural equation modelling

Structural equation modelling (SEM), can be viewed as an extension of confirmatory factor analysis, in which we also model causal relationships between latent variables. This allows us to test relationships between different abstract concepts.

In a SEM, we distinguish between two types of latent variables:

- **Endogenous variables:** latent variables that are (partially) determined or explained by other latent variables in the model. The unexplained part is represented by an error term. Usually denoted η.
- **Exogenous variables:** latent variables that are not determined or explained by the other latent variables in the model. Usually denoted ξ (in line with the notation we used for CFA).

10.3.1   Fitting a SEM

Let's look at an example. We'll use the `PoliticalDemocracy` dataset from `lavaan`, known from the classic SEM textbook by Bollen (1989):

```
library(lavaan)
?PoliticalDemocracy
View(PoliticalDemocracy)
```

The dataset concerns industrialisation and democracy in 75 developing countries in the 1960s. From subject-matter theory, there are three latent variables:

- **Industrialisation in 1960** ξ_1: associated with x_1, x_2, and x_3,
- **Democracy in 1960** η_1: affected by ξ_1 and associated with y_1, y_2, y_3, and y_4,
- **Democracy in 1965** η_2: affected by ξ_1 and η_1, and associated with y_5, y_6, y_7, and y_8.

ξ_1 is exogenous, whereas η_1 and η_2 are endogenous. The theory is that industrialisation drives democracy in developing countries, i.e., that there is a causal relationship.

In addition to the above, it is believed that some of the observed variables are correlated, for instance y_1 (expert ratings of the freedom of the press in 1960) and y_5 (expert ratings of the freedom of the press in 1965). The model is summarised in Figure 10.2.

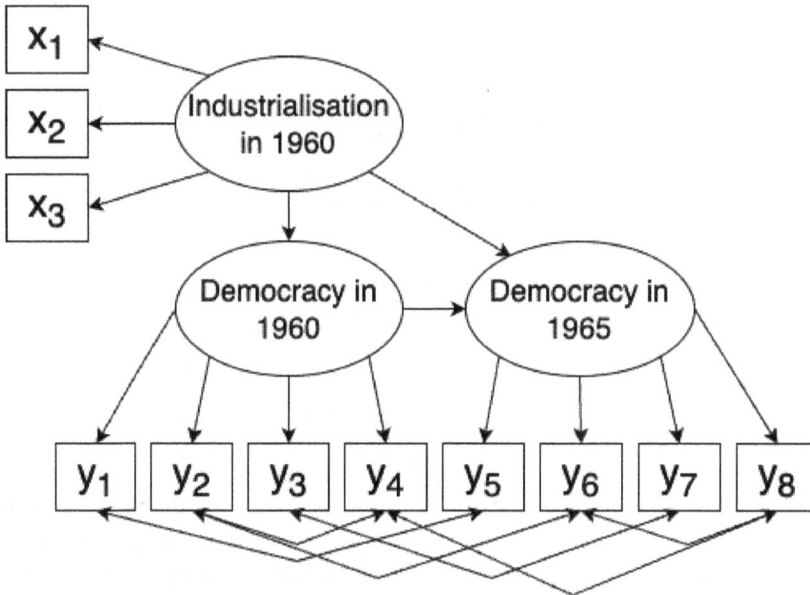

FIGURE 10.2 Structural equation model for the industrialisation and democracy model.

To specify the model for use with `lavaan`, we must specify three things. First, measurements models that describe the hypothesised relationships between latent variables and observed variables. These are written in the same way as for CFA, e.g., `"Ind1960 =~ x1 + x2 + x3"`. Next, we need to specify latent variable regressions, describing relations between exogeneous and endogenous variables. These are written like regression formulas, e.g., `"Dem1965 ~ Ind1960 + Dem1960"`. Finally, we specify the hypothesised correlations between observed variables, written with double tilde symbols: `"y1 ~~ y5"`. The whole model then becomes:

```
model <- "
  # Measurement model:
  Ind1960 =~ x1 + x2 + x3
  Dem1960 =~ y1 + y2 + y3 + y4
  Dem1965 =~ y5 + y6 + y7 + y8

  # Latent variable regressions:
  Dem1960 ~ Ind1960
  Dem1965 ~ Ind1960 + Dem1960

  # Residual correlations:
  y1 ~~ y5
  y2 ~~ y4
  y2 ~~ y6
  y3 ~~ y7
  y4 ~~ y8
  y6 ~~ y8"
```

To fit the SEM specified above, we use `sem`:

```
m <- sem(model, data = PoliticalDemocracy)
summary(m)
```

The associations between the latent variables and the observed variables are all significant, as is the association between industrialisation in 1960 and 1965. The relations between the latent variables are also significant. However, some of the stipulated correlations between observed variables are not.

The output is very similar to that from `cfa`. In fact, `cfa` is a wrapper-function for `sem`, and the two have many arguments in common. We could for instance add `std.lv = TRUE` to the `sem` call to fix the variances of the latent variables.

```
m <- sem(model, data = PoliticalDemocracy,
         std.lv = TRUE)
summary(m)
```

Note that this changes not only the estimated coefficients, but also the p-values! The associations between the latent variables and the observed variables are still significant, as is the association between industrialisation in 1960 and 1965. The relations between democracy in 1965 and the other latent variables are however not significant.

The results from the hypotheses tests in a SEM can change if we change what parameters we keep fixed. This is unlikely to happen for large sample sizes, but when we have a complex model and a small sample size, as in this example, it is not uncommon. We should consider it a sign that our sample size is too small to test this model.

The p-values presented in the table are based on a normality assumption. If we like, we can use the bootstrap to compute p-values instead. To run the model again using the bootstrap with 1,000 bootstrap replicates to compute p-values, we do the following (which may take a minute to run):

```
m <- sem(model, data = PoliticalDemocracy,
        se = "bootstrap", bootstrap = 1000)
summary(m)
```

10.3.2   Assessing and plotting the model

Model fit is assessed using the same measures as for CFA (see Section 10.2.3):

```
summary(m, fit.measures = TRUE)
```

In this case, the χ^2 test and the CFI, TLI, RMSEA, and SRMR measures all indicate that the model is a good fit.

Just as for CFA, we can use `semPaths` to plot the path diagram for the model:

```
library(semPlot)
semPaths(m, exoVar = FALSE, residuals = FALSE)
```

$\sim$

Exercise 10.4. Consider the `HolzingerSwineford1939` data that we studied in Section 10.2. Assume that theory stipulates that there is an additional exogenous latent variable, "Mental maturity", measured by the observed variables `ageyr` and `sex`, that causes the endogenous variables "visual", "textual", and "speed".

Evaluate this hypothesised dependence structure by running a SEM. Plot the associated path diagram.

Exercise 10.5. Download the `treegrowth.csv` dataset from the book's web page[4]. It contains simulated data inspired by Liu *et al.* (2016): five soil measurements, $x_1, \ldots, x_5$, and four measurements of tree growth, $y_1, \ldots, y_5$. It is hypothesised that the soil measurements relate to a latent variable describing environmental conditions, which affects a latent variable describing plant traits. The latter is associated with the tree growth measurements.

Evaluate this hypothesised dependence structure by running a SEM. Plot the associated path diagram.

10.4   Mediation and moderation in regression models

For hundreds of years, scurvy was the bane of many a sailor, killing at least two million between the 16th and 19th century (Drymon, 2008). In one of the first-ever controlled clinical trials, the Scottish physician James Lind showed that scurvy could be treated and prevented by eating citrus fruit.

But what was the mechanism involved in this?

---
[4]http://www.modernstatisticswithr.com/data.zip

In some cases, there is causal relationship between an explanatory variable and a response variable, but not a *direct* causal relationship. Instead, the explanatory variable affects a third variable, which in turn has an effect on the response variable.

Citrus and scurvy is an example of this. Eating citrus does not directly prevent scurvy, but it will raise your vitamin C levels, which in turn prevents scurvy. In this case, we say that vitamin C level is a *mediator variable* that *mediates* the effect of citrus-eating on scurvy.

An explanatory variable can have a *direct effect* (which does not act through the mediator) on the response variable, an *indirect effect* (which acts through the mediator), or both. The sum of the direct effect and the indirect effect is called the *total effect* of the explanatory variable. The purpose of dividing the total effect into these two parts is to understand the mechanism behind a known causal relationship, e.g., to understand *how* eating citrus prevents scurvy.

Mediation analysis is concerned with finding if a third variable (e.g., vitamin C levels) is involved in the mechanism through which a variable affects another. There should always be strong theoretical support for choosing a particular mediator variable; the analysis is used to confirm the theory, not to generate it. In addition to this, the following assumptions must be met:

- **Correctly specified model:** the functional relationships between the variables are correct and don't interact with each other.
- **Independent and homoscedastic errors:** the error terms should not be correlated with each other or the explanatory variables and should all have the same variance.
- **No omitted variables:** the model accounts for all variables influencing the response variable. This is almost never true in practice – all models are wrong, but some are useful, as the saying goes[5]. For a mediation model to be useful, it is important that the effects of the omitted variables are negligible. In the case of vitamin C and scurvy, for instance, vitamin C uptake is affected by certain medical conditions, and if an effect like this is large enough and the conditions common enough, leaving it out from the model can skew the results substantially.

10.4.1 Fitting a mediation model

There are several packages that can be used for mediation analyses in R. We'll use `lavaan` and the SEM framework – the formula language of which can also be used when all variables in the model have been observed. Our examples will use the following dataset from the `psych` package:

```
library(psych)
?Tal_Or
View(Tal_Or)
```

The data comes from an experiment in which people's reactions to a news story were measured, along with their presumptions about the influence of the media, under two experimental conditions. The effect of the experimental condition is thought to be mediated by the presumed media influence. We visualise this using a diagram, as in Figure 10.3.

---

[5]This aphorism is usually attributed to George Box, one of the greats of applied statistics. His 2013 memoir *An Accidental Statistician* is a must-read for those interested in the history of statistics.

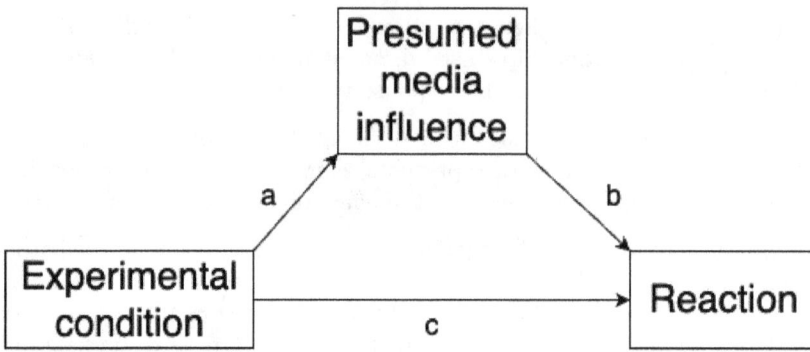

FIGURE 10.3 Mediation model for the news story reaction model.

Let x denote the experimental condition, m denote the presumed media influence, and y denote the reaction. Our mediation model is then as follows:

$$m = a \cdot x + \epsilon_1$$
$$y = b \cdot m + c \cdot x + \epsilon_2$$

The direct effect of x is c, the indirect effect is $a \cdot b$, and the total effect is $a \cdot b + c$.

First, we specify the model in a string. Then, we fit the model using sem, computing p-values using the bootstrap, and print the results using summary:

```
library(lavaan)
model <- "# Model:
          pmi ~ a*cond
          reaction ~ b*pmi + c*cond

          # Indirect and total effects:
          indirect_effect := a*b
          total_effect := indirect_effect + c"
m <- sem(model,
         data = Tal_Or,
         se = "bootstrap",
         bootstrap = 1000)
summary(m)
```

The most interesting part of the output is the first and last tables, showing estimates and p-values for the indirect and total effects (the output below has been cropped):

```
Regressions:
                 Estimate  Std.Err  z-value  P(>|z|)
  pmi ~
    cond      (a)  0.477    0.237    2.010    0.044
  reaction ~
    pmi       (b)  0.506    0.078    6.471    0.000
    cond      (c)  0.254    0.260    0.977    0.329
```

```
Defined Parameters:
                 Estimate  Std.Err  z-value  P(>|z|)
  indirect_effect   0.241    0.133    1.816    0.069
  total_effect      0.496    0.268    1.853    0.064
```

Changing from experimental condition 0 to 1 increases the presumed media influence by 0.477 units. This effect is significant ($p = 0.044$ in my run). A one unit increase in presumed media influence increases the reaction score by 0.506 units (also significant, $p < 0.001$). However, the indirect effect of the experimental condition is not significant ($p = 0.069$). Neither is the direct effect ($p = 0.329$) or the total effect (0.064). In conclusion, there is no statistical evidence of the experimental condition affecting the reaction score, neither directly nor indirectly.

~

Exercise 10.6. Download the `treegrowth.csv` dataset from the book's web page[6]. The variable x_1 describes calcium levels in soil, while x_4 contains pH measurements. It is thought that the effect of x_1 on plant height y_1 is mediated by x_4. Test this hypothesis.

10.4.2   Mediation with confounders

In mediation analysis, *confounders* are variables that potentially affect the mediator variable, the outcome, or both. Ignoring such variables can bias the estimates. In our example with presumed media influence, two potential confounders are gender and age. Adding them to our diagram, we get Figure 10.4.

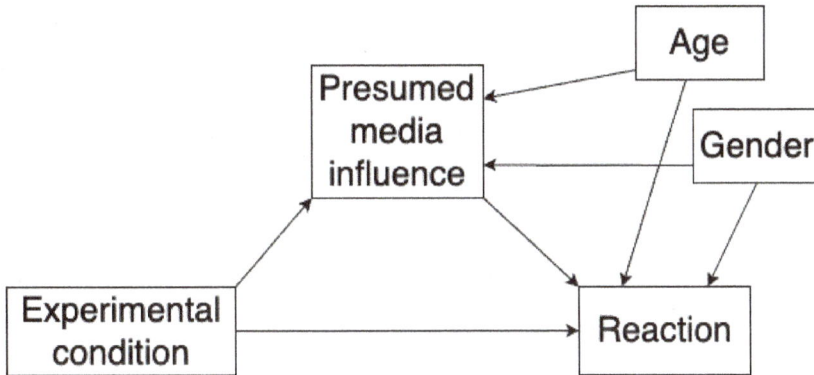

FIGURE 10.4 Mediation model with two confounders.

Letting x_1 denote the experimental condition, x_2 denote gender and x_3 denote age, our model is now:

$$m = a_1 \cdot x_1 + a_2 \cdot x_2 + a_3 \cdot x_3 + \epsilon_1$$
$$y = b \cdot m + c_1 \cdot x + c_2 \cdot x_2 + c_3 \cdot x_3 + \epsilon_2$$

The direct effect of x_1 is c_1, the indirect effect is $a_1 \cdot b$, and the total effect is $a_1 \cdot b + c$. All three effects are now adjusted for gender and age. We fit the model as follows:

---

[6]http://www.modernstatisticswithr.com/data.zip

```
model <- "# Model:
        pmi ~ a1*cond + a2*gender + a3*age
        reaction ~ b*pmi + c1*cond + c2*gender + c3*age

        # Indirect and total effects:
        indirect_effect := a1*b
        total_effect := indirect_effect + c1"
m <- sem(model,
        data = Tal_Or,
        se = "bootstrap",
        bootstrap = 1000)
summary(m)
```

10.4.3   Moderation

Mediation models often also include moderation effects. A *moderator* is a variable that modifies the strength of the relationship between two other variables, by strengthening, weakening, or even reversing the dependence. A visual representation of moderation is shown in Figure 10.5.

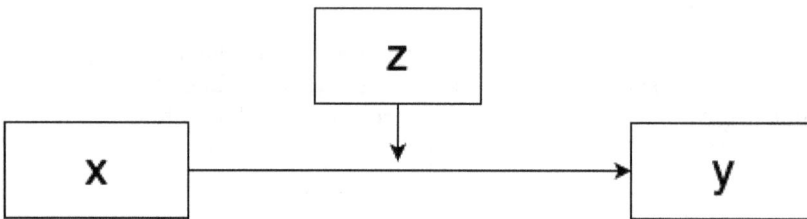

FIGURE 10.5 Model with moderation.

If x is an explanatory variable, z is a moderator, and y is the response variable, then a linear model including a moderation effect would look like this:

$$y = \beta_0 + \beta_1 x + \beta_2 z + \beta_{12} xz.$$

You probably already know how to run a moderation analysis, even if the term "moderation" is unfamiliar to you. The effect of a moderator variable is namely described by the *interaction term* β_{12}, just as in Sections 8.1.3 and 8.2.2. If β_{12} is significantly non-zero, then the moderation effect is significant. If β_1 and β_{12} have the same sign, z strengthens the dependence between x and y. If they have opposite signs, z weakens the dependence (and, if $|\beta_{12}| > |\beta_1|$, reverses it).

If both variables are numeric, it is recommended to centre them; see Section 8.2.2 for more on this.

10.4.4   Mediated moderation and moderated mediation

Mediated moderation occurs when there is moderation of both the total and indirect effects of the explanatory variable. Returning to our example with presumed media influence, we'll consider a model where a moderator variable, age, affects the paths from the explanatory variable, as shown in Figure 10.6:

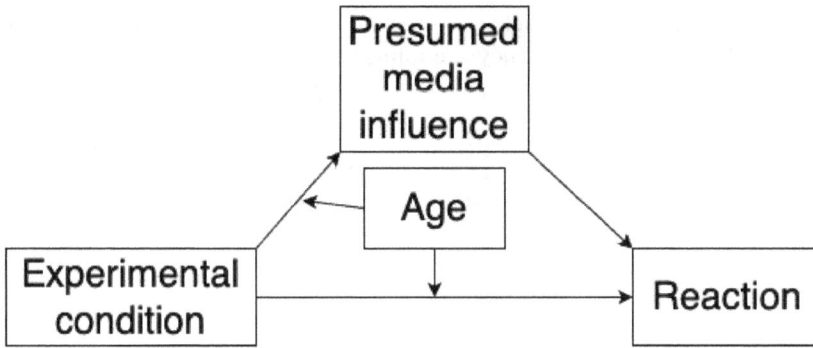

FIGURE 10.6 Model with mediated moderation.

```
model <- "# Model:
        pmi ~ a1*cond + a2*age + a12*cond:age
        reaction ~ b*pmi + c1*cond + c2*age + c13*cond:age

        # Moderated effects:
        mediated_moderation := a12*b
        total_moderated_effect := mediated_moderation + c13"
m <- sem(model,
        data = Tal_Or,
        se = "bootstrap",
        bootstrap = 1000)
summary(m)
```

Here, neither the mediated moderation nor the total moderated effect is significant.

Moderated mediation occurs when a moderator affects either the effect a of the explanatory variable on the mediator, or the effect b of the mediator on the response variable. Note that this is different from a model with confounders: confounders add new coefficients to the models, whereas a moderator adds an interaction term affecting a and/or b.

Let's say that we want to test whether age moderates the effect of the mediator on the response variable, as illustrated in Figure 10.7:

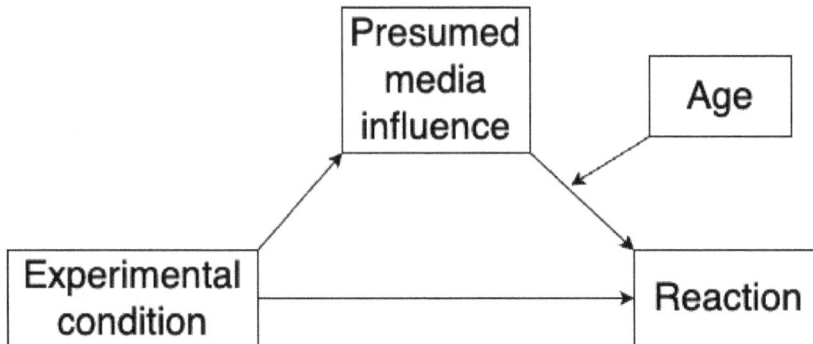

FIGURE 10.7 Model with moderated mediation.

To test this, we add age and an interaction between the moderator and age to the regression formula. First, we centre both variables, as they are numeric.

Without pipes:

```
Tal_Or$pmi <- scale(Tal_Or$pmi, scale = FALSE)
Tal_Or$age <- scale(Tal_Or$age, scale = FALSE)
```

With pipes:

```
library(dplyr)
Tal_Or |> mutate(pmi = scale(pmi, scale = FALSE),
                 age = scale(age, scale = FALSE)) -> Tal_Or
```

Next, we fit the model:

```
model <- "# Model:
          pmi ~ a1*cond
          reaction ~ b*pmi + c1*cond + c2*age + bc*pmi:age

          # Moderated effect:
          moderated_mediation := a1*bc"
m <- sem(model,
         data = Tal_Or,
         se = "bootstrap",
         bootstrap = 1000)
summary(m)
```

In this example, the moderated mediation effect is not significant.

$$\sim$$

Exercise 10.7. Download the `treegrowth.csv` dataset from the book's web page[7]. The variable x_1 describes calcium levels in soil, x_2 describes nitrate levels, and x_4 contains pH measurements. It is thought that the effect of x_1 on plant height y_1 is mediated by x_4, and that x_2 moderates the relations between x_1 and x_4, and x_1 and y_1. Test this hypothesis of mediated moderation.

---

[7]http://www.modernstatisticswithr.com/data.zip

11

Predictive modelling and machine learning

In predictive modelling, we fit statistical models that use historical data to make predictions about future (or unknown) outcomes. This practice is a cornerstone of modern statistics and includes methods ranging from classical parametric linear regression to black-box machine learning models.

After reading this chapter, you will be able to use R to:

- Fit predictive models for regression and classification,
- Evaluate predictive models,
- Use cross-validation and the bootstrap for out-of-sample evaluations,
- Handle imbalanced classes in classification problems,
- Fit regularised (and possibly also generalised) linear models, e.g., using the lasso,
- Fit a number of machine learning models, including k-nearest neighbours (kNN), decision trees, random forests, and boosted trees, and
- Make forecasts based on time series data.

11.1   Evaluating predictive models

In many ways, modern predictive modelling differs from the more traditional inference problems that we studied in the previous chapter. The goal of predictive modelling is (usually) not to test whether some variable affects another or to study causal relationships. Instead, our only goal is to make good predictions. It is little surprise then that the tools we use to evaluate predictive models differ from those used to evaluate models used for other purposes, like hypothesis testing. In this section, we will have a look at how to evaluate predictive models.

The terminology used in predictive modelling differs a little from that used in traditional statistics. For instance, explanatory variables are often called *features* or *predictors*, and predictive modelling is often referred to as *supervised learning*. We will stick with the terms used in Chapter 7, to keep the terminology consistent within the book.

Predictive models can be divided into two categories:

- *Regression*, where we want to make predictions for a numeric variable,
- *Classification*, where we want to make predictions for a categorical variable.

There are many similarities between these two, but we need to use different measures when evaluating their predictive performance. Let's start with models for numeric predictions, i.e., regression models.

11.1.1   Evaluating regression models

Let's return to the `mtcars` data that we studied in Section 8.1. There, we fitted a linear model to explain the fuel consumption of cars:

```
m <- lm(mpg ~ ., data = mtcars)
```

(Recall that the formula `mpg ~ .` means that all variables in the dataset, except `mpg`, are used as explanatory variables in the model.)

A number of measures of how well the model fits the data have been proposed. Without going into details (it will soon be apparent why), we can mention examples like the coefficient of determination R^2, and information criteria like AIC and BIC. All of these are straightforward to compute for our model:

```
summary(m)$r.squared      # R^2
summary(m)$adj.r.squared  # Adjusted R^2
AIC(m)                    # AIC
BIC(m)                    # BIC
```

R^2 is a popular tool for assessing model fit, with values close to 1 indicating a good fit and values close to 0 indicating a poor fit (i.e., that most of the variation in the data isn't accounted for).

It is nice if our model fits the data well, but what really matters in predictive modelling is how close the predictions from the model are to the truth. We therefore need ways to measure the distance between predicted values and observed values – ways to measure the size of the average prediction error. A common measure is the root mean square error (RMSE). Given n observations $y_1, y_2, \ldots, y_n$ for which our model makes the predictions $\hat{y}_1, \ldots, \hat{y}_n$, this is defined as

$$RMSE = \sqrt{\frac{\sum_{i=1}^{n}(\hat{y}_i - y_i)^2}{n}},$$

that is, as the named implies, the square root of the mean of the squared errors $(\hat{y}_i - y_i)^2$.

Another common measure is the mean absolute error (MAE):

$$MAE = \frac{\sum_{i=1}^{n}|\hat{y}_i - y_i|}{n}.$$

Let's compare the predicted values $\hat{y}_i$ to the observed values y_i for our `mtcars` model `m`:

```
rmse <- sqrt(mean((predict(m) - mtcars$mpg)^2))
mae <- mean(abs(predict(m) - mtcars$mpg))
rmse; mae
```

There is a problem with this computation, and it is a big one. What we just computed was the difference between predicted values and observed values *for the sample that was used to fit the model*. This doesn't necessarily tell us anything about how well the model will fare when used to make predictions about new observations. It is, for instance, entirely possible that our model has *overfitted* to the sample, and essentially has learned the examples therein by heart, ignoring the general patterns that we were trying to model. This would lead to a small $RMSE$ and MAE, and a high R^2, but would render the model useless for predictive purposes.

All the computations that we've just done – R^2, AIC, BIC, $RMSE$, and MAE – were examples of *in-sample evaluations* of our model. There are a number of problems associated with in-sample evaluations, all of which have been known for a long time; see, e.g., Picard & Cook (1984). In general, they tend to be overly optimistic and overestimate how well the model will perform for new data. It is about time that we got rid of them for good.

A fundamental principle of predictive modelling is that the model chiefly should be judged on how well it makes predictions for new data. To evaluate its performance, we therefore need to carry out some form of *out-of-sample evaluation*, i.e., to use the model to make predictions for new data (that weren't used to fit the model). We can then compare those predictions to the actual observed values for those data, and, e.g., compute the $RMSE$ or MAE to measure the size of the average prediction error. Out-of-sample evaluations, when done right, are less overoptimistic than in-sample evaluations, and they are also better in the sense that they actually measure the right thing.

$\sim$

Exercise 11.1. To see that a high R^2 and low p-values say very little about the predictive performance of a model, consider the following dataset with 30 randomly generated observations of four variables:

```
exdata <- data.frame(x1 = c(0.87, -1.03, 0.02, -0.25, -1.09, 0.74,
          0.09, -1.64, -0.32, -0.33, 1.40, 0.29, -0.71, 1.36, 0.64,
          -0.78, -0.58, 0.67, -0.90, -1.52, -0.11, -0.65, 0.04,
          -0.72, 1.71, -1.58, -1.76, 2.10, 0.81, -0.30),
          x2 = c(1.38, 0.14, 1.46, 0.27, -1.02, -1.94, 0.12, -0.64,
          0.64, -0.39, 0.28, 0.50, -1.29, 0.52, 0.28, 0.23, 0.05,
          3.10, 0.84, -0.66, -1.35, -0.06, -0.66, 0.40, -0.23,
          -0.97, -0.78, 0.38, 0.49, 0.21),
          x3 = c(1, 0, 0, 0, 0, 1, 0, 0, 0, 1, 0, 0, 0, 0, 1, 1, 0,
          1, 0, 0, 0, 0, 1, 0, 0, 0, 1, 1, 0, 1),
          y = c(3.47, -0.80, 4.57, 0.16, -1.77, -6.84, 1.28, -0.52,
          1.00, -2.50, -1.99, 1.13, -4.26, 1.16, -0.69, 0.89, -1.01,
          7.56, 2.33, 0.36, -1.11, -0.53, -1.44, -0.43, 0.69, -2.30,
          -3.55, 0.99, -0.50, -1.67))
```

1. The true relationship between the variables, used to generate the y variables, is $y = 2x_1 - x_2 + x_3 \cdot x_2$. Plot the y values in the data against this expected value. Does a linear model seem appropriate?
2. Fit a linear regression model with x1, x2, and x3 as explanatory variables (without any interactions) using the first 20 observations of the data. Do the p-values and R^2 indicate a good fit?
3. Make predictions for the remaining 10 observations. Are the predictions accurate?
4. A common (mal)practice is to remove explanatory variables that aren't significant from a linear model (see Section 8.2.7 for some comments on this). Remove any variables from the regression model with a p-value above 0.05, and refit the model using the first 20 observations. Do the p-values and R^2 indicate a good fit? Do the predictions for the remaining 10 observations improve?

5. Finally, fit a model with x1, x2, and x3*x2 as explanatory variables (i.e., a correctly specified model) to the first 20 observations. Do the predictions for the remaining 10 observations improve?

11.1.2   Test-training splits

In some cases, our data is naturally separated into two sets, one of which can be used to fit a model and the other to evaluate it. A common example of this is when data has been collected during two distinct time periods, and the older data is used to fit a model that is evaluated on the newer data, to see if historical data can be used to predict the future.

In most cases though, we don't have that luxury. A popular alternative is to artificially create two sets by randomly withdrawing a part of the data, 10% or 20%, say, which can be used for evaluation. In machine learning lingo, model fitting is known as *training* and model evaluation as *testing*. The set used for training (fitting) the model is therefore often referred to as the *training data*, and the set used for testing (evaluating) the model is known as the *test data*.

Let's try this out with the mtcars data. We'll use 80% of the data for fitting our model and 20% for evaluating it.

```
# Set the sizes of the test and training samples.
# We use 20% of the data for testing:
n <- nrow(mtcars)
ntest <- round(0.2*n)
ntrain <- n - ntest

# Split the data into two sets:
train_rows <- sample(1:n, ntrain)
mtcars_train <- mtcars[train_rows,]
mtcars_test <- mtcars[-train_rows,]
```

In this case, our training set consists of 26 observations and our test set of 6 observations. Let's fit the model using the training set and use the test set for evaluation:

```
# Fit model to training set:
m <- lm(mpg ~ ., data = mtcars_train)

# Evaluate on test set:
rmse <- sqrt(mean((predict(m, mtcars_test) - mtcars_test$mpg)^2))
mae <- mean(abs(predict(m, mtcars_test) - mtcars_test$mpg))
rmse; mae
```

Because of the small sample sizes here, the results can vary a lot if you rerun the two code chunks above several times (try it!). When I ran them 10 times, the *RMSE* varied between 1.8 and 7.6 – quite a difference on the scale of mpg! This problem is usually not as pronounced if you have larger sample sizes, but even for fairly large datasets, there can be a lot of variability depending on how the data happens to be split. It is not uncommon to get a "lucky" or "unlucky" test set that either overestimates or underestimates the model's performance.

In general, I'd therefore recommend that you only use test-training splits of your data as a last resort (and only use them with sample sizes of 10,000 or more). Better tools are available in the form of the bootstrap and its darling cousin, cross-validation.

11.1.3 Leave-one-out cross-validation and `caret`

The idea behind cross-validation is similar to that behind test-training splitting of the data. We partition the data into several sets and use one of them for evaluation. The key difference is that in a cross-validation we partition the data into more than two sets and use all of them (one-by-one) for evaluation.

To begin with, we split the data into k sets, where k is equal to or less than the number of observations n. We then put the first set aside, to use for evaluation, and fit the model to the remaining $k-1$ sets. The model predictions are then evaluated on the first set. Next, we put the first set back among the others and remove the second set to use that for evaluation. And so on. This means that we fit k models to k different (albeit similar) training sets, and evaluate them on k test sets (none of which are used for fitting the model that is evaluated on them).

The most basic form of cross-validation is leave-one-out cross-validation (LOOCV), where $k = n$ so that each observation is its own set. For each observation, we fit a model using all other observations, and then we compare the prediction of that model to the actual value of the observation. We can do this using a `for` loop (Section 6.4.1) as follows:

```
# Leave-one-out cross-validation:
pred <- vector("numeric", nrow(mtcars))
for(i in 1:nrow(mtcars))
{
    # Fit model to all observations except observation i:
    m <- lm(mpg ~ ., data = mtcars[-i,])

    # Make a prediction for observation i:
    pred[i] <- predict(m, mtcars[i,])
}

# Evaluate predictions:
rmse <- sqrt(mean((pred - mtcars$mpg)^2))
mae <- mean(abs(pred - mtcars$mpg))
rmse; mae
```

We will use cross-validation a lot, and so it is nice not to have to write a lot of code each time we want to do it. To that end, we'll install the `caret` package, which not only lets us do cross-validation but also acts as a wrapper for a large number of packages for predictive models. That means that we won't have to learn a ton of functions to be able to fit different types of models. Instead, we just have to learn a few functions from `caret`. Let's install the package and some of the packages it needs to function fully:

```
install.packages("caret", dependencies = TRUE)
```

Now, let's see how we can use `caret` to fit a linear regression model and evaluate it using cross-validation. The two main functions used for this are `trainControl`, which we use to say that we want to perform a LOOCV (`method = "LOOCV"`) and `train`, where we state the model formula and specify that we want to use `lm` for fitting the model:

```
library(caret)
tc <- trainControl(method = "LOOCV")
m <- train(mpg ~ .,
           data = mtcars,
           method = "lm",
           trControl = tc)
```

`train` has now done several things in parallel. First of all, it has fitted a linear model to the entire dataset. To see the results of the linear model we can use `summary`, just as if we'd fitted it with `lm`:

```
summary(m)
```

Many, but not all, functions that we would apply to an object fitted using `lm` still work fine with a linear model fitted using `train`, including `predict`. Others, like `coef` and `confint` no longer work (or work differently) – but that is not that big a problem. We only use `train` when we are fitting a linear regression model with the intent of using it for prediction – and in such cases, we are typically not interested in the values of the model coefficients or their confidence intervals. If we need them, we can always refit the model using `lm`.

What makes `train` great is that `m` also contains information about the predictive performance of the model, computed, in this case, using LOOCV:

```
# Print a summary of the cross-validation:
m

# Extract the measures:
m$results
```

~

Exercise 11.2. Download the `estates.xlsx` data from the book's web page[1]. It describes the selling prices (in thousands of SEK) of houses in and near Uppsala, Sweden, along with a number of variables describing the location, size, and standard of the house.

Fit a linear regression model to the data, with `selling_price` as the response variable and the remaining variables as explanatory variables. Perform an out-of-sample evaluation of your model. What are the $RMSE$ and MAE? Do the prediction errors seem acceptable?

11.1.4   *k*-fold cross-validation

LOOCV is a very good way of performing out-of-sample evaluation of your model. It can however become overoptimistic if you have "twinned" or duplicated data in your sample, i.e., observations that are identical or nearly identical (in which case the model for all intents and purposes already has "seen" the observation for which it is making a prediction). It can

---

[1]http://www.modernstatisticswithr.com/data.zip

also be quite slow if you have a large dataset, as you need to fit n different models, each using a lot of data.

A much faster option is k-fold cross-validation, which is the name for cross-validation where k is lower than n – usually much lower, with $k = 10$ being a common choice. To run a 10-fold cross-validation with `caret`, we change the arguments of `trainControl`, and then we run `train` exactly as before:

```
tc <- trainControl(method = "cv" , number = 10)
m <- train(mpg ~ .,
           data = mtcars,
           method = "lm",
           trControl = tc)

m
```

As with test-training splitting, the results from a k-fold cross-validation will vary each time it is run (unless $k = n$). To reduce the variance of the estimates of the prediction error, we can repeat the cross-validation procedure multiple times, and average the errors from all runs. This is known as a repeated k-fold cross-validation. To run one hundred 10-fold cross-validations, we change the settings in `trainControl` as follows:

```
tc <- trainControl(method = "repeatedcv",
                   number = 10, repeats = 100)
m <- train(mpg ~ .,
           data = mtcars,
           method = "lm",
           trControl = tc)

m
```

Repeated k-fold cross-validations are more computer-intensive than simple k-fold cross-validations, but in return the estimates of the average prediction error are much more stable.

Which type of cross-validation to use for different problems remains an open question. Several studies (e.g., Zhang & Yang (2015), and the references therein) indicate that in most settings larger k is better (with LOOCV being the best), but there are exceptions to this rule – e.g., when you have a lot of twinned data. This is in contrast to an older belief that a high k leads to estimates with high variances, tracing its roots back to a largely unsubstantiated claim in Efron (1983), which you still can see repeated in many books. When n is very large, the difference between different k is typically negligible.

A downside to k-fold cross-validation is that the model is fitted using $\frac{k-1}{k}n$ observations instead of n. If n is small, this can lead to models that are noticeably worse than the model fitted using n observations. LOOCV is the best choice in such cases, as it uses $n - 1$ observations (so, almost n) when fitting the models. On the other hand, there is also the computational aspect – LOOCV is simply not computationally feasible for large datasets with numerically complex models. In summary, my recommendation is to use LOOCV when possible, particularly for smaller datasets, and to use repeated 10-fold cross-validation otherwise. For very large datasets, or toy examples, you can resort to a simple 10-fold cross-validation (which still is a better option than test-training splitting).

~

Exercise 11.3. Return to the `estates.xlsx` data from the previous exercise. Refit your linear model, but this time:

  1. Use 10-fold cross-validation for the evaluation. Run it several times and check the MAE. How much does the MAE vary between runs?

  2. Run repeated 10-fold cross-validations a few times. How much does the MAE vary between runs?

11.1.5   Twinned observations

If you want to use LOOCV but are concerned about twinned observations, you can use `duplicated`, which returns a `logical` vector showing which rows are duplicates of previous rows. However, it will not find near-duplicates. Let's try it on the `diamonds` data from `ggplot2`:

```
library(ggplot2)
# Are there twinned observations?
duplicated(diamonds)

# Count the number of duplicates:
sum(duplicated(diamonds))

# Show the duplicates:
diamonds[which(duplicated(diamonds)),]
```

If you plan on using LOOCV, you may want to remove duplicates. We saw how to do this in Section 5.8.2:

With `data.table`:

With `dplyr`:

```
library(data.table)
diamonds <- as.data.table(diamonds)
unique(diamonds)
```

```
library(dplyr)
diamonds |> distinct()
```

11.1.6   Bootstrapping

An alternative to cross-validation is to draw bootstrap samples, some of which are used to fit models, and some to evaluate them. This has the benefit that the models are fitted to n observations instead of $\frac{k-1}{k}n$ observations. This is in fact the default method in `trainControl`. To use it for our `mtcars` model, with 999 bootstrap samples, we run the following:

```
library(caret)
tc <- trainControl(method = "boot",
                   number = 999)
m <- train(mpg ~ .,
```

```
            data = mtcars,
            method = "lm",
            trControl = tc)
```

```
m
m$results
```

~

Exercise 11.4. Return to the `estates.xlsx` data from the previous exercise. Refit your linear model, but this time use the bootstrap to evaluate the model. Run it several times and check the MAE. How much does the MAE vary between runs?

11.1.7 Evaluating classification models

Classification models, or classifiers, differ from regression models in that they aim to predict which *class* (category) an observation belongs to, rather than to predict a number. Because the target variable, the class, is categorical, it would make little sense to use measures like $RMSE$ and MAE to evaluate the performance of a classifier. Instead, we will use other measures that are better suited to this type of problem.

To begin with, though, we'll revisit the `wine` data that we studied in Section 8.3. It contains characteristics of wines that belong to either of two classes: white and red. Let's create the dataset:

```
# Import data about white and red wines:
white <- read.csv("https://tinyurl.com/winedata1",
                  sep = ";")
red <- read.csv("https://tinyurl.com/winedata2",
                sep = ";")

# Add a type variable:
white$type <- "white"
red$type <- "red"

# Merge the datasets:
wine <- rbind(white, red)
wine$type <- factor(wine$type)

# Check the result:
summary(wine)
```

In Section 8.3, we fitted a logistic regression model to the data using `glm`:

```
m <- glm(type ~ pH + alcohol, data = wine, family = binomial)
summary(m)
```

Logistic regression models are regression models, because they give us a numeric output: class probabilities. These probabilities, however, can be used for classification – we can for instance classify a wine as being red if the predicted probability that it is red is at least 0.5. We can therefore use logistic regression as a classifier and refer to it as such; although, we should bear in mind that it actually is more than that[2].

We can use `caret` and `train` to fit the same logistic regression model and use cross-validation or the bootstrap to evaluate it. We should supply the arguments `method = "glm"` and `family = "binomial"` to `train` to specify that we want a logistic regression model. Let's do that and run a repeated 10-fold cross-validation of the model – this takes longer to run than our `mtcars` example because the dataset is larger:

```
library(caret)
tc <- trainControl(method = "repeatedcv",
                   number = 10, repeats = 100)
m <- train(type ~ pH + alcohol,
           data = wine,
           trControl = tc,
           method = "glm",
           family = "binomial")


m
```

The summary reports two figures from the cross-validation:

- *Accuracy*: the proportion of correctly classified observations,
- *Cohen's kappa*: a measure combining the observed accuracy with the accuracy expected under random guessing (which is related to the balance between the two classes in the sample).

We mentioned a little earlier that we can use logistic regression for classification by, for instance, classifying a wine as being red if the predicted probability that it is red is at least 0.5. It is of course possible to use another threshold as well and classify wines as being red if the probability is at least 0.2, 0.3333, or 0.62. When setting this threshold, there is a tradeoff between the occurrence of what is known as *false negatives* and *false positives*. Imagine that we have two classes (white and red), and that we label one of them as negative (white) and one as positive (red). Then:

- A *false negative* is a positive (red) observation incorrectly classified as negative (white),
- A *false positive* is a negative (white) observation incorrectly classified as positive (red).

In the `wine` example, there is little difference between these types of errors. But in other examples, the distinction is an important one. Imagine for instance that we, based on some data, want to classify patients as being sick (positive) or healthy (negative). In that case, it might be much worse to get a false negative (patient won't get the treatment they need) than a false positive (patient will have to run a few more tests). For any given threshold, we can compute two measures of the frequency of these types of errors:

- *Sensitivity* or *true positive rate*: proportion of positive observations that are correctly classified as being positive,
- *Specificity* or *true negative rate*: proportion of negative observations that are correctly classified as being negative.

---

[2]Many, but not all, classifiers also output predicted class probabilities. The distinction between regression models and classifiers is blurry at best.

If we increase the threshold for at what probability a wine is classified as being red (positive), then the sensitivity will increase, but the specificity will decrease. And if we lower the threshold, the sensitivity will decrease while the specificity increases.

It would make sense to try several different thresholds, to see for which threshold we get a good compromise between sensitivity and specificity. We will use the MLeval package to visualise the result of this comparison, so let's install that:

```
install.packages("MLeval")
```

Sensitivity and specificity are usually visualised using receiver operation characteristic curves (ROC curves). We'll plot such a curve for our wine model. The function evalm from MLeval can be used to collect the data that we need from the cross-validations of a model m created using train. To use it, we need to set savePredictions = TRUE and classProbs = TRUE in trainControl:

```
tc <- trainControl(method = "repeatedcv",
                   number = 10, repeats = 100,
                   savePredictions = TRUE,
                   classProbs = TRUE)

m <- train(type ~ pH + alcohol,
           data = wine,
           trControl = tc,
           method = "glm",
           family = "binomial")

library(MLeval)
plots <- evalm(m)

# ROC:
plots$roc
```

The x-axis shows the *false positive rate* of the classifier (which is 1 minus the specificity – we'd like this to be as low as possible) and the y-axis shows the corresponding sensitivity of the classifier (we'd like this to be as high as possible). The red line shows the false positive rate and sensitivity of our classifier, with each point on the line corresponding to a different threshold. The grey line shows the performance of a classifier that is no better than random guessing – ideally, we want the red line to be much higher than that.

The beauty of the ROC curve is that it gives us a visual summary of how the classifier performs for all possible thresholds. It is instrumental if we want to compare two or more classifiers, as you will do in Exercise 11.5.

The legend shows a summary measure, AUC, the area under the ROC curve. An AUC of 0.5 means that the classifier is no better than random guessing, and an AUC of 1 means that the model always makes correct predictions for all thresholds. Getting an AUC that is lower than 0.5, meaning that the classifier is *worse* than random guessing, is exceedingly rare and can be a sign of some error in the model fitting.

evalm also computes a 95% confidence interval for the AUC, which can be obtained as follows:

```
plots$optres[[1]][13,]
```

Another very important plot provided by `evalm` is the *calibration curve*. It shows how well calibrated the model is. If the model is well calibrated, then the predicted probabilities should be close to the true frequencies. As an example, this means that among wines for which the predicted probability of the wine being red is about 20%, 20% should actually be red. For a well-calibrated model, the red curve should closely follow the grey line in the plot:

```
# Calibration curve:
plots$cc
```

Our model doesn't appear to be that well calibrated, meaning that we can't really trust its predicted probabilities.

If we just want to quickly print the AUC without plotting the ROC curves, we can set `summaryFunction = twoClassSummary` in `trainControl`, after which the AUC will be printed instead of accuracy and Cohen's kappa (although it is erroneously called ROC instead of AUC). The sensitivity and specificity for the 0.5 threshold are also printed:

```
tc <- trainControl(method = "repeatedcv",
                   number = 10, repeats = 100,
                   summaryFunction = twoClassSummary,
                   savePredictions = TRUE,
                   classProbs = TRUE)

m <- train(type ~ pH + alcohol,
           data = wine,
           trControl = tc,
           method = "glm",
           family = "binomial",
           metric = "ROC")
m
```

$\sim$

Exercise 11.5. Fit a second logistic regression model, `m2`, to the `wine` data, that also includes `fixed.acidity` and `residual.sugar` as explanatory variables. You can then run

```
library(MLeval)
plots <- evalm(list(m, m2),
               gnames = c("Model 1", "Model 2"))
```

to create ROC curves and calibration plots for both models. Compare their curves. Is the new model better than the simpler model?

11.1.8 Visualising decision boundaries

For models with two explanatory variables, the *decision boundaries* of a classifier can easily be visualised. These show the different regions of the sample space that the classifier associates with the different classes. Let's look at an example of this using the model m fitted to the wine data at the end of the previous section. We'll create a grid of points using expand.grid and make predictions for each of them (i.e., classify each of them). We can then use geom_contour to draw the decision boundaries:

```
contour_data <- expand.grid(
  pH = seq(min(wine$pH), max(wine$pH), length = 500),
  alcohol = seq(min(wine$alcohol), max(wine$alcohol), length = 500))

predictions <- data.frame(contour_data,
                          type = as.numeric(predict(m, contour_data)))

library(ggplot2)
ggplot(wine, aes(pH, alcohol, colour = type)) +
    geom_point(size = 2) +
    stat_contour(aes(x = pH, y = alcohol, z = type),
                 data = predictions, colour = "black")
```

In this case, points to the left of the black line are classified as white, and points to the right of the line are classified as red. It is clear from the plot (both from the point clouds and from the decision boundaries) that the model won't work very well, as many wines will be misclassified.

11.2   Ethical issues in predictive modelling

Even when they are used for the best of intents, predictive models can inadvertently create injustice and bias and lead to discrimination. This is particularly so for models that, in one way or another, make predictions about people. Real-world examples include facial recognition systems that perform worse for people with darker skin (Buolamwini & Gebru, 2018) and recruitment models that are biased against women (Dastin, 2018).

A common issue that can cause this type of problem is difficult-to-spot biases in the training data. If female applicants have been less likely to get a job at a company in the past, then a recruitment model built on data from that company will likely also become biased against women. It can be problematic to simply take data from the past and to consider it as the "ground-truth" when building models.

Similarly, predictive models can create situations where people are prevented from improving their circumstances, and for instance are stopped from getting out of poverty because they are poor. As an example, if people from a certain (poor) zip code historically often have defaulted on their loans, then a predictive model determining who should be granted a student loan may reject an applicant from that area solely on those grounds, even though they otherwise might be an ideal candidate for a loan (which would have allowed them to get an education and a better-paying job). Finally, in extreme cases, predictive models can be used by authoritarian governments to track and target dissidents in a bid to block democracy and human rights.

When working on a predictive model, you should always keep these risks in mind and ask yourself some questions. How will your model be used, and by whom? Are there hidden biases in the training data? Are the predictions good enough, and if they aren't, what could be the consequences for people who get erroneous predictions? Are the predictions good enough for all groups of people, or does the model have worse performance for some groups? Will the predictions improve fairness or cement structural unfairness that was implicitly incorporated in the training data?

$$\sim$$

Exercise 11.6. *Discuss the following.* You are working for a company that tracks the behaviour of online users using cookies. The users have all agreed to be tracked by clicking on an "Accept all cookies" button, but most can be expected not to have read the terms and conditions involved. You analyse information from the cookies, consisting of data about more or less all parts of the users' digital lives, to serve targeted ads to the users. Is this acceptable? Does the accuracy of your targeting models affect your answer? What if the ads are relevant to the user 99% of the time? What if they only are relevant 1% of the time?

Exercise 11.7. *Discuss the following.* You work for a company that has developed a facial recognition system. In a final trial before releasing your product, you discover that your system performs poorly for people over the age of 70 (the accuracy is 99% for people below 70 and 65% for people above 70). Should you release your system without making any changes to it? Does your answer depend on how it will be used? What if it is used instead of keycards to access offices? What if it is used to unlock smartphones? What if it is used for ID controls at voting stations? What if it is used for payments?

Exercise 11.8. *Discuss the following.* Imagine a model that predicts how likely it is that a suspect committed a crime that they are accused of, and that said model is used in courts of law. The model is described as being faster, fairer, and more impartial than human judges. It is a highly complex black-box machine learning model built on data from previous trials. It uses hundreds of variables, and so it isn't possible to explain why it gives a particular prediction for a specific individual. The model makes correct predictions 99% of the time. Is using such a model in the judicial system acceptable? What if an innocent person is predicted by the model to be guilty, without an explanation of why it found them to be guilty? What if the model makes correct predictions 90% or 99.99% of the time? Are there things that the model shouldn't be allowed to take into account, such as skin colour or income? If so, how can you make sure that such variables aren't implicitly incorporated into the training data?

11.3   Challenges in predictive modelling

There are a number of challenges that often come up in predictive modelling projects. In this section we'll briefly discuss some of them.

11.3.1   Handling class imbalance

Imbalanced data, where the proportions of different classes differ a lot, are common in practice. In some areas, such as the study of rare diseases, such datasets are inherent to the field. Class imbalance can cause problems for many classifiers, as they tend to become prone to classify too many observations as belonging to the more common class.

One way to mitigate this problem is to use *down-sampling* and *up-sampling* when fitting the model. In down-sampling, only a (random) subset of the observations from the larger class are used for fitting the model, so that the number of cases from each class becomes balanced. In up-sampling, the number of observations in the smaller class are artificially increased by resampling, also to achieve balance. These methods are only used when fitting the model, to avoid problems with the model overfitting to the class imbalance.

To illustrate the need and use for these methods, let's create a more imbalanced version of the `wine` data:

```
# Create imbalanced wine data:
wine_imb <- wine[1:5000,]

# Check class balance:
table(wine_imb$type)
```

Next, we fit three logistic models – one the usual way, one with down-sampling, and one with up-sampling. We'll use 10-fold cross-validation to evaluate their performance.

```
library(caret)

# Fit a model the usual way:
tc <- trainControl(method = "cv" , number = 10,
                   savePredictions = TRUE,
                   classProbs = TRUE)
m1 <- train(type ~ pH + alcohol,
            data = wine_imb,
            trControl = tc,
            method = "glm",
            family = "binomial")

# Fit with down-sampling:
tc <- trainControl(method = "cv" , number = 10,
                   savePredictions = TRUE,
                   classProbs = TRUE,
                   sampling = "down")
m2 <- train(type ~ pH + alcohol,
            data = wine_imb,
            trControl = tc,
            method = "glm",
            family = "binomial")

# Fit with up-sampling:
tc <- trainControl(method = "cv" , number = 10,
                   savePredictions = TRUE,
                   classProbs = TRUE,
                   sampling = "up")
m3 <- train(type ~ pH + alcohol,
            data = wine_imb,
            trControl = tc,
```

```
            method = "glm",
            family = "binomial")
```

Looking at the accuracy of the three models, m1 seems to be the winner:

```
m1$results
m2$results
m3$results
```

Bear in mind, though, that the accuracy can be very high when you have imbalanced classes, even if your model has overfitted and always predicts that all observations belong to the same class. Perhaps ROC curves will paint a different picture?

```
library(MLeval)
plots <- evalm(list(m1, m2, m3),
               gnames = c("Imbalanced data",
                          "Down-sampling",
                          "Up-sampling"))
```

The three models have virtually identical performance in terms of AUC, so thus far there doesn't seem to be an advantage to using down-sampling or up-sampling.

Now, let's make predictions for all the red wines that the models haven't seen in the training data. What are the predicted probabilities of them being red, for each model?

```
# Number of red wines:
size <- length(5001:nrow(wine))

# Collect the predicted probabilities in a data frame:
red_preds <- data.frame(pred = c(
          predict(m1, wine[5001:nrow(wine),], type = "prob")[, 1],
          predict(m2, wine[5001:nrow(wine),], type = "prob")[, 1],
          predict(m3, wine[5001:nrow(wine),], type = "prob")[, 1]),
          method = rep(c("Standard",
                         "Down-sampling",
                         "Up-sampling"),
                     c(size, size, size)))

# Plot the distributions of the predicted probabilities:
library(ggplot2)
ggplot(red_preds, aes(pred, colour = method)) +
     geom_density()
```

When the model is fitted using the standard methods, almost all red wines get very low predicted probabilities of being red. This isn't the case for the models that used down-sampling and up-sampling, meaning that m2 and m3 are much better at correctly classifying red wines. Note that we couldn't see any differences between the models in the ROC curves, but that there are huge differences between them when they are applied to new data.

Problems related to class imbalance can be difficult to detect, so always be careful when working with imbalanced data.

11.3.2 Assessing variable importance

caret contains a function called varImp that can be used to assess the relative importance of different variables in a model. dotPlot can then be used to plot the results:

```r
library(caret)
tc <- trainControl(method = "LOOCV")
m <- train(mpg ~ .,
           data = mtcars,
           method = "lm",
           trControl = tc)


varImp(m)            # Numeric summary
dotPlot(varImp(m))   # Graphical summary
```

Getting a measure of variable importance sounds really good – it can be useful to know which variables influence the model the most. Unfortunately, varImp uses a nonsensical importance measure: the *t*-statistics of the coefficients of the linear model. In essence, this means that variables with a lower p-value are assigned higher importance. But the p-value is *not* a measure of effect size, nor the predictive importance of a variable (see, e.g., Wasserstein & Lazar (2016)). I strongly advise against using varImp for linear models.

There are other options for computing variable importance for linear and generalised linear models, for instance in the relaimpo package, but mostly these rely on in-sample metrics like R^2. Since our interest is in the predictive performance of our model, we are chiefly interested in how much the different variables *affect the predictions*. In Section 11.5.2 we will see an example of such an evaluation, for another type of model.

11.3.3 Extrapolation

It is always dangerous to use a predictive model with data that comes from outside the range of the variables in the training data. We'll use bacteria.csv as an example of that – download that file from the books' web page[3] and set file_path to its path. The data has two variables, Time and OD. The first describes the time of a measurement, and the second describes the optical density (OD) of a container with bacteria. The more the bacteria grow, the greater the OD. First, let's load and plot the data:

```r
# Read and format data:
bacteria <- read.csv(file_path)
bacteria$Time <- as.POSIXct(bacteria$Time, format = "%H:%M:%S")

# Plot the bacterial growth:
library(ggplot2)
ggplot(bacteria, aes(Time, OD)) +
      geom_line()
```

---

[3]http://www.modernstatisticswithr.com/data.zip

Now, let's fit a linear model to data from hours 3-6, during which the bacteria are in their exponential phase, where they grow faster:

```
# Fit model:
m <- lm(OD ~ Time, data = bacteria[45:90,])

# Plot fitted model:
ggplot(bacteria, aes(Time, OD)) +
      geom_line() +
      geom_abline(aes(intercept = coef(m)[1], slope = coef(m)[2]),
                  colour = "red")
```

The model fits the data that it's been fitted to extremely well, but it does very poorly outside this interval. It overestimates the future growth and underestimates the previous OD.

In this example, we had access to data from outside the range used for fitting the model, which allowed us to see that the model performs poorly outside the original data range. In most cases, however, we do not have access to such data. When extrapolating outside the range of the training data, there is always a risk that the patterns governing the phenomena we are studying are completely different, and it is important to be aware of this.

11.3.4 Missing data and imputation

The estates.xlsx data that you studied in Exercise 11.2 contained a lot of missing data, and as a consequence, you had to remove a lot of rows from the dataset. Another option is do what we did in Section 8.7 and use *imputation*, i.e., add artificially generated observations in place of the missing values. This allows you to use the entire dataset – even those observations where some variables are missing. caret has functions for doing this, using methods that are based on some of the machine learning models that we will look at in Section 11.5.

To see an example of imputation, let's create some missing values in mtcars:

```
mtcars_missing <- mtcars
rows <- sample(1:nrow(mtcars), 5)
cols <- sample(1:ncol(mtcars), 2)
mtcars_missing[rows, cols] <- NA
mtcars_missing
```

If we try to fit a model to this data, we'll get an error message about NA values:

```
library(caret)
tc <- trainControl(method = "repeatedcv",
                   number = 10, repeats = 100)
m <- train(mpg ~ .,
           data = mtcars_missing,
           method = "lm",
           trControl = tc)
```

By adding preProcess = "knnImpute" and na.action = na.pass to train we can use the observations that are the most similar to those with missing values to impute data:

```
library(caret)
tc <- trainControl(method = "repeatedcv",
                   number = 10, repeats = 100)
m <- train(mpg ~ .,
           data = mtcars_missing,
           method = "lm",
           trControl = tc,
           preProcess = "knnImpute",
           na.action = na.pass)

m$results
```

You can compare the results obtained for this model to those obtained using the complete dataset:

```
m <- train(mpg ~ .,
           data = mtcars,
           method = "lm",
           trControl = tc)

m$results
```

Here, these are probably pretty close (we didn't have a lot of missing data, after all), but not identical.

11.3.5 Endless waiting

Comparing many different models can take a lot of time, especially if you are working with large datasets. Waiting for the results can seem to take forever. Fortuitously, modern computers have processing units, CPUs, that can perform multiple computations in parallel using different *cores* or *threads*. This can significantly speed up model fitting, as for instance it allows us to fit the same model to different subsets in a cross-validation in parallel, i.e., at the same time.

In Section 12.2 you'll learn how to perform any type of computation in parallel. However, caret is so simple to run in parallel that we'll have a quick look at that right away. We'll use the foreach, parallel, and doParallel packages, so let's install them:

```
install.packages(c("foreach", "parallel", "doParallel"))
```

The number of cores available on your machine determines how many processes can be run in parallel. To see how many you have, use detectCores:

```
library(parallel)
detectCores()
```

You should avoid the temptation of using all available cores for your parallel computation – you'll always need to reserve at least one for running RStudio and other applications.

To enable parallel computations, we use `registerDoParallel` to *register* the parallel backend to be used. Here is an example where we create three workers (and so use three cores in parallel[4]):

```
library(doParallel)
registerDoParallel(3)
```

After this, it will likely take less time to fit your `caret` models, as model fitting now will be performed using parallel computations on three cores. That means that you'll spend less time waiting and more time modelling. Hurrah! One word of warning though: parallel computations require more memory, so you may run into problems with RAM if you are working on very large datasets.

11.3.6   Overfitting to the test set

Although out-of-sample evaluations are better than in-sample evaluations of predictive models, they are not without risks. Many practitioners like to fit several different models to the same dataset, and then compare their performance (indeed, we ourselves have done and will continue to do so!). When doing this, there is a risk that we overfit our models to the data used for the evaluation. The risk is greater when using test-training splits but is not non-existent for cross-validation and bootstrapping. An interesting example of this phenomenon is presented by Recht et al. (2019), who show that the celebrated image classifiers trained on a dataset known as ImageNet perform significantly worse when used on new data.

When building predictive models that will be used in a real setting, it is a good practice to collect an additional *evaluation set* that is used to verify that the model still works well when faced with new data, and that wasn't part of the model fitting or the model testing. If your model performs worse than expected on the evaluation set, it is a sign that you've overfitted your model to the test set.

Apart from testing so many models that one happens to perform well on the test data (thus overfitting), there are several mistakes that can lead to overfitting. One example is data leakage, where part of the test data "leaks" into the training set. This can happen in several ways: maybe you include an explanatory variable that is a function of the response variable (e.g., price per square metre when trying to predict housing prices), or maybe you have twinned or duplicate observations in your data. Another example is to not include all steps of the modelling in the evaluation, for instance by first using the entire dataset to select which variables to include, and then using cross-validation to assess the performance of the model. If you use the data for variable selection, then that needs to be a part of your cross-validation as well.

In contrast to much of traditional statistics, out-of-sample evaluations are example-based. We must be aware that what worked at one point won't necessarily work in the future. It is entirely possible that the phenomenon that we are modelling is non-stationary, meaning that the patterns in the training data differ from the patterns in future data. In that case, our model can be overfitted in the sense that it describes patterns that no longer are valid. It is therefore important to not only validate a predictive model once, but to return to it at a later point to check that it still performs as expected. Model evaluation is a task that lasts as long as the model is in use.

---

[4]If your CPU has three or fewer cores, you should lower this number.

11.4 Regularised regression models

The standard method used for fitting linear models, ordinary least squares (OLS), can be shown to yield the best unbiased estimator of the regression coefficients (under certain assumptions). But what if we are willing to use estimators that are *biased*? A common way of measuring the performance of an estimator is the mean squared error (MSE). If $\hat{\theta}$ is an estimator of a parameter θ, then

$$MSE(\theta) = E((\hat{\theta} - \theta)^2) = Bias^2(\hat{\theta}) + Var(\hat{\theta}),$$

which is known as the *bias-variance decomposition* of the MSE. This means that if increasing the bias allows us to decrease the variance, it is possible to obtain an estimator with a lower MSE than what is possible for unbiased estimators.

Regularised regression models are linear or generalised linear models in which a small (typically) bias is introduced in the model fitting. Often this can lead to models with better predictive performance. Moreover, it turns out that this also allows us to fit models in situations where it wouldn't be possible to fit ordinary (generalised) linear models, for example when the number of variables is greater than the sample size.

To introduce the bias, we add a *penalty* term to the *loss function* used to fit the regression model. In the case of linear regression, the usual loss function is the squared ℓ_2 norm, meaning that we seek the estimates β_i that minimise

$$\sum_{i=1}^{n}(y_i - \beta_0 - \beta_1 x_{i1} - \beta_2 x_{i2} - \cdots - \beta_p x_{ip})^2.$$

When fitting a regularised regression model, we instead seek the $\beta = (\beta_1, \ldots, \beta_p)$ that minimise

$$\sum_{i=1}^{n}(y_i - \beta_0 - \beta_1 x_{i1} - \beta_2 x_{i2} - \cdots - \beta_p x_{ip})^2 + p(\beta, \lambda),$$

for some penalty function $p(\beta, \lambda)$. The penalty function increases the "cost" of having large β_i, which causes the estimates to "shrink" towards 0. λ is a *shrinkage parameter* used to control the strength of the shrinkage – the larger λ is, the greater the shrinkage. It is usually chosen using cross-validation.

Regularised regression models are not invariant under linear rescalings of the explanatory variables, meaning that if a variable is multiplied by some number a, then this can change the fit of the entire model in an arbitrary way. For that reason, it is widely agreed that the explanatory variables should be standardised to have mean 0 and variance 1 before fitting a regularised regression model. Fortunately, the functions that we will use for fitting these models do that for us, so that we don't have to worry about standardisation. Moreover, they then rescale the model coefficients to be on the original scale, to facilitate interpretation of the model. We can therefore interpret the regression coefficients in the same way as we would for any other regression model.

In this section, we'll look at how to use regularised regression in practice. Further mathematical details are deferred to Section 14.5. We will make use of model-fitting functions from the `glmnet` package, so let's start by installing that:

```
install.packages("glmnet")
```

We will use the `mtcars` data to illustrate regularised regression. We'll begin by once again fitting an ordinary linear regression model to the data:

```
library(caret)
tc <- trainControl(method = "LOOCV")
m1 <- train(mpg ~ .,
            data = mtcars,
            method = "lm",
            trControl = tc)

summary(m1)
```

11.4.1   Ridge regression

The first regularised model that we will consider is *ridge regression* (Hoerl & Kennard, 1970), for which the penalty function is $p(\beta, \lambda) = \lambda \sum_{j=1}^{p} \beta_i^2$. We will fit such a model to the `mtcars` data using `train`. LOOCV will be used, both for evaluating the model and for finding the best choice of the shrinkage parameter λ. This process is often called hyperparameter[5] *tuning* – we tune the hyperparameter λ until we get a good model.

```
library(caret)
# Fit ridge regression:
tc <- trainControl(method = "LOOCV")
m2 <- train(mpg ~ .,
            data = mtcars,
            method = "glmnet",
            tuneGrid = expand.grid(alpha = 0,
                                   lambda = seq(0, 10, 0.1)),
            metric = "RMSE",
            trControl = tc)
```

In the `tuneGrid` setting of `train` we specified that values of λ in the interval $[0, 10]$ should be evaluated. When we print the `m` object, we will see $RMSE$ and MAE of the models for different values of λ (with $\lambda = 0$ being ordinary non-regularised linear regression):

```
# Print the results:
m2

# Plot the results:
library(ggplot2)
ggplot(m2, metric = "RMSE")
ggplot(m2, metric = "MAE")
```

To only print the results for the best model, we can use:

---

[5]Parameters like λ that describe "settings" used for the method rather than parts of the model are often referred to as *hyperparameters*.

```
m2$results[which(m2$results$lambda == m2$finalModel$lambdaOpt),]
```

Note that the $RMSE$ is substantially lower than that for the ordinary linear regression (m1).

In the metric setting of train, we said that we wanted $RMSE$ to be used to determine which value of λ gives the best model. To get the coefficients of the model with the best choice of λ, we use coef as follows:

```
# Check the coefficients of the best model:
coef(m2$finalModel, m2$finalModel$lambdaOpt)
```

Comparing these coefficients to those from the ordinary linear regression (summary(m1)), we see that the coefficients of the two models actually differ quite a lot.

If we want to use our ridge regression model for prediction, it is straightforward to do so using predict(m), as predict automatically uses the best model for prediction.

It is also possible to choose λ without specifying the region in which to search for the best λ, i.e., without providing a tuneGrid argument. In this case, some (arbitrarily chosen) default values will be used instead:

```
m2 <- train(mpg ~ .,
            data = mtcars,
            method = "glmnet",
            metric =  "RMSE",
            trControl = tc)
m2
```

$\sim$

Exercise 11.9. Return to the estates.xlsx data from Exercise 11.2. Refit your linear model, but this time use ridge regression instead. Do the $RMSE$ and MAE improve?

Exercise 11.10. Return to the wine data from Exercise 11.5. Fitting the models below will take a few minutes, so be prepared to wait for a little while.

1. Fit a logistic ridge regression model to the data (make sure to add family = "binomial" so that you actually fit a logistic model and not a linear model), using all variables in the dataset (except type) as explanatory variables. Use five-fold cross-validation for choosing λ and evaluating the model (other options are too computer-intensive). What metric is used when finding the optimal λ?

2. Set summaryFunction = twoClassSummary in trainControl and metric = "ROC" in train and refit the model using AUC to find the optimal λ. Does the choice of λ change, for this particular dataset?

11.4.2 The lasso

The next regularised regression model that we will consider is the *lasso* (Tibshirani, 1996), for which $p(\beta, \lambda) = \lambda \sum_{j=1}^{p} |\beta_i|$. This is an interesting model because it simultaneously performs estimation and *variable selection*, by completely removing some variables from the model. This is particularly useful if we have a large number of variables, in which case the lasso may create a simpler model while maintaining high predictive accuracy. Let's fit a lasso model to our data, using MAE to select the best λ:

```
library(caret)
tc <- trainControl(method = "LOOCV")
m3 <- train(mpg ~ .,
            data = mtcars,
            method = "glmnet",
            tuneGrid = expand.grid(alpha = 1,
                                   lambda = seq(0, 10, 0.1)),
            metric = "MAE",
            trControl = tc)

# Plot the results:
library(ggplot2)
ggplot(m3, metric = "RMSE")
ggplot(m3, metric = "MAE")

# Results for the best model:
m3$results[which(m3$results$lambda == m3$finalModel$lambdaOpt),]

# Coefficients for the best model:
coef(m3$finalModel, m3$finalModel$lambdaOpt)
```

The variables that were removed from the model are marked by points (.) in the list of coefficients. The $RMSE$ is comparable to that from the ridge regression and is better than that for the ordinary linear regression, but the number of variables used is fewer. The lasso model is more parsimonious, and therefore it is easier to interpret (and present to your boss/client/supervisor/colleagues!).

If you only wish to extract the names of the variables with non-zero coefficients from the lasso model (i.e., a list of the variables retained in the variable selection), you can do so using the code below. This can be useful if you have a large number of variables and quickly want to check which have non-zero coefficients:

```
rownames(coef(m3$finalModel, m3$finalModel$lambdaOpt))[
        coef(m3$finalModel, m3$finalModel$lambdaOpt)[,1]!= 0]
```

~

Exercise 11.11. Return to the `estates.xlsx` data from Exercise 11.2. Refit your linear model, but this time use the lasso instead. Do the $RMSE$ and MAE improve?

Exercise 11.12. To see how the lasso handles variable selection, simulate a dataset where only the first 5 out of 200 explanatory variables are correlated with the response variable:

```
n <- 100 # Number of observations
p <- 200 # Number of variables
# Simulate explanatory variables:
x <- matrix(rnorm(n*p), n, p)
# Simulate the response variable:
y <- 2*x[,1] + x[,2] - 3*x[,3] + 0.5*x[,4] + 0.25*x[,5] + rnorm(n)
# Collect the simulated data in a data frame:
simulated_data <- data.frame(y, x)
```

1. Fit a linear model to the data (using the model formula y ~ .). What happens?

2. Fit a lasso model to this data. Does it select the correct variables? What if you repeat the simulation several times, or if you change the values of n and p?

11.4.3 Elastic net

A third option is the *elastic net* (Zou & Hastie, 2005), which essentially is a compromise between ridge regression and the lasso. Its penalty function is $p(\beta, \lambda, \alpha) = \lambda\left(\alpha \sum_{j=1}^{p} |\beta_i| + (1-\alpha) \sum_{j=1}^{p} \beta_i^2\right)$, with $0 \leq \alpha \leq 1$. $\alpha = 0$ yields the ridge estimator, $\alpha = 1$ yields the lasso, and α between 0 and 1 yields a combination of both. When fitting an elastic net model, we search for an optimal choice of α, along with the choice of λ_i. To fit such a model, we can run the following:

```
library(caret)
tc <- trainControl(method = "LOOCV")
m4 <- train(mpg ~ .,
            data = mtcars,
            method = "glmnet",
            tuneGrid = expand.grid(alpha = seq(0, 1, 0.1),
                                   lambda = seq(0, 10, 0.1)),
            metric = "RMSE",
            trControl = tc)

# Print best choices of alpha and lambda:
m4$bestTune

# Print the RMSE and MAE for the best model:
m4$results[which(rownames(m4$results) == rownames(m4$bestTune)),]

# Print the coefficients of the best model:
coef(m4$finalModel, m4$bestTune$lambda, m4$bestTune$alpha)
```

In this example, the ridge regression happened to yield the best fit, in terms of the cross-validation *RMSE*.

~

Exercise 11.13. Return to the `estates.xlsx` data from Exercise 11.2. Refit your linear model, but this time use the elastic net instead. Does the $RMSE$ and MAE improve?

11.4.4 Choosing the best model

So far, we have used the values of λ and α that give the best results according to a performance metric, such as $RMSE$ or AUC. However, it is often the case that we can find a more parsimonious, i.e., simpler, model with almost as good performance. Such models can sometimes be preferable, because of their relative simplicity. Using those models can also reduce the risk of overfitting. `caret` has two functions that can be used for this:

- `oneSE`, which follows a rule-of-thumb from Breiman et al. (1984), which states that the simplest model within one standard error of the model with the best performance should be chosen,
- `tolerance`, which chooses the simplest model that has a performance within (by default) 1.5% of the model with the best performance.

Neither of these can be used with LOOCV, but work for other cross-validation schemes and the bootstrap.

We can set the rule for selecting the "best" model using the argument `selectionFunction` in `trainControl`. By default, it uses a function called `best` that simply extracts the model with the best performance. Here are some examples for the lasso:

```
library(caret)
# Choose the best model (this is the default!):
  tc <- trainControl(method = "repeatedcv",
                     number = 10, repeats = 100)
  m3 <- train(mpg ~ .,
             data = mtcars,
             method = "glmnet",
             tuneGrid = expand.grid(alpha = 1,
                                    lambda = seq(0, 10, 0.1)),
             metric = "RMSE",
             trControl = tc)

  # Print the best model:
  m3$bestTune
  coef(m3$finalModel, m3$finalModel$lambdaOpt)

# Choose model using oneSE:
  tc <- trainControl(method = "repeatedcv",
                     number = 10, repeats = 100,
                     selectionFunction = "oneSE")
  m3 <- train(mpg ~ .,
             data = mtcars,
             method = "glmnet",
             tuneGrid = expand.grid(alpha = 1,
                                    lambda = seq(0, 10, 0.1)),
             trControl = tc)
```

```
# Print the "best" model (according to the oneSE rule):
m3$bestTune
coef(m3$finalModel, m3$finalModel$lambdaOpt)
```

In this example, the difference between the models is small – and it usually is. In some cases, using `oneSE` or `tolerance` leads to a model that has better performance on new data, but in other cases the model that has the best performance in the evaluation also has the best performance for new data.

11.4.5 Regularised mixed models

`caret` does not handle regularisation of (generalised) linear mixed models. If you want to work with such models, you'll therefore need a package that provides functions for this:

```
install.packages("glmmLasso")
```

Regularised mixed models are strange birds. Mixed models are primarily used for inference about the fixed effects, whereas regularisation primarily is used for predictive purposes. The two don't really seem to match. They can, however, be very useful if our main interest is *estimation* rather than prediction or hypothesis testing, where regularisation can help decrease overfitting. Similarly, it is not uncommon for linear mixed models to be numerically unstable, with the model fitting sometimes failing to converge. In such situations, a regularised LMM will often work better. Let's study an example concerning football (soccer) teams, from Groll & Tutz (2014), that shows how to incorporate random effects and the lasso in the same model:

```
library(glmmLasso)

data(soccer)
?soccer
View(soccer)
```

We want to model the point totals for these football teams. We suspect that variables like `transfer.spendings` can affect the performance of a team:

```
ggplot(soccer, aes(transfer.spendings, points, colour = team)) +
    geom_point() +
    geom_smooth(method = "lm", colour = "black", se = FALSE)
```

Moreover, it also seems likely that other non-quantitative variables also affect the performance, which could cause the teams to all have different intercepts. Let's plot them side-by-side:

```
library(ggplot2)
ggplot(soccer, aes(transfer.spendings, points, colour = team)) +
    geom_point() +
    theme(legend.position = "none") +
    facet_wrap(~ team, nrow = 3)
```

When we model the point totals, it seems reasonable to include a random intercept for `team`. We'll also include other fixed effects describing the crowd capacity of the teams' stadiums, and their playing style (e.g. ball possession and number of yellow cards).

The `glmmLasso` functions won't automatically centre and scale the data for us, which you'll recall is recommended to do before fitting a regularised regression model. We'll create a copy of the data with centred and scaled numeric explanatory variables:

```
soccer_scaled <- soccer
soccer_scaled[, c(4:16)] <- scale(soccer_scaled[, c(4:16)],
                           center = TRUE,
                           scale = TRUE)
```

Next, we'll run a `for` loop to find the best λ. Because we are interested in fitting a model to this particular dataset rather than making predictions, we will use an in-sample measure of model fit, *BIC*, to compare the different values of λ. The code below is partially adapted from `demo("glmmLasso-soccer")`:

```
# Number of effects used in model:
params <- 10

# Set parameters for optimisation:
lambda <- seq(500, 0, by = -5)
BIC_vec <- rep(Inf, length(lambda))
m_list <- list()
Delta_start <- as.matrix(t(rep(0, params + 23)))
Q_start <- 0.1

# Search for optimal lambda:
pbar <- txtProgressBar(min = 0, max = length(lambda), style = 3)
for(j in 1:length(lambda))
{
  setTxtProgressBar(pbar, j)

  m <- glmmLasso(points ~ 1 + transfer.spendings +
                 transfer.receits +
                 ave.unfair.score +
                 tackles  +
                 yellow.card +
                 sold.out +
                 ball.possession +
                 capacity +
                 ave.attend,
                 rnd = list(team =~ 1),
                 family = poisson(link = log),
                 data = soccer_scaled,
                 lambda = lambda[j],
                 switch.NR = FALSE,
                 final.re = TRUE,
                 control = list(start = Delta_start[j,],
```

```
                    q_start = Q_start[j]))

  BIC_vec[j] <- m$bic
  Delta_start <- rbind(Delta_start, m$Deltamatrix[m$conv.step,])
  Q_start <- c(Q_start,m$Q_long[[m$conv.step + 1]])
  m_list[[j]] <- m
}
close(pbar)

# Print the optimal model:
opt_m <- m_list[[which.min(BIC_vec)]]
summary(opt_m)
```

Don't pay any attention to the p-values in the summary table. Variable selection can affect p-values in all sorts of strange ways, and because we've used the lasso to select what variables to include, the p-values presented here are no longer valid.

Note that the coefficients printed by the code above are on the scale of the standardised data. To make them possible to interpret, let's finish by transforming them back to the original scale of the variables:

```
sds <- sqrt(diag(cov(soccer[, c(4:16)])))
sd_table <- data.frame(1/sds)
sd_table["(Intercept)",] <- 1
coef(opt_m) * sd_table[names(coef(opt_m)),]
```

11.5 Machine learning models

In this section we will have a look at the smorgasbord of machine learning models that can be used for predictive modelling. Some of these models differ from more traditional regression models in that they are black-box models, meaning that we don't always know what's going on inside the fitted model. This is in contrast to, e.g., linear regression, where we can look at and try to interpret the β coefficients. Another difference is that these models have been developed solely for prediction, and so they often lack some of the tools that we associate with traditional regression models, like confidence intervals and p-values.

Because we use `caret` for the model fitting, fitting a new type of model mostly amounts to changing the `method` argument in `train`. But please note that I wrote *mostly* – there are a few other differences, e.g., in the preprocessing of the data to which you need to pay attention. We'll point these out as we go.

11.5.1 Decision trees

Decision trees are a class of models that can be used for both classification and regression. Their use is perhaps best illustrated by an example, so let's fit a decision tree to the `estates` data from Exercise 11.2. We set `file_path` to the path to `estates.xlsx` and import and clean the data as before:

```
library(openxlsx)
estates <- read.xlsx(file_path)
estates <- na.omit(estates)
```

Next, we fit a decision tree by setting `method = "rpart"`[6], which uses functions from the rpart package to fit the tree:

```
library(caret)
tc <- trainControl(method = "LOOCV")

m <- train(selling_price ~ .,
           data = estates,
           trControl = tc,
           method = "rpart",
           tuneGrid = expand.grid(cp = 0))

m
```

So, what is this? We can plot the resulting decision tree using the `rpart.plot` package, so let's install and use that:

```
install.packages("rpart.plot")

library(rpart.plot)
prp(m$finalModel)
```

What we see in Figure 11.1 *is* our machine learning model – our decision tree. When it is used for prediction, the new observation is fed to the top of the tree, where a question about the new observation is asked: "is `tax_value` < 1610"? If the answer is *yes*, the observation continues down the line to the left, to the next question. If the answer is *no*, it continues down the line to the right, to the question "is `tax_value` < 2720", and so on. After a number of questions, the observation reaches a circle – a so-called *leaf node*, with a number in it. This number is the predicted selling price of the house, which is based on observations in the training data that belong to the same leaf. When the tree is used for classification, the predicted probability of class A is the proportion of observations from the training data in the leaf that belong to class A.

`prp` has a number of parameters that lets us control what our tree plot looks like. `box.palette`, `shadow.col`, `nn`, `type`, `extra`, and `cex` are all useful – read the documentation for `prp` to see how they affect the plot:

```
prp(m$finalModel,
    box.palette = "RdBu",
    shadow.col = "gray",
    nn = TRUE,
    type = 3,
```

---

[6]The name `rpart` may seem cryptic: it is an abbreviation for Recursive Partitioning and Regression Trees, which is a type of decision trees.

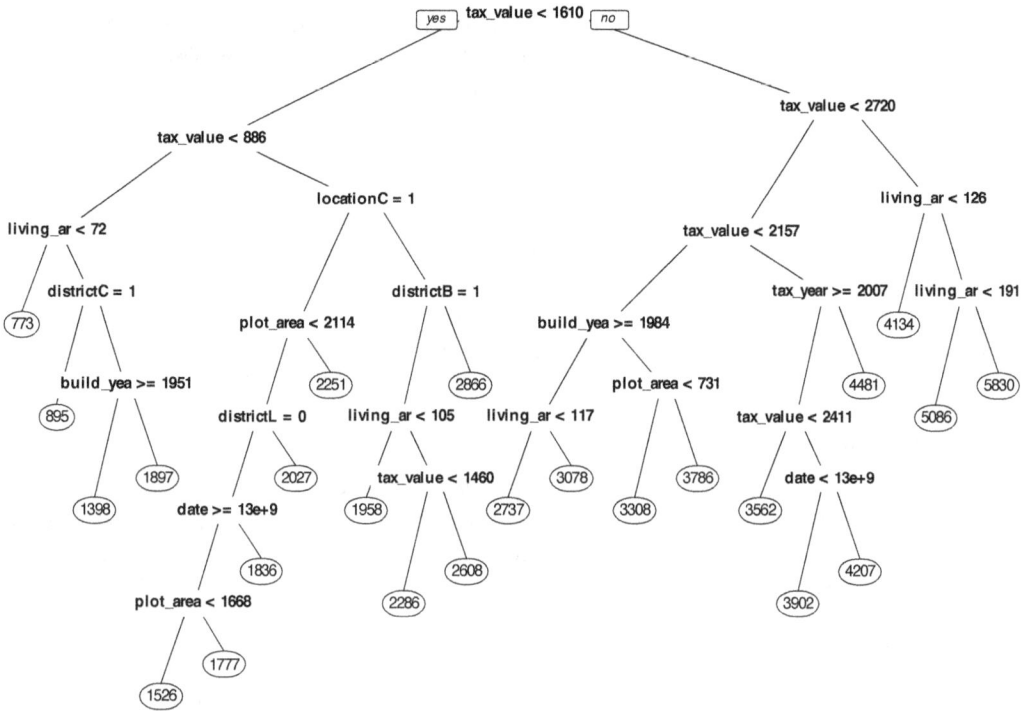

FIGURE 11.1 Decision tree for predicting real estate prices.

```
      extra = 1,
      cex = 0.75)
```

When fitting the model, rpart builds the tree from the top down. At each split, it tries to find a question that will separate subgroups in the data as much as possible. There is no need to standardise the data (in fact, this won't change the shape of the tree at all).

~

Exercise 11.14. Fit a classification tree model to the wine data, using pH, alcohol, fixed.acidity, and residual.sugar as explanatory variables. Evaluate its *AUC* using repeated 10-fold cross-validation.

    1. Plot the resulting decision tree. It is too large to be easily understandable and needs to be *pruned*. This is done using the parameter cp. Try increasing the value of cp in tuneGrid = expand.grid(cp = 0) to different values between 0 and 1. What happens with the tree?

    2. Use tuneGrid = expand.grid(cp = seq(0, 0.01, 0.001)) to find an optimal choice of cp. What is the result?

Exercise 11.15. Fit a regression tree model to the bacteria.csv data to see how OD changes with Time, using the data from observations 45 to 90 of the data frame, as in the example in Section 11.3.3. Then make predictions for all observations in the dataset. Plot the actual OD values along with your predictions. Does the model extrapolate well?

Exercise 11.16. Fit a classification tree model to the `seeds` data from Section 4.11, using `Variety` as the response variable and `Kernel_length` and `Compactness` as explanatory variables. Plot the resulting decision boundaries, as in Section 11.1.8. Do they seem reasonable to you?

11.5.2 Random forests

Random forest (Breiman, 2001) is an *ensemble method*, which means that it is based on combining multiple predictive models. In this case, it is a combination of multiple decision trees that have been built using different subsets of the data. Each tree is fitted to a bootstrap sample of the data (a procedure known as *bagging*), and at each split only a random subset of the explanatory variables are used. The predictions from these trees are then averaged to obtain a single prediction. While the individual trees in the forest tend to have rather poor performance, the random forest itself often performs better than a single decision tree fitted to all of the data using all variables.

To fit a random forest to the `estates` data (loaded in the same way as in Section 11.5.1), we set `method = "rf"`, which will let us do the fitting using functions from the `randomForest` package. The random forest has a parameter called `mtry` that determines the number of randomly selected explanatory variables. As a rule-of-thumb, `mtry` close to $\sqrt{p}$, where p is the number of explanatory variables in your data, is usually a good choice. When trying to find the best choice for `mtry` I recommend trying some values close to that.

For the `estates` data we have 11 explanatory variables, and so a value of `mtry` close to $\sqrt{11} \approx 3$ could be a good choice. Let's try a few different values with a 10-fold cross-validation:

```
library(caret)
tc <- trainControl(method = "cv",
                   number = 10)


m <- train(selling_price ~ .,
         data = estates,
         trControl = tc,
         method = "rf",
         tuneGrid = expand.grid(mtry = 2:4))


m
```

In my run, an `mtry` equal to 4 gave the best results. Let's try larger values as well, just to see if that gives a better model:

```
m <- train(selling_price ~ .,
         data = estates,
         trControl = tc,
         method = "rf",
         tuneGrid = expand.grid(mtry = 4:10))


m
```

We can visually inspect the impact of `mtry` by plotting `m`:

```
ggplot(m)
```

For this data, a value of `mtry` that is a little larger than what usually is recommended seems to give the best results. It was a good thing that we didn't just blindly go with the rule-of-thumb but instead tried a few different values.

Random forests have a built-in variable importance measure, which is based on measuring how much worse the model fares when the values of each variable are permuted. This is a much more sensible measure of variable importance than that presented in Section 11.3.2. The importance values are reported on a relative scale, with the value for the most important variable always being 100. Let's have a look:

```
dotPlot(varImp(m))
```

~

Exercise 11.17. Fit a decision tree model and a random forest to the `wine` data, using all variables (except `type`) as explanatory variables. Evaluate their performance using 10-fold cross-validation. Which model has the best performance?

Exercise 11.18. Fit a random forest to the `bacteria.csv` data to see how `OD` changes with `Time`, using the data from observations 45 to 90 of the data frame, as in the example in Section 11.3.3. Then make predictions for all observations in the dataset. Plot the actual OD values along with your predictions. Does the model extrapolate well?

Exercise 11.19. Fit a random forest model to the `seeds` data from Section 4.11, using `Variety` as the response variable and `Kernel_length` and `Compactness` as explanatory variables. Plot the resulting decision boundaries, as in Section 11.1.8. Do they seem reasonable to you?

11.5.3 Boosted trees

Another useful class of ensemble method that relies on combining decision trees is *boosted trees*. Several different versions are available; we'll use a version called Stochastic Gradient Boosting (Friedman, 2002), which is available through the `gbm` package. Let's start by installing that:

```
install.packages("gbm")
```

The decision trees in the ensemble are built sequentially, with each new tree giving more weight to observations for which the previous trees performed poorly. This process is known as *boosting*.

When fitting a boosted trees model in `caret`, we set `method = "gbm"`. There are four parameters that we can use to find a better fit. The two most important are `interaction.depth`, which determines the maximum tree depth (values greater than $\sqrt{p}$, where p is the number of explanatory variables in your data, are discouraged) and `n.trees`, which specifies the number of trees to fit (also known as the number of boosting iterations). Both can have a large impact on the model fit. Let's try a few values with the `estates` data (loaded in the same way as in Section 11.5.1):

```
library(caret)
tc <- trainControl(method = "cv",
                   number = 10)

m <- train(selling_price ~ .,
           data = estates,
           trControl = tc,
           method = "gbm",
           tuneGrid = expand.grid(
               interaction.depth = 1:3,
               n.trees = seq(20, 200, 10),
               shrinkage = 0.1,
               n.minobsinnode = 10),
           verbose = FALSE)
```

```
m
```

The setting `verbose = FALSE` is used to stop `gbm` from printing details about each fitted tree. We can plot the model performance for different settings:

```
ggplot(m)
```

As you can see, using more trees (a higher number of boosting iterations) seems to lead to a better model. However, if we use too many trees, the model usually overfits, leading to a worse performance in the evaluation:

```
m <- train(selling_price ~ .,
           data = estates,
           trControl = tc,
           method = "gbm",
           tuneGrid = expand.grid(
               interaction.depth = 1:3,
               n.trees = seq(25, 500, 25),
               shrinkage = 0.1,
               n.minobsinnode = 10),
           verbose = FALSE)
```

```
ggplot(m)
```

A table and plot of variable importance is given by `summary`:

```
summary(m)
```

In many problems, boosted trees are among the best-performing models. They do, however, require a lot of tuning, which can be time-consuming, both in terms of how long it takes to run the tuning and in terms of how much time you have to spend fiddling with the different parameters. Several different implementations of boosted trees are available in `caret`. A good alternative to `gbm` is `xgbTree` from the `xgboost` package. I've chosen not to use that for the examples here, as it often is slower to train due to having a larger number of hyperparameters (which in return makes it even more flexible!).

~

Exercise 11.20. Fit a boosted trees model to the `wine` data, using all variables (except `type`) as explanatory variables. Evaluate its performance using repeated 10-fold cross-validation. What is the best *AUC* that you can get by tuning the model parameters?

Exercise 11.21. Fit a boosted trees regression model to the `bacteria.csv` data to see how `OD` changes with `Time`, using the data from observations 45 to 90 of the data frame, as in the example in Section 11.3.3. Then make predictions for all observations in the dataset. Plot the actual OD values along with your predictions. Does the model extrapolate well?

Exercise 11.22. Fit a boosted trees model to the `seeds` data from Section 4.11, using `Variety` as the response variable and `Kernel_length` and `Compactness` as explanatory variables. Plot the resulting decision boundaries, as in Section 11.1.8. Do they seem reasonable to you?

11.5.4 Model trees

A downside to all the tree-based models that we've seen so far is their inability to extrapolate when the explanatory variables of a new observation are outside the range in the training data. You've seen this, e.g., in Exercise 11.15. Methods based on *model trees* solve this problem by fitting, e.g., a linear model in each leaf node of the decision tree. Ordinary decision trees fit regression models that are piecewise constant, while model trees utilising linear regression fit regression models that are piecewise linear.

The model trees that we'll now have a look at aren't available in `caret`, meaning that we can't use its functions for evaluating models using cross-validations. We can, however, still perform cross-validation using a `for` loop, as we did in the beginning of Section 11.1.3. Model trees are available through the `partykit` package, which we'll install next. We'll also install `ggparty`, which contains tools for creating good-looking plots of model trees:

```
install.packages(c("partykit", "ggparty"))
```

The model trees in `partykit` differ from classical decision trees not only in how the nodes are treated, but also in how the splits are determined; see Zeileis et al. (2008) for details. To illustrate their use, we'll return to the `estates` data. The model formula for model trees has two parts. The first specifies the response variable and what variables to use for the linear models in the nodes, and the second part specifies what variables to use for the splits. In our example, we'll use `living_area` as the sole explanatory variable in our linear models, and `location`, `build_year`, `tax_value`, and `plot_area` for the splits (in this particular example, there is no overlap between the variables used for the linear models and the variables used for the splits, but it is perfectly fine to have an overlap if you like!).

As in Section 11.5.1, we set `file_path` to the path to `estates.xlsx` and import and clean the data. We can then fit a model tree with linear regressions in the nodes using `lmtree`:

```
library(openxlsx)
estates <- read.xlsx(file_path)
```

```
estates <- na.omit(estates)

# Make location a factor variable:
estates$location <- factor(estates$location)

# Fit model tree:
library(partykit)
m <- lmtree(selling_price ~ living_area | location + build_year +
                                          tax_value + plot_area,
            data = estates)
```

Next, we plot the resulting tree – make sure that you enlarge your Plot panel so that you can see the linear models fitted in each node:

```
library(ggparty)
autoplot(m)
```

By adding additional arguments to `lmtree`, we can control, e.g., the amount of pruning. You can find a list of all the available arguments by having a look at `?mob_control`. To do automated likelihood-based pruning, we can use `prune = "AIC"` or `prune = "BIC"`, which yields a slightly shorter tree:

```
m <- lmtree(selling_price ~ living_area | location + build_year +
                                          tax_value + plot_area,
            data = estates,
            prune = "BIC")
```

```
autoplot(m)
```

As per usual, we can use `predict` to make predictions from our model. Similar to how we used `lmtree` above, we can use `glmtree` to fit a logistic regression in each node, which can be useful for classification problems. We can also fit Poisson regressions in the nodes using `glmtree`, creating more flexible Poisson regression models. For more information on how you can control how model trees are plotted using `ggparty`, have a look at `vignette("ggparty-graphic-partying")`.

~

Exercise 11.23. In this exercise, you will fit model trees to the `bacteria.csv` data to see how OD changes with Time.

    1. Fit a model tree and a decision tree, using the data from observations 45 to 90 of the data frame, as in the example in Section 11.3.3. Then make predictions for all observations in the dataset. Plot the actual OD values along with your predictions. Do the models extrapolate well?

    2. Now, fit a model tree and a decision tree using the data from observations 20 to 120 of the data frame. Then make predictions for all observations in the dataset. Does this improve the models' ability to extrapolate?

11.5.5 Discriminant analysis

In *linear discriminant analysis* (LDA), prior knowledge about how common different classes are is used to classify new observations using *Bayes' theorem*. It relies on the assumption that the data from each class is generated by a multivariate normal distribution, and that all classes share a common covariance matrix. The resulting decision boundary is a hyperplane.

As part of fitting the model, LDA creates linear combinations of the explanatory variables, which are used for separating different classes. These can be used both for classification and as a supervised alternative to principal components analysis (PCA, Section 4.11).

LDA does not require any tuning. It does, however, allow you to specify prior class probabilities if you like, using the `prior` argument, allowing for Bayesian classification. If you don't provide a prior, the class proportions in the training data will be used instead. Here is an example using the `wine` data from Section 11.1.7:

```r
library(caret)
tc <- trainControl(method = "repeatedcv",
                   number = 10, repeats = 100,
                   summaryFunction = twoClassSummary,
                   savePredictions = TRUE,
                   classProbs = TRUE)

# Without the use of a prior:
# Prior probability of a red wine is 0.25 (i.e., the
# proportion of red wines in the dataset).
m_no_prior <- train(type ~  pH + alcohol + fixed.acidity +
                      residual.sugar,
           data = wine,
           trControl = tc,
           method = "lda",
           metric = "ROC")

# With a prior:
# Prior probability of a red wine is set to be 0.5.
m_with_prior <- train(type ~  pH + alcohol + fixed.acidity +
                        residual.sugar,
           data = wine,
           trControl = tc,
           method = "lda",
           metric = "ROC",
           prior = c(0.5, 0.5))

m_no_prior
m_with_prior
```

As I mentioned, LDA can also be used as an alternative to PCA, which we studied in Section 4.11. Let's have a look at the `seeds` data that we used in that section:

```r
# The data is downloaded from the UCI Machine Learning Repository:
# http://archive.ics.uci.edu/ml/datasets/seeds
```

```
seeds <- read.table("https://tinyurl.com/seedsdata",
        col.names = c("Area", "Perimeter", "Compactness",
        "Kernel_length", "Kernel_width", "Asymmetry",
        "Groove_length", "Variety"))
seeds$Variety <- factor(seeds$Variety)
```

When `caret` fits an LDA, it uses the `lda` function from the MASS package, which uses the same syntax as `lm`. If we use `lda` directly, without involving `caret`, we can extract the scores (linear combinations of variables) for all observations. We can then plot these, to get something similar to a plot of the first two principal components. There is a difference though – PCA seeks to create new variables that summarise as much as possible of the variance in the data, whereas LDA seeks to create new variables that can be used to discriminate between pre-specified groups.

```
# Run an LDA:
library(MASS)
m <- lda(Variety ~ ., data = seeds)

# Save the LDA scores:
lda_preds <- data.frame(Type = seeds$Variety,
                        Score = predict(m)$x)
View(lda_preds)
# There are 3 varieties of seeds. LDA creates 1 less new variable
# than the number of categories - so 2 in this case. We can
# therefore visualise these using a simple scatterplot.

# Plot the two LDA scores for each observation to get a visual
# representation of the data:
library(ggplot2)
ggplot(lda_preds, aes(Score.LD1, Score.LD2, colour = Type)) +
    geom_point()
```

~

Exercise 11.24. An alternative to linear discriminant analysis is *quadratic discriminant analysis* (QDA). This is closely related to LDA, the difference being that we no longer assume that the classes have equal covariance matrices. The resulting decision boundaries are quadratic (i.e., non-linear). Run a QDA on the `wine` data, by using `method = "qda"` in `train`.

Exercise 11.25. Fit an LDA classifier and a QDA classifier to the `seeds` data from Section 4.11, using `Variety` as the response variable and `Kernel_length` and `Compactness` as explanatory variables. Plot the resulting decision boundaries, as in Section 11.1.8. Do they seem reasonable to you?

Exercise 11.26. An even more flexible version of discriminant analysis is *mixture discriminant analysis* (MDA), which uses normal mixture distributions for classification. That way, we no longer have to rely on the assumption of normality. It is available through the mda package and can be used in `train` with 'method = "mda"'. Fit an MDA classifier to the seeds data from Section 4.11, using `Variety` as the response variable and `Kernel_length` and `Compactness` as explanatory variables. Plot the resulting decision boundaries, as in Section 11.1.8. Do they seem reasonable to you?

11.5.6 Support vector machines

Support vector machines (SVM), is a flexible class of methods for classification and regression. Like LDA, they rely on hyperplanes to separate classes. Unlike LDA, however, more weight is put to points close to the border between classes. Moreover, the data is projected into a higher-dimensional space, with the intention of creating a projection that yields a good separation between classes. Several different projection methods can be used, typically represented by *kernels* – functions that measure the inner product in these high-dimensional spaces.

Despite the fancy mathematics, using SVMs is not that difficult. With `caret`, we can fit SVMs with many different types of kernels using the `kernlab` package. Let's install it:

```
install.packages("kernlab")
```

The simplest SVM uses a linear kernel, creating a linear classification that is reminiscent of LDA. Let's look at an example using the `wine` data from Section 11.1.7. The parameter C is a regularisation parameter:

```
library(caret)
tc <- trainControl(method = "cv",
                   number = 10,
                   summaryFunction = twoClassSummary,
                   savePredictions = TRUE,
                   classProbs = TRUE)

m <- train(type ~ pH + alcohol + fixed.acidity + residual.sugar,
           data = wine,
           trControl = tc,
           method = "svmLinear",
           tuneGrid = expand.grid(C = c(0.5, 1, 2)),
           metric = "ROC")
```

There are a number of other non-linear kernels that can be used, with different hyperparameters that can be tuned. Without going into details about the different kernels, some important examples are:

- `method = "svmPoly"`: polynomial kernel. The tuning parameters are `degree` (the polynomial degree, e.g., 3 for a cubic polynomial), `scale` (scale) and `C` (regularisation).
- `method = "svmRadialCost"`: radial basis/Gaussian kernel. The only tuning parameter is `C` (regularisation).
- `method = "svmRadialSigma"`: radial basis/Gaussian kernel with tuning of σ. The tuning parameters are `C` (regularisation) and `sigma` (σ).
- `method = "svmSpectrumString"`: spectrum string kernel. The tuning parameters are `C` (regularisation) and `length` (length).

~

Exercise 11.27. Fit an SVM to the `wine` data, using all variables (except `type`) as explanatory variables, using a kernel of your choice. Evaluate its performance using repeated 10-fold cross-validation. What is the best *AUC* that you can get by tuning the model parameters?

Exercise 11.28. In this exercise, you will fit SVM regression models to the `bacteria.csv` data to see how `OD` changes with `Time`.

> 1. Fit an SVM, using the data from observations 45 to 90 of the data frame, as in the example in Section 11.3.3. Then make predictions for all observations in the dataset. Plot the actual OD values along with your predictions. Does the model extrapolate well?

> 2. Now, fit an SVM using the data from observations 20 to 120 of the data frame. Then make predictions for all observations in the dataset. Does this improve the model's ability to extrapolate?

Exercise 11.29. Fit SVM classifiers with different kernels to the `seeds` data from Section 4.11, using `Variety` as the response variable and `Kernel_length` and `Compactness` as explanatory variables. Plot the resulting decision boundaries, as in Section 11.1.8. Do they seem reasonable to you?

11.5.7   Nearest neighbours classifiers

In classification problems with numeric explanatory variables, a natural approach to finding the class of a new observation is to look at the classes of neighbouring observations, i.e., of observations that are "close" to it in some sense. This requires a distance measure, to measure how close observations are. A kNN classifier classifies the new observations by letting the k Nearest Neighbours – the k points that are the closest to the observation – "vote" about the class of the new observation. As an example, if $k = 3$, two of the three closest neighbours belong to class A, and one of the three closest neighbours belongs to class B, then the new observation will be classified as A. If we like, we can also use the proportion of different classes among the nearest neighbours to get predicted probabilities of the classes (in our example: 2/3 for A, 1/3 for B).

What makes kNN appealing is that it doesn't require a complicated model; instead, we simply compare observations to each other. A major downside is that we have to compute the distance between each new observation and all observations in the training data, which can be time-consuming if you have large datasets. Moreover, we consequently have to store the training data indefinitely, as it is used each time we use the model for prediction. This can cause problems, e.g., if the data are of a kind that falls under the European General Data Protection Regulation (GDPR), which limits how long data can be stored, and for what purpose.

A common choice of distance measure, which is the default when we set `method = "knn"` in `train`, is the common Euclidean distance. We need to take care to standardise our variables before using it, as variables with a high variance otherwise automatically will contribute more to the Euclidean distance. Unlike in regularised regression, `caret` does *not* do this for us. Instead, we must provide the argument `preProcess = c("center", "scale")` to `train`.

An important choice in kNN is what value to use for the parameter k. If k is too small, we use too little information, and if k is too large, the classifier will become prone to classify all observations as belonging to the most common class in the training data. k is usually chosen using cross-validation or bootstrapping. To have `caret` find a good choice of k for us (as we did with λ in regularised regression models), we use the argument `tuneLength` in train, e.g., `tuneLength = 15` to try 15 different values of k.

By now, I think you've seen enough examples of how to fit models in `caret` that you can figure out how to fit a model with `knn` on your own (using the information above, of course). In the next exercise, you will give kNN a go, using the `wine` data.

~

Exercise 11.30. Fit a kNN classification model to the `wine` data, using `pH`, `alcohol`, `fixed.acidity`, and `residual.sugar` as explanatory variables. Evaluate its performance using 10-fold cross-validation, using AUC to choose the best k. Is it better than the logistic regression models that you fitted in Exercise 11.5?

Exercise 11.31. Fit a kNN classifier to the `seeds` data from Section 4.11, using `Variety` as the response variable and `Kernel_length` and `Compactness` as explanatory variables. Plot the resulting decision boundaries, as in Section 11.1.8. Do they seem reasonable to you?

11.6 Forecasting time series

A time series, as those we studied in Section 4.7, is a series of observations sorted in time order. The goal of time series analysis is to model temporal patterns in data. This allows us to take correlations between observations into account (today's stock prices are correlated to yesterday's), to capture seasonal patterns (ice cream sales always increase during the summer), and to incorporate those into predictions, or forecasts, for the future. This section acts as a brief introduction to how this can be done.

11.6.1 Decomposition

In Section 4.7.5 we saw how time series can be *decomposed* into three components:

- *Seasonal* component, describing recurring seasonal patterns,
- *Trend* component, describing a trend over time,
- *Remainder* component, describing random variation.

Let's have a quick look at how to do this in R, using the `a10` data from `fpp2`:

```
library(forecast)
library(ggplot2)
library(fpp2)
?a10
autoplot(a10)
```

The `stl` function uses repeated LOESS smoothing to decompose the series. The `s.window` parameter lets us set the length of the season in the data. We can set it to `"periodic"` to have `stl` find the periodicity of the data automatically:

```
autoplot(stl(a10, s.window = "periodic"))
```

We can access the different parts of the decomposition as follows:

```
a10_stl <- stl(a10, s.window = "periodic")
a10_stl$time.series[,"seasonal"]
a10_stl$time.series[,"trend"]
a10_stl$time.series[,"remainder"]
```

When modelling time series data, we usually want to remove the seasonal component, as it makes the data structure too complicated. We can then add it back when we use the model for forecasting. We'll see how to do that in the following sections.

11.6.2   Forecasting using ARIMA models

The `forecast` package contains a large number of useful methods for fitting time series models. Among them is `auto.arima` which can be used to fit autoregressive integrated moving average (*ARIMA*) models to time series data. ARIMA models are a flexible class of models that can capture many different types of temporal correlations and patterns. `auto.arima` helps us select a model that seems appropriate based on historical data, using an in-sample criterion, a version of *AIC*, for model selection.

`stlm` can be used to fit a model after removing the seasonal component, and then automatically add it back again when using it for a forecast. The `modelfunction` argument lets us specify what model to fit. Let's use `auto.arima` for model fitting through `stlm`:

```
library(forecast)
library(fpp2)

# Fit the model after removing the seasonal component:
tsmod <- stlm(a10, s.window = "periodic", modelfunction = auto.arima)
```

For model diagnostics, we can use `checkresiduals` to check whether the residuals from the model look like white noise (i.e., look normal):

```
# Check model diagnostics:
checkresiduals(tsmod)
```

In this case, the variance of the series seems to increase with time, which the model fails to capture. We therefore see more large residuals than what is expected under the model.

Nevertheless, let's see how we can make a forecast for the next 24 months. The function for this is the aptly named `forecast`:

```
# Plot the forecast (with the seasonal component added back)
# for the next 24 months:
forecast(tsmod, h = 24)
```

```
# Plot the forecast along with the original data:
autoplot(forecast(tsmod, h = 24))
```

In addition to the forecasted curve, `forecast` also provides prediction intervals. By default, these are based on an asymptotic approximation. To obtain bootstrap prediction intervals instead, we can add `bootstrap = TRUE` to `forecast`:

```
autoplot(forecast(tsmod, h = 24, bootstrap = TRUE))
```

The `forecast` package is designed to work well with pipes. To fit a model using `stlm` and `auto.arima` and then plot the forecast, we could have used:

```
a10 |> stlm(s.window = "periodic", modelfunction = auto.arima) |>
        forecast(h = 24, bootstrap = TRUE) |> autoplot()
```

It is also possible to incorporate seasonal effects into ARIMA models by adding seasonal terms to the model. `auto.arima` will do this for us if we apply it directly to the data:

```
a10 |> auto.arima() |>
        forecast(h = 24, bootstrap = TRUE) |> autoplot()
```

For this data, the forecasts from the two approaches are very similar.

In Section 11.3 we mentioned that a common reason for predictive models failing in practical applications is that many processes are non-stationary, meaning that their patterns change over time. ARIMA model are designed to handle some types of non-stationarity, which can make them particularly useful for modelling such processes.

∼

Exercise 11.32. Return to the `writing` dataset from the `fma` package that we studied in Exercise 4.15. Remove the seasonal component. Fit an ARIMA model to the data and use it to plot a forecast for the next three years, with the seasonal component added back and with bootstrap prediction intervals.

11.7 Deploying models

The process of making a prediction model available to other users or systems, for instance by running them on a server, is known as *deployment*. In addition to the need for continuous model evaluation, mentioned in Section 11.3.6, you will also need to check that your R code works as intended in the environment you deploy your model. For instance, if you developed your model using R 4.1 and then run it on a server running R 3.6 with out-of-date versions of the packages you used, there is a risk that some of the functions that you use behave differently from what you expected. Maybe something that should be a `factor` variable becomes a `character` variable, which breaks that part of your code where you use `levels`. A lot of the time, small changes are enough to make the code work in the new environment (add a line that converts the variable to a `factor`), but sometimes large changes can be needed. Likewise, you must check that the model still works after the software is updated on the server.

11.7.1 Creating APIs with `plumber`

An Application Programming Interface (API) is an interface that lets other systems access your R code – which is exactly what you want when you're ready to deploy your model. By using the `plumber` package to create an API (or a REST API, to be more specific), you can let other systems (web page, Java script, Python script, and so on) access your model. Those systems can call your model, sending some input, and then receive its output in different formats, e.g., a JSON list, a csv file, or an image.

We'll illustrate how this works with a simple example. First, let's install `plumber`:

```
install.packages("plumber")
```

Next, assume that we've fitted a model (we'll use the linear regression model for `mtcars` that we've used several times before). We can use this model to make predictions:

```
m <- lm(mpg ~ hp + wt, data = mtcars)

predict(m, newdata = data.frame(hp = 150, wt = 2))
```

We would like to make these predictions available to other systems. That is, we'd like to allow other systems to send values of `hp` and `wt` to our model and get predictions in return. To do so, we start by writing a function for the predictions:

```
m <- lm(mpg ~ hp + wt, data = mtcars)

predictions <- function(hp, wt)
{
    predict(m, newdata = data.frame(hp = hp, wt = wt))
}

predictions(150, 2)
```

To make this accessible to other systems, we save this function in a script called `mtcarsAPI.R` (make sure to save it in your working directory), which looks as follows:

```
# Fit the model:
m <- lm(mpg ~ hp + wt, data = mtcars)

#* Return the prediction:
#* @param hp
#* @param wt
#* @post /predictions
function(hp, wt)
{
```

```
        predict(m, newdata = data.frame(hp = as.numeric(hp),
                                         wt = as.numeric(wt)))
}
```

The only changes that we have made are some additional special comments (#*), which specify what input is expected (parameters `hp` and `wt`) and that the function is called `predictions`. `plumber` uses this information to create the API. The functions made available in an API are referred to as *endpoints*.

To make the function available to other systems, we run `pr` as follows:

```
library(plumber)
pr("mtcarsAPI.R") |> pr_run(port = 8000)
```

The function will now be available on port 8000 of your computer. To access it, you can open your browser and go to the following URL:

- `http://localhost:8000/predictions?hp=150&wt=2`

Try changing the values of `hp` and `wt` and see how the returned value changes.

That's it! As long as you leave your R session running with `plumber`, other systems will be able to access the model using the URL. Typically, you would run this on a server and not on your personal computer.

11.7.2 Different types of output

You won't always want to return a number. Maybe you want to use R to create a plot, send a file, or print some text. Here is an example of an R script, which we'll save as `exampleAPI.R`, that returns different types of output – an image, a text, and a downloadable csv file:

```
#* Plot some random numbers
#* param n The number of points to plot
#* @serializer png
#* @get /plot
function(n = 15) {
  x <- rnorm(as.numeric(n))
  y <- rnorm(as.numeric(n))
  plot(x, y, col = 2, pch = 16)
}

#* Print a message
#* @param name Your name
#* @get /message
function(name = "") {
  list(message = paste("Hello", name, "- I'm happy to see you!"))
}

#* Download the mtcars data as a csv file
#* @serializer csv
#* @get /download
```

```
function() {
  mtcars
}
```

After you've saved the file in your working directory, run the following to create the API:

```
library(plumber)
pr("mtcarsAPI.R") |> pr_run(port = 8000)
```

You can now try the different endpoints:

- `http://localhost:8000/plot`
- `http://localhost:8000/plot?n=50`
- `http://localhost:8000/message?name=Oskar`
- `http://localhost:8000/download`

We've only scratched the surface of `plumber`'s capabilities here. A more thorough guide can be found on the official `plumber` web page at: https://www.rplumber.io/

12

Advanced topics

This chapter contains brief descriptions of more advanced uses of R. First, we cover more details surrounding packages. We then deal with two topics that are important for computational speed: parallelisation and matrix operations. Finally, there are some tips for how to play well with others (which in this case means using R in combination with programming languages like Python and C++).

After reading this chapter, you will know how to:

- Update and remove R packages,
- Install R packages from other repositories than CRAN,
- Run computations in parallel,
- Perform matrix computations using R, and
- Integrate R with other programming languages.

12.1 More on packages

12.1.1 Loading and auto-installing packages

The best way to load R packages is usually to use `library`, as we've done in the examples in this book. If the package that you're trying to load isn't installed, this will return an error message:

```
library("theWrongPackageName")
```

Alternatively, you can use `require` to load packages. This will only display a warning but won't cause your code to stop executing, which usually would be a problem if the rest of the code depends on the package[1]!

However, `require` also returns a `logical`: `TRUE` if the package is installed, and `FALSE` otherwise. This is useful if you want to load a package and automatically install it if it doesn't exist.

To load the `beepr` package, and install it if it doesn't already exist, we can use `require` inside an `if` condition, as in the code chunk below. If the package exists, the package will be loaded (by `require`) and `TRUE` will be returned, and otherwise `FALSE` will be returned. By using `!` to turn `FALSE` into `TRUE` and vice versa, we can make R install the package if it is missing:

```
if(!require("beepr")) { install.packages("beepr"); library(beepr) }
beep(4)
```

---

[1]And why else would you load it...?

423

12.1.2   Updating R and your packages

You can download new versions of R and RStudio following the same steps as in Section 2.1. On Windows, you can have multiple versions of R installed simultaneously.

To update a specific R package, you can use `install.packages`. For instance, to update the `beepr` package, you'd run:

```
install.packages("beepr")
```

If you make a major update of R, you may have to update most or all of your packages. To update all your packages, you simply run `update.packages()`. If this fails, you can try the following instead:

```
pkgs <- installed.packages()
pkgs <- pkgs[is.na(pkgs[, "Priority"]), 1]
install.packages(pkgs)
```

12.1.3   Alternative repositories

In addition to CRAN, two important sources for R packages are Bioconductor, which contains a large number of packages for bioinformatics, and GitHub, where many developers post development versions of their R packages (which often contain functions and features not yet included in the version of the package that has been posted on CRAN).

To install packages from GitHub, you need the `devtools` package. You can install it using:

```
install.packages("devtools")
```

If you for instance want to install the development version of `dplyr` (which you can find at https://github.com/tidyverse/dplyr), you can then run the following:

```
library(devtools)
install_github("tidyverse/dplyr")
```

Using development versions of packages can be great, because it gives you the most up-to-date version of packages. Bear in mind that they are development versions though, which means that they can be less stable and have more bugs.

To install packages from Bioconductor, you can start by running this code chunk, which installs the `BiocManager` package that is used to install Bioconductor packages:

```
install.packages("BiocManager")
# Install core packages:
library(BiocManager)
install()
```

You can have a look at the list of packages at:

https://www.bioconductor.org/packages/release/BiocViews.html#___Software

If you for instance find the `affyio` package interesting, you can then install it using:

```
library(BiocManager)
install("affyio")
```

12.1.4 Removing packages

This is probably not something that you'll find yourself doing often, but if you need to uninstall a package, you can do so using `remove.packages`. Perhaps you've installed the development version of a package and want to remove it, so that you can install the stable version again? If you for instance want to uninstall the `beepr` package[2], you'd run the following:

```
remove.packages("beepr")
```

12.2 Speeding up computations with parallelisation

Modern computers have CPUs with multiple cores and threads, which allows us to speed up computations by performing them in parallel. Some functions in R do this by default, but far from all do. In this section, we'll have a look at how to run parallel versions of `for` loops and functionals.

12.2.1 Parallelising `for` loops

First, we'll have a look at how to parallelise a `for` loop. We'll use the `foreach`, `parallel`, and `doParallel` packages, so let's install them if you haven't already:

```
install.packages(c("foreach", "parallel", "doParallel"))
```

To see how many cores are available on your machine, you can use `detectCores`:

```
library(parallel)
detectCores()
```

It is unwise to use all available cores for your parallel computation – you'll always need to reserve at least one for running RStudio and other applications.

To run the steps of a `for` loop in parallel, we must first use `registerDoParallel` to *register* the parallel backend to be used. Here is an example where we create three workers (and so use three cores in parallel[3]) using `registerDoParallel`. When we then use `foreach` to create a `for` loop, these three workers will execute different steps of the loop in parallel. Note that this wouldn't work if each step of the loop depended on output from the previous step.

---

[2]Why though?!

[3]If your CPU has three or fewer cores, you should lower this number.

foreach returns the output created at the end of each step of the loop in a `list` (Section 5.2):

```
library(doParallel)
registerDoParallel(3)

loop_output <- foreach(i = 1:9) %dopar%
{
    i^2
}

loop_output
unlist(loop_output) # Convert the list to a vector
```

If the output created at the end of each iteration is a vector, we can collect the output in a `matrix` object as follows:

```
library(doParallel)
registerDoParallel(3)

loop_output <- foreach(i = 1:9) %dopar%
{
    c(i, i^2)
}

loop_output
matrix(unlist(loop_output), 9, 2, byrow = TRUE)
```

If you have nested loops, you should run the outer loop in parallel, but not the inner loops. The reason for this is that parallelisation only really helps if each step of the loop takes a comparatively long time to run. In fact, there is a small overhead cost associated with assigning different iterations to different cores, meaning that parallel loops can be slower than regular loops if each iteration runs quickly.

An example where each step often takes a while to run is simulation studies. Let's rewrite the simulation we used to compute the type I error rates of different versions of the t-test in Section 7.2.2 using a parallel `for` loop instead. First, we define the function as in Section 7.2.2 (minus the progress bar):

```
# Load package used for permutation t-test:
library(MKinfer)

# Create a function for running the simulation:
simulate_type_I <- function(n1, n2, distr, level = 0.05, B = 999,
                            alternative = "two.sided", ...)
{
    # Create a data frame to store the results in:
    p_values <- data.frame(p_t_test = rep(NA, B),
                           p_perm_t_test = rep(NA, B),
                           p_wilcoxon = rep(NA, B))
```

```r
    for(i in 1:B)
    {
            # Generate data:
            x <- distr(n1, ...)
            y <- distr(n2, ...)

            # Compute p-values:
            p_values[i, 1] <- t.test(x, y,
                                alternative = alternative)$p.value
            p_values[i, 2] <- perm.t.test(x, y,
                                alternative = alternative,
                                R = 999)$perm.p.value
            p_values[i, 3] <- wilcox.test(x, y,
                                alternative = alternative)$p.value
    }

    # Return the type I error rates:
    return(colMeans(p_values < level))
}
```

Next, we create a parallel version:

```r
# Register parallel backend:
library(doParallel)
registerDoParallel(3)

# Create a function for running the simulation in parallel:
simulate_type_I_parallel <- function(n1, n2, distr, level = 0.05,
                            B = 999,
                            alternative = "two.sided", ...)
{

        results <- foreach(i = 1:B)  %dopar%
        {
                # Generate data:
                x <- distr(n1, ...)
                y <- distr(n2, ...)

                # Compute p-values:
                p_val1 <- t.test(x, y,
                                alternative = alternative)$p.value
                p_val2 <- perm.t.test(x, y,
                                alternative = alternative,
                                R = 999)$perm.p.value
                p_val3 <- wilcox.test(x, y,
                                alternative = alternative)$p.value

                # Return vector with p-values:
                c(p_val1, p_val2, p_val3)
```

```
    }

    # Each element of the results list is now a vector
    # with three elements.
    # Turn the list into a matrix:
    p_values <- matrix(unlist(results), B, 3, byrow = TRUE)

    # Return the type I error rates:
    return(colMeans(p_values < level))
}
```

We can now compare how long the two functions take to run using the tools from Section 6.6 (we'll not use mark in this case, as it requires both functions to yield identical output, which won't be the case for a simulation):

```
time1 <- system.time(simulate_type_I(20, 20, rlnorm,
                                      B = 999, sdlog = 3))
time2 <- system.time(simulate_type_I_parallel(20, 20, rlnorm,
                                               B = 999, sdlog = 3))

# Compare results:
time1; time2; time2/time1
```

As you can see, the parallel function is considerably faster. If you have more cores, you can try increasing the value in `registerDoParallel` and see how that affects the results.

12.2.2 Parallelising functionals

The `parallel` package contains parallelised versions of the `apply` family of functions, with names like `parApply`, `parLapply`, and `mclapply`. Which of these you should use depends in part on your operating system, as different operating systems handle multicore computations differently. Here is the first example from Section 6.5.3, run in parallel with three workers:

```
# Non-parallel version:
lapply(airquality, function(x) { (x-mean(x))/sd(x) })

# Parallel version for Linux/Mac:
library(parallel)
mclapply(airquality, function(x) { (x-mean(x))/sd(x) },
         mc.cores = 3)

# Parallel version for Windows (a little slower):
library(parallel)
myCluster <- makeCluster(3)
parLapply(myCluster, airquality, function(x) { (x-mean(x))/sd(x) })
stopCluster(myCluster)
```

Similarly, the `furrr` package lets us run `purrr` functionals in parallel. It relies on a package called `future`. Let's install them both:

```r
install.packages(c("future", "furrr"))
```

To run functionals in parallel, we load the `furrr` package and use `plan` to set the number of parallel workers:

```r
library(furrr)
# Use 3 workers:
plan(multisession, workers = 3)
```

We can then run parallel versions of functions like `map` and `imap`, by using functions from `furrr` with the same names, only with `future_` added at the beginning. Here is the first example from Section 6.5.3, run in parallel:

```r
library(magrittr)
airquality %>% future_map(~(.-mean(.))/sd(.))
```

Just as for `for` loops, parallelisation of functionals only really helps if each iteration of the functional takes a comparatively long time to run (and so there is no benefit to using parallelisation in this particular example).

12.3   Linear algebra and matrices

Linear algebra is the beating heart of many statistical methods. R has a wide range of functions for creating and manipulating matrices, and doing matrix algebra. In this section, we'll have a look at some of them.

12.3.1   Creating matrices

To create a `matrix` object, we can use the `matrix` function. It always coerces all elements to be of the same type (Section 5.1):

```r
# Create a 3x2 matrix, one column at a time:
matrix(c(2, -1, 3, 1, -2, 4), 3, 2)

# Create a 3x2 matrix, one row at a time:
# (No real need to include line breaks in the vector with
# the values, but I like to do so to see what the matrix
# will look like!)
matrix(c(2, -1,
         3, 1,
         -2, 4), 3, 2, byrow = TRUE)
```

Matrix operations require the dimension of the matrices involved to match. To check the dimension of a `matrix`, we can use `dim`:

```
A <- matrix(c(2, -1, 3, 1, -2, 4), 3, 2)
dim(A)
```

To create a unit matrix (all 1s) or a zero matrix (all 0s), we use `matrix` with a single value in the first argument:

```
# Create a 3x3 unit matrix:
matrix(1, 3, 3)
```

```
# Create a 2x3 zero matrix:
matrix(0, 2, 3)
```

The `diag` function has three uses. First, it can be used to create a diagonal matrix (if we supply a vector as input). Second, it can be used to create an identity matrix (if we supply a single number as input). Third, it can be used to extract the diagonal from a square matrix (if we supply a matrix as input). Let's give it a go:

```
# Create a diagonal matrix with 2, 4, 6 along the diagonal:
diag(c(2, 4, 6))
```

```
# Create a 9x9 identity matrix:
diag(9)
```

```
# Create a square matrix and then extract its diagonal:
A <- matrix(1:9, 3, 3)
A
diag(A)
```

Similarly, we can use `lower.tri` and `upper.tri` to extract a matrix of `logical` values, describing the location of the lower and upper triangular part of a matrix:

```
# Create a matrix_
A <- matrix(1:9, 3, 3)
A
```

```
# Which are the elements in the lower triangular part?
lower.tri(A)
A[lower.tri(A)]
```

```
# Set the lower triangular part to 0:
A[lower.tri(A)] <- 0
A
```

To transpose a matrix, use `t`:

```
t(A)
```

Matrices can be combined using `cbind` and `rbind`:

```r
A <- matrix(c(1:3, 3:1, 2, 1, 3), 3, 3, byrow = TRUE)   # 3x3
B <- matrix(c(2, -1, 3, 1, -2, 4), 3, 2)                # 3x2

# Add B to the right of A:
cbind(A, B)

# Add the transpose of B below A:
rbind(A, t(B))

# Adding B below A doesn't work, because the dimensions
# don't match:
rbind(A, B)
```

12.3.2 Sparse matrices

The `Matrix` package contains functions for creating and speeding up computations with sparse matrices (i.e., matrices with lots of 0s), as well as for creating matrices with particular structures. You likely already have it installed, as many other packages rely on it. `Matrix` distinguishes between sparse and dense matrices:

```r
# Load or/and install Matrix:
if(!require("Matrix")) { install.packages("Matrix"); library(Matrix) }

# Create a dense 8x8 matrix using the Matrix package:
A <- Matrix(1:64, 8, 8)

# Create a copy and randomly replace 40 elements by 0:
B <- A
B[sample(1:64, 40)] <- 0
B

# Store B as a sparse matrix instead:
B <- as(B, "sparseMatrix")
B
```

To visualise the structure of a sparse matrix, we can use `image`:

```r
image(B)
```

An example of a slightly larger, 72×72 sparse matrix is given by `CAex`:

```r
data(CAex)
CAex
image(CAex)
```

`Matrix` contains additional classes for, e.g., symmetric sparse matrices and triangular matrices. See `vignette("Introduction", "Matrix")` for further details.

12.3.3   Matrix operations

In this section, we'll use the following matrices and vectors to show how to perform various matrix operations:

```
# Matrices:
A <- matrix(c(1:3, 3:1, 2, 1, 3), 3, 3, byrow = TRUE)  # 3x3
B <- matrix(c(2, -1, 3, 1, -2, 4), 3, 2)               # 3x2
C <- matrix(c(4, 1, 1, 2), 2, 2)           # Symmetric 2x2

# Vectors:
a <- 1:9                   # Length 9
b <- c(2, -1, 3, 1, -2, 4) # Length 6
d <- 9:1                   # Length 9
y <- 4:6                   # Length 3
```

To perform element-wise addition and subtraction with matrices, use + and -:

```
A + A
A - t(A)
```

To perform element-wise multiplication, use *:

```
2 * A  # Multiply all elements by 2
A * A  # Square all elements
```

To perform matrix multiplication, use %*%. Remember that matrix multiplication is non-commutative, and so the order of the matrices is important:

```
A %*% B  # A is 3x3, B is 3x2
B %*% C  # B is 3x2, C is 2x2
B %*% A  # Won't work, because B is 3x2 and A 3x3!
```

Given the vectors a, b, and d defined above, we can compute the outer product $a \otimes b$ using %o% and the dot product $a \cdot d$ by using %*% and t in the right manner:

```
a %o% b  # Outer product
a %*% t(b) # Alternative way of getting the outer product
t(a) %*% d # Dot product
```

To find the inverse of a square matrix, we can use solve. To find the generalised Moore-Penrose inverse of any matrix, we can use ginv from MASS:

```
solve(A)
solve(B)  # Doesn't work because B isn't square

library(MASS)
ginv(A)  # Same as solve(A), because A is non-singular and square
ginv(B)
```

`solve` can also be used to solve equations systems. To solve the equation $Ax = y$:

```
solve(A, y)
```

The eigenvalues and eigenvectors of a square matrix can be found using `eigen`:

```
eigen(A)
eigen(A)$values    # Eigenvalues only
eigen(A)$vectors   # Eigenvectors only
```

The singular value decomposition, QR decomposition, and the Choleski factorisation of a matrix are computed as follows:

```
svd(A)
qr(A)
chol(C)
```

`qr` also provides the rank of the matrix:

```
qr(A)$rank
qr(B)$rank
```

Finally, you can get the determinant[4] of a matrix using `det`:

```
det(A)
```

As a P.S., I'll also mention the `matlab` package, which contains functions for running computations using MATLAB®-like function calls. This is useful if you want to reuse MATLAB code in R without translating it row-by-row. Incidentally, this also brings us nicely into the next section.

12.4   Integration with other programming languages

R is great for a lot of things, but it is obviously not the best choice for every task. There are a number of packages that can be used to harvest the power of other languages, or to integrate your R code with code that you or others have developed in other programming languages. In this section, we'll mention a few of them.

12.4.1   Integration with C++

C++ is commonly used to speed up functions, for instance involving loops that can't be vectorised or parallelised due to dependencies between different iterations. The `Rcpp` package (Eddelbuettel & Balamuta, 2018) allows you to easily call C++ functions from R, as well as calling R functions from C++. See `vignette("Rcpp-introduction", "Rcpp")` for details.

---

[4]Do you *really* need it, though?

An important difference between R and C++ that you should be aware of is that the indexing of vectors (and similar objects) in C++ starts with 0. So the first element of the vector is element 0, the second is element 1, and so forth. Bear this in mind if you pass a vector and a list of indices to C++ functions.

12.4.2   Integration with Python

The `reticulate` package can be used to call Python functions from R. See

```
vignette("calling_python", "reticulate")
```

for some examples.

Some care has to be taken when sending data back and forth between R and Python. In R `NA` is used to represent missing data and `NaN` (not a number) is used to represent things that should be numbers but aren't (e.g., the result of computing `0/0`). Perfectly reasonable! However, for reasons unknown to humanity, popular Python packages like Pandas, NumPy and SciKit-Learn use `NaN` instead of `NA` to represent missing data – but only for `double` (`numeric`) variables. `integer` and `logical` variables have no way to represent missing data in Pandas. Tread gently if there are `NA` or `NaN` values in your data.

Like in C++, the indexing of vectors (and similar objects) in Python starts with 0.

12.4.3   Integration with Tensorflow and PyTorch

Tensorflow, Keras, and PyTorch are popular frameworks for deep learning. To use Tensorflow or Keras with R, you can use the `keras` package. See `vignette("index", "keras")` for an introduction and Chollet & Allaire (2022) for a thorough treatise. Similarly, to use PyTorch with R, use the `torch` package. In both cases, it can take some tampering to get the frameworks to run on a GPU.

12.4.4   Integration with Spark

If you need to process large datasets using Spark, you can do so from R using the `sparklyr` package. It can be used both with local and cloud clusters, and (as the name seems to imply) it is easy to integrate with `dplyr`.

13

Debugging

In Section 2.18, I gave some general advice about what to do when there is an error in your R code:

1. Read the error message carefully and try to decipher it. Have you seen it before? Does it point to a particular variable or function? Check Section 13.2 of this book, which deals with common error messages in R.

2. Check your code. Have you misspelt any variable or function names? Are there missing brackets, strange commas, or invalid characters?

3. Copy the error message and do a web search using the message as your search term. It is more than likely that somebody else has encountered the same problem, and that you can find a solution to it online. This is a great shortcut for finding solutions to your problem. In fact, **this may well be the single most important tip in this entire book**.

4. Read the documentation for the function causing the error message, and look at some examples of how to use it (both in the documentation and online, e.g., in blog posts). Have you used it correctly?

5. Use the debugging tools presented in Chapter 13, or try to simplify the example that you are working with (e.g., removing parts of the analysis or the data) and see if that removes the problem.

6. If you still can't find a solution, post a question at a site like Stack Overflow[1] or the RStudio community forums[2]. Make sure to post your code and describe the context in which the error message appears. If at all possible, post a reproducible example, i.e., a piece of code that others can run, that causes the error message. This will make it a lot easier for others to help you.

The debugging tools mentioned in point 5 above are an important part of your toolbox, particularly if you're doing more advanced programming with R.

In this chapter you will learn how to:

- Debug R code,
- Recognise and resolve common errors in R code, and
- Interpret and resolve common warning messages in R.

---

[1] https://stackoverflow.com/
[2] https://community.rstudio.com/

13.1   Debugging

Debugging is the process of finding and removing bugs in your scripts. R and RStudio have several functions that can be used for this purpose. We'll have a closer look at some of them here.

13.1.1   Find out where the error occured with `traceback`

If a function returns an error, it is not always clear *where* exactly the error occurred. Let's say that we want to compute the correlation between two variables, but we have forgotten to assign values to the variables:

```
cor(variable1, variable2)
```

The resulting error message is:

```
> cor(variable1, variable2)
Error in is.data.frame(y) : object 'variable2' not found
```

Why is the function `is.data.frame` throwing an error? We were using `cor`, not `is.data.frame`!

Functions often make calls to other functions, which in turn make calls to other functions, and so on. When you get an error message, the error could have taken place in any one of these functions. To find out in which function the error occurred, you can run `traceback`, which shows the sequence of calls that lead to the error:

```
traceback()
```

Which in this case will yield the output:

```
> traceback()
2: is.data.frame(y)
1: cor(variable1, variable2)
```

What this tells you is that `cor` makes a call to `is.data.frame`, and that is where the error occurs. This can help you understand why a function that you weren't aware you were calling (`is.data.frame` in this case) is throwing an error, but won't tell you *why* there was an error. To find out, you can use `debug`, which we'll discuss next.

As a side note, if you'd like to know why and when `cor` called `is.data.frame` you can print the code for `cor` in the Console by typing the function name without parentheses:

```
cor
```

Reading the output, you can see that it makes a call to `is.data.frame` on the 10th line:

```
 1  function (x, y = NULL, use = "everything", method = c("pearson",
 2      "kendall", "spearman"))
 3  {
 4      na.method <- pmatch(use, c("all.obs", "complete.obs",
 5                                  "pairwise.complete.obs",
 6          "everything", "na.or.complete"))
 7      if (is.na(na.method))
 8          stop("invalid 'use' argument")
 9      method <- match.arg(method)
10      if (is.data.frame(y))
11          y <- as.matrix(y)
```

...

13.1.2 Interactive debugging of functions with `debug`

If you are looking for an error in a script, you can simply run the script one line at a time until the error occurs, to find out where the error is. But what if the error is inside of a function, as in the example above?

Once you know in which function the error occurs, you can have a look inside it using debug. debug takes a function name as input, and the next time you run that function, an interactive debugger starts, allowing you to step through the function one line at a time. That way, you can find out exactly where in the function the error occurs. We'll illustrate its use with a custom function:

```
transform_number <- function(x)
{
  square <- x^2
  if(x >= 0) { logx <- log(x) } else { stop("x must be positive") }
  if(x >= 0) { sqrtx <- sqrt(x) } else { stop("x must be positive") }
  return(c(x.squared = square, log = logx, sqrt = sqrtx))
}
```

The function appears to work just fine:

```
transform_number(2)
transform_number(-1)
```

However, if we input an NA, an error occurs:

```
transform_number(NA)
```

We now run debug:

```
debug(transform_number)
transform_number(NA)
```

Two things happen. First, a tab with the code for transform_number opens. Second, a *browser* is initialised in the Console panel. This allows you to step through the code, by typing one of the following and pressing Enter:

- n to run the next line,
- c to run the function until it finishes or an error occurs,
- a variable name to see the current value of that variable (useful for checking that variables have the intended values),
- Q to quit the browser and stop the debugging.

If you either use n a few times, or c, you can see that the error occurs on line number 4 of the function:

```
if(x >= 0) { logx <- log(x) } else { stop("x must be positive") }
```

Because this function was so short, you could probably see that already, but for longer and more complex functions, debug is an excellent way to find out where exactly the error occurs.

The browser will continue to open for debugging each time transform_number is run. To turn it off, use undebug:

```
undebug(transform_number)
```

13.1.3 Investigate the environment with recover

By default, R prints an error message, returns to the global environment, and stops the execution when an error occurs. You can use recover to change this behaviour so that R stays in the environment where the error occurred. This allows you to investigate that environment, e.g., to see if any variables have been assigned the wrong values.

```
transform_number(NA)
recover()
```

This gives you the same list of function calls as traceback (called the *function stack*), and you can select which of these you'd like to investigate (in this case there is only one, which you access by writing 1 and pressing Enter). The environment for that call shows up in the Environment panel, which in this case shows you that the local variable x has been assigned the value NA (which is what causes an error when the condition x >= 0 is checked).

13.2 Common error messages

Some errors are more frequent than others. Below is a list of some of the most common ones, along with explanations of what they mean, and how to resolve them.

13.2.1 +

If there is a + sign at the beginning of the last line in the Console, and it seems that your code doesn't run, that is likely due to missing brackets or quotes. Here is an example where a bracket is missing:

```
> 1 + 2*(3 + 2
+
```

Type) in the Console to finish the expression, and your code will run. The same problem can occur if a quote is missing:

```
> myString <- "Good things come in threes
+
```

Type " in the Console to finish the expression, and your code will run.

13.2.2 could not find function

This error message appears when you try to use a function that doesn't exist. Here is an example:

```
> age <- c(28, 48, 47, 71, 22, 80, 48, 30, 31)
> means(age)
Error in means(age) : could not find function "means"
```

This error is either due to a misspelling (in which case you should fix the spelling) or due to attempting to use a function from a package that hasn't been loaded (in which case you should load the package using library(package_name)). If you are unsure which package the function belongs to, doing a quick web search for "R function_name" usually does the trick.

13.2.3 object not found

R throws this error message if we attempt to use a variable that does not exist:

```
> name_of_a_variable_that_doesnt_exist + 1 * pi^2
Error: object 'name_of_a_variable_that_doesnt_exist' not found
```

This error may be due to a spelling error, so check the spelling of the variable name. It is also commonly encountered if you return to an old R script and try to run just a part of it – if the variable is created on an earlier line that hasn't been run, R won't find it because it hasn't been created yet.

13.2.4 cannot open the connection and No such file or directory

This error message appears when you try to load a file that doesn't exist:

```
> read.csv("not-a-real-file-name.csv")
Error in file(file, "rt") : cannot open the connection
In addition: Warning message:
In file(file, "rt") :
  cannot open file 'not-a-real-file-name.csv': No such file or
  directory
```

Check the spelling of the file name, and that you have given the correct path to it (see Section 2.15). If you are unsure about the path, you can use

```
read.csv(file.choose())
```

to interactively search for the file in question.

13.2.5 invalid 'description' argument

When you try to import data from an Excel file, you can run into error messages like:

```
Error in file(con, "r") : invalid 'description' argument
In addition: Warning message:
In unzip(xlsxFile, exdir = xmlDir) : error 1 in extracting from zip file
```

and

```
Error: Evaluation error: zip file 'C:\Users\mans\Data\some_file.xlsx' cannot be
opened.
```

These usually appear if you have the file open in Excel *at the same time* that you're trying to import data from it in R. Excel temporarily locks the file so that R can't open it. Close Excel and then import the data.

13.2.6 missing value where TRUE/FALSE needed

This message appears when a condition in a conditional statement evaluates to NA. Here is an example:

```
x <- c(8, 5, 9, NA)
for(i in seq_along(x))
{
    if(x[i] > 7) { cat(i, "\n") }
}
```

which yields:

```
> x <- c(8, 5, 9, NA)
> for(i in seq_along(x))
+ {
+     if(x[i] > 7) { cat(i, "\n") }
+ }
1
3
Error in if (x[i] > 7) { : missing value where TRUE/FALSE needed
```

The error occurs when i is 4, because the expression x[i] > 7 becomes NA > 7, which evaluates to NA. if statements require that the condition evaluates to either TRUE or FALSE. When this error occurs, you should investigate why you get an NA instead.

13.2.7   unexpected '=' in ...

This message indicates that you have an assignment happening in the wrong place. You probably meant to use == to check for equality, but accidentally wrote = instead, as in this example:

```
x <- c(8, 5, 9, NA)
for(i in seq_along(x))
{
    if(x[i] = 5) { cat(i, "\n") }
}
```

which yields:

```
> x <- c(8, 5, 9, NA)
> for(i in seq_along(x))
+ {
+     if(x[i] = 5) { cat(i, "\n") }
Error: unexpected '=' in:
"{
    if(x[i] ="
> }
Error: unexpected '}' in "}"
```

Replace the = by == and your code should run as intended. If you really intended to assign a value to a variable inside the if condition, you should probably rethink that.

13.2.8   attempt to apply non-function

This error occurs when you put parentheses after something that isn't a function. It is easy to make that mistake, e.g., when doing a mathematical computation.

```
> 1+2(2+3)
Error: attempt to apply non-function
```

In this case, we need to put a multiplication symbol * between 2 and (to make the code run:

```
> 1+2*(2+3)
[1] 11
```

13.2.9 undefined columns selected

If you try to select a column that doesn't exist from a data frame, this message will be printed. Let's start by defining an example data frame:

```
age <- c(28, 48, 47, 71, 22, 80, 48, 30, 31)
purchase <- c(20, 59, 2, 12, 22, 160, 34, 34, 29)
bookstore <- data.frame(age, purchase)
```

If we attempt to access the third column of the data, we get the error message:

```
> bookstore[,3]
Error in `[.data.frame`(bookstore, , 3) : undefined columns selected
```

Check that you really have the correct column number. It is common to get this error if you have removed columns from your data.

13.2.10 subscript out of bounds

This error message is similar to the last example above but occurs if you try to access the column in another way:

```
> bookstore[[3]]
Error in .subset2(x, i, exact = exact) : subscript out of bounds
```

Check that you really have the correct column number. It is common to get this error if you have removed columns from your data, or if you are running a for loop accessing element [i, j] of your data frame, where either i or j is greater than the number of rows and columns of your data.

13.2.11 Object of type 'closure' is not subsettable

This error occurs when you use square brackets [ ] directly after a function:

```
> x <- c(8, 5, 9, NA)
> sqrt[x]
Error in sqrt[x] : object of type 'closure' is not subsettable
```

You probably meant to use parentheses ( ) instead. Or perhaps you wanted to use the square brackets on the object returned by the function:

```
> sqrt(x)[2]
[1] 2.236068
```

13.2.12   `$ operator is invalid for atomic vectors`

This message is printed when you try to use the $ operator with an object that isn't a `list` or a data frame, for instance with a vector. Even though the elements in a vector can be named, you cannot access them using $:

```
> x <- c(a = 2, b = 3)
> x
a b
2 3
> x$a
Error in x$a : $ operator is invalid for atomic vectors
```

If you need to access the element named a, you can do so using bracket notation:

```
> x["a"]
a
2
```

Or use a data frame instead:

```
> x <- data.frame(a = 2, b = 3)
> x$a
[1] 2
```

13.2.13   `(list) object cannot be coerced to type 'double'`

This error occurs when you try to convert the elements of a `list` to `numeric`. First, we create a `list`:

```
x <- list(a = c("1", "2", "3"),
          b = c("1", "4", "1889"))
```

If we now try to apply `as.numeric`, we get the error:

```
> as.numeric(x)
Error: 'list' object cannot be coerced to type 'double'
```

You can apply `unlist` to collapse the `list` to a vector:

```
as.numeric(unlist(x))
```

You can also use `lapply` (see Section 6.5):

```
lapply(x, as.numeric)
```

13.2.14 arguments imply differing number of rows

This message is printed when you try to create a data frame with different numbers of rows for different columns, like in this example, where a has three rows and b has four:

```
> x <- data.frame(a = 1:3, b = 6:9)
Error in data.frame(a = 1:3, b = 6:9) :
  arguments imply differing number of rows: 3, 4
```

If you really need to create an object with different numbers of rows for different columns, create a list instead:

```
x <- list(a = 1:3, b = 6:9)
```

13.2.15 non-numeric argument to a binary operator

This error occurs when you try to use mathematical operators with non-numerical variables. For instance, it occurs if you try to add character variables:

```
> "Hello" + "World"
Error in "Hello" + "world" : non-numeric argument to binary operator
```

If you want to combine character variables, use paste instead:

```
paste("Hello", "world")
```

13.2.16 non-numeric argument to mathematical function

This error message is similar to the previous one, and it appears when you try to apply a mathematical function, like log or exp to non-numerical variables:

```
> log("1")
Error in log("1") : non-numeric argument to mathematical function
```

Make sure that the data you are inputting doesn't contain character variables.

13.2.17 cannot allocate vector of size ...

This message is shown when you're trying to create an object that would require more RAM than is available. You can try to free up RAM by closing other programs and removing data that you don't need using rm (see Section 5.14). Also check your code so that you don't make copies of your data, which takes up more RAM. Replacing base R and dplyr code for data wrangling with data.table code can also help, as data.table uses considerably less RAM for most tasks.

13.2.18 Error in plot.new() : figure margins too large

This error occurs when your Plot panel (or file, if you are saving your plot as a graphics file) is too small to fit the graphic that you're trying to create. Enlarge your Plot panel (or increase the size of the graphics file) and run the code again.

13.2.19 Error in .Call.graphics(C_palette2, .Call(C_palette2, NULL)) : invalid graphics state

This error can happen when you create plots with ggplot2. You can usually solve it by running dev.off() to close the previous plot window. In rare cases, you may have to reinstall ggplot2 (see Section 12.1).

13.2.20 Error in select(...) : unused argument (...)

This error occurs when you attempt to use select from dplyr after loading the MASS package, which also has a function called select. To solve it, specify that you wish to use select from dplyr:

```
airquality |> dplyr::select(Ozone)
```

13.3 Common warning messages

13.3.1 replacement has ... rows ...

This occurs when you try to assign values to rows in a data frame, but the object you are assigning to them has a different number of rows. Here is an example:

```
> x <- data.frame(a = 1:3, b = 6:8)
> y <- data.frame(a = 4:5, b = 10:11)
> x[3,] <- y
Warning messages:
1: In `[<-.data.frame`(`*tmp*`, 3, , value = list(a = 4:5, b = 10:11)) :
  replacement element 1 has 2 rows to replace 1 rows
2: In `[<-.data.frame`(`*tmp*`, 3, , value = list(a = 4:5, b = 10:11)) :
  replacement element 2 has 2 rows to replace 1 rows
```

You can fix this, e.g., by changing the number of rows to place the data in:

```
x[3:4,] <- y
```

13.3.2 the condition has length > 1 and only the first element will be used

This warning is thrown when the condition in a conditional statement is a vector rather than a single value. Here is an example:

```
> x <- 1:3
> if(x == 2) { cat("Two!") }
Warning message:
In if (x == 2) { :
  the condition has length > 1 and only the first element will be used
```

Only the first element of the vector is used for evaluating the condition. See if you can change the condition so that it doesn't evaluate to a vector. If you actually want to evaluate the condition for all elements of the vector, either collapse it using any or all or wrap it in a loop:

```
x <- 1:3
if(any(x == 2)) { cat("Two!") }

for(i in seq_along(x))
{
    if(x[i] == 2) { cat("Two!") }
}
```

13.3.3 number of items to replace is not a multiple of replacement length

This error occurs when you try to assign too many values to too short a vector. Here is an example:

```
> x <- c(8, 5, 9, NA)
> x[4] <- c(5, 7)
Warning message:
In x[4] <- c(5, 7) :
  number of items to replace is not a multiple of replacement length
```

Don't try to squeeze more values than can fit into a single element! Instead, do something like this:

```
x[4:5] <- c(5, 7)
```

13.3.4 longer object length is not a multiple of shorter object length

This warning is printed, e.g., when you try to add two vectors of different lengths together. If you add two vectors of equal lengths, everything is fine:

```
a <- c(1, 2, 3)
b <- c(4, 5, 6)
a + b
```

R does *element-wise* addition, i.e., adds the first element of a to the first element of b, and so on.

But what happens if we try to add two vectors of different lengths together?

```
a <- c(1, 2, 3)
b <- c(4, 5, 6, 7)
a + b
```

This yields the following warning message:

```
> a + b
[1] 5 7 9 8
Warning message:
In a + b : longer object length is not a multiple of shorter object length
```

R *recycles* the numbers in a in the addition, so that the first element of a is added to the fourth element of b. Was that really what you wanted? Maybe. But probably not.

13.3.5 NAs introduced by coercion

This warning is thrown when you try to convert something that cannot be converted to another data type:

```
> as.numeric("two")
[1] NA
Warning message:
NAs introduced by coercion
```

You can try using gsub to manually replace values instead:

```
x <- c("one", "two")
x <- gsub("one", 1, x)
as.numeric(x)
```

13.3.6 package is not available (for R version x.x.x)

This warning message (which perhaps should be an error message rather than a warning) occurs when you try to install a package that isn't available for the version of R that you are using.

```
> install.packages("great_name_for_a_package")
Installing package into '/home/mans/R/x86_64-pc-linux-gnu-library/4.4'
(as 'lib' is unspecified)
Warning in install.packages :
  package 'great_name_for_a_package' is not available (for R version
  4.4.0)
```

This can be either due to the fact that you've misspelt the package name or that the package isn't available for your version of R, either because you are using an out-of-date version or because the package was developed for an older version of R. In the former case, consider updating to a newer version of R. In the latter case, if you really need the package, you can find and download older version of R at R-project.org[3].

On Windows, it is relatively easy to have multiple versions of R installed side-by-side (in RStudio, you can choose which version to use through the menus, by going to *Tools > Global options* and choosing *R version*. On a Mac, you can use the R Switch app available from: https://rud.is/rswitch/

---

[3]https://cran.r-project.org/bin/windows/base/old/

14

Mathematical appendix

This chapter contains remarks regarding the mathematical background of certain methods and concepts encountered in Chapters 7-11. Sections 14.2 and 14.3 consist of reworked materials from Thulin (2014b). Most of this chapter assumes some familiarity with mathematical statistics, on the level of Casella & Berger (2002) or Liero & Zwanzig (2012).

14.1 Bootstrap confidence intervals

We wish to construct a confidence interval for a parameter θ based on a statistic t. Let t_{obs} be the value of the statistic in the original sample, t_i^* be a bootstrap replicate of the statistic, for $i = 1, 2, \ldots, B$, and t^* be the mean of the statistic among the bootstrap replicates. Let se^* be the standard error of the bootstrap estimate, and $b^* = t^* - t_{obs}$ be the bias of the bootstrap estimate. For a confidence level $1 - \alpha$ ($\alpha = 0.05$ being a common choice), let $z_{\alpha/2}$ be the $1 - \frac{\alpha}{2}$ quantile of the standard normal distribution (with $z_{0.025} = 1.9599\ldots$). Moreover, let $\theta_{\alpha/2}$ be the $1 - \frac{\alpha}{2}$-quantile of the bootstrap distribution of the t_i^*'s.

The *bootstrap normal confidence interval* is

$$t_{obs} - b^* \pm z_{\alpha/2} \cdot se^*.$$

The *bootstrap basic confidence interval* is

$$\left(2t_{obs} - \theta_{\alpha/2}, 2t_{obs} - \theta_{1-\alpha/2} \right).$$

The *bootstrap percentile confidence interval* is

$$\left(\theta_{1-\alpha/2}, \theta_{\alpha/2} \right).$$

For the *bootstrap BCa confidence interval*, let

$$\hat{z} = \Theta^{-1}\left(\frac{\#\{t_i^* < t_{obs}\}}{B} \right),$$

where Θ is the cumulative distribution function for the normal distribution. Let $t_{(-i)}^*$ be the mean of the bootstrap replicates after deleting the i:th replicate, and define the acceleration term

$$\hat{a} = \frac{\sum_{i=1}^n (t^* - t_{(-i)}^*)}{6 \left(\sum_{i=1}^n (t^* - t_{(-i)}^*)^2 \right)^{3/2}}.$$

Finally, let

$$\alpha_1 = \Theta\left(\hat{z} + \frac{\hat{z} + z_{1-\alpha/2}}{1 - \hat{a}(\hat{z} + z_{1-\alpha/2})}\right)$$

and

$$\alpha_2 = \Theta\left(\hat{z} + \frac{\hat{z} + z_{\alpha/2}}{1 - \hat{a}(\hat{z} + z_{\alpha/2})}\right).$$

Then the confidence interval is

$$\left(\theta_{\alpha_1}, \theta_{\alpha_2}\right).$$

For the *studentised bootstrap confidence interval*, we additionally have an estimate se_t^* for the standard error of the statistic. Moreover, we compute $q_i = \frac{t_i^* - t_{obs}}{se_t^*}$ for each bootstrap replicate, and define $q_{\alpha/2}$ as the $1 - \frac{\alpha}{2}$-quantile of the bootstrap distribution of q_i's. The confidence interval is then

$$\left(t_{obs} - se_t^* \cdot q_{\alpha}/2, t_{obs} + se_t^* \cdot q_{1-\alpha/2}\right).$$

14.2 The equivalence between confidence intervals and hypothesis tests

Let θ be an unknown parameter in the parameter space $\Theta \subseteq \mathbb{R}$, and let the sample $\mathbf{x} = (x_1, \ldots, x_n) \in \mathcal{X}^n \subseteq \mathbb{R}^n$ be a realisation of the random variable $\mathbf{X} = (X_1, \ldots, X_n)$. In frequentist statistics, there is a fundamental connection between interval estimation and point-null hypothesis testing of θ, which we describe next. We define a confidence interval $I_\alpha(\mathbf{X})$ as a random interval such that its *coverage probability*

$$P_\theta(\theta \in I_\alpha(\mathbf{X})) = 1 - \alpha \qquad \text{for all } \alpha \in (0, 1).$$

Consider a two-sided test of the point-null hypothesis $H_0(\theta_0) : \theta = \theta_0$ against the alternative $H_1(\theta_0) : \theta \neq \theta_0$. Let $\lambda(\theta_0, \mathbf{x})$ denote the p-value of the test. For any $\alpha \in (0, 1)$, $H_0(\theta_0)$ is rejected at the level α if $\lambda(\theta_0, x) \leq \alpha$. The level α *rejection region* is the set of $\mathbf{x}$ which lead to the rejection of $H_0(\theta_0)$:

$$R_\alpha(\theta_0) = \{\mathbf{x} \in \mathbb{R}^n : \lambda(\theta_0, \mathbf{x}) \leq \alpha\}.$$

Now, consider a family of two-sided tests with p-values $\lambda(\theta, \mathbf{x})$, for $\theta \in \Theta$. For such a family we can define an *inverted rejection region*

$$Q_\alpha(\mathbf{x}) = \{\theta \in \Theta : \lambda(\theta, \mathbf{x}) \leq \alpha\}.$$

For any fixed θ_0, $H_0(\theta_0)$ is rejected if $\mathbf{x} \in R_\alpha(\theta_0)$, which happens if and only if $\theta_0 \in Q_\alpha(\mathbf{x})$, that is,

$$\mathbf{x} \in R_\alpha(\theta_0) \Leftrightarrow \theta_0 \in Q_\alpha(\mathbf{x}).$$

If the test is based on a test statistic with a completely specified absolutely continuous null distribution, then $\lambda(\theta_0, \mathbf{X}) \sim U(0,1)$ under $H_0(\theta_0)$ (Liero & Zwanzig, 2012). Then

$$P_{\theta_0}(\mathbf{X} \in R_\alpha(\theta_0)) = P_{\theta_0}(\lambda(\theta_0, \mathbf{X}) \leq \alpha) = \alpha.$$

Since this holds for any $\theta_0 \in \Theta$ and since the equivalence relation $\mathbf{x} \in R_\alpha(\theta_0) \Leftrightarrow \theta_0 \in Q_\alpha(\mathbf{x})$ implies that

$$P_{\theta_0}(\mathbf{X} \in R_\alpha(\theta_0)) = P_{\theta_0}(\theta_0 \in Q_\alpha(\mathbf{X})),$$

it follows that the random set $Q_\alpha(\mathbf{x})$ always covers the true parameter θ_0 with probability α. Consequently, letting $Q_\alpha^C(\mathbf{x})$ denote the complement of $Q_\alpha(\mathbf{x})$, for all $\theta_0 \in \Theta$ we have

$$P_{\theta_0}(\theta_0 \in Q_\alpha^C(\mathbf{X})) = 1 - \alpha,$$

meaning that the complement of the inverted rejection region is a $1 - \alpha$ confidence interval for θ. This equivalence between a family of tests and a confidence interval $I_\alpha(\mathbf{x}) = Q_\alpha^C(\mathbf{x})$, illustrated in Figure 14.1, provides a simple way of constructing confidence intervals through test inversion, and vice versa.

The figure shows the rejection regions and confidence intervals corresponding to the z-test for a normal mean, for different null means θ and different sample means $\bar{x}$, with $\sigma = 1$. $H_0(\theta)$ is rejected if $(\bar{x}, \theta)$ is in the shaded light grey region. Shown in dark grey is the rejection region $R_{0.05}(-0.9) = (-\infty, -1.52) \cup (-0.281, \infty)$ and the confidence interval $I_{0.05}(1/2) = Q_{0.05}^C(1/2) = (-0.120, 1.120)$.

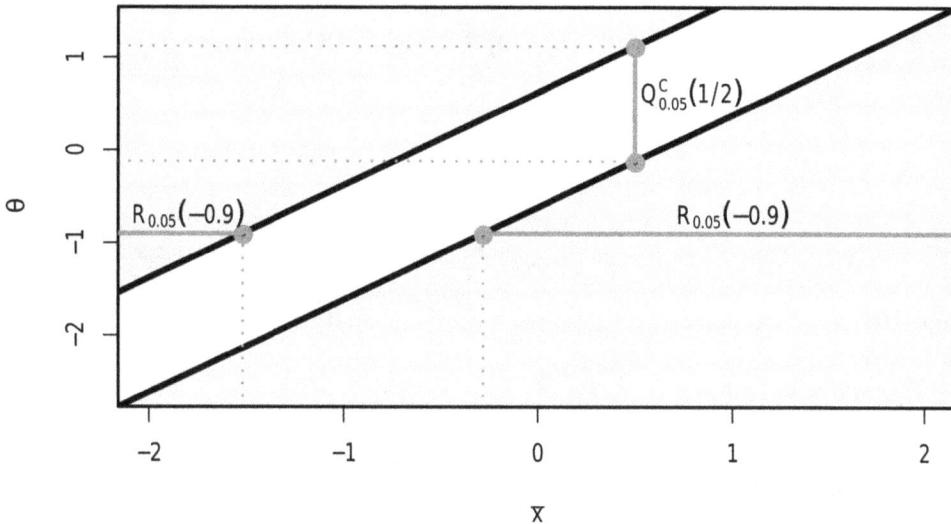

FIGURE 14.1 Equivalence between confidence intervals and hypothesis tests.

14.3 Two types of p-values

The `symmetric` argument in `perm.t.test` and `boot.t.test` controls how the p-values of the test are computed. In most cases, the difference is not that large:

```
library(MKinfer)
library(ggplot2)
boot.t.test(sleep_total ~ vore, data =
            subset(msleep, vore == "carni" | vore == "herbi"),
          symmetric = FALSE)

boot.t.test(sleep_total ~ vore, data =
            subset(msleep, vore == "carni" | vore == "herbi"),
          symmetric = TRUE)
```

In other cases, the choice matters more. Below, we will discuss the difference between the two approaches.

Let $T(\mathbf{X})$ be a test statistic on which a two-sided test of the point-null hypothesis that $\theta = \theta_0$ is based, and let $\lambda(\theta_0, \mathbf{x})$ denote its p-value. Assume for simplicity that $T(\mathbf{x}) < 0$ implies that $\theta < \theta_0$ and that $T(\mathbf{x}) > 0$ implies that $\theta > \theta_0$. We'll call the symmetric = FALSE scenario the *twice-the-smaller-tail* approach to computing p-values. In it, the first step is to check whether $T(\mathbf{x}) < 0$ or $T(\mathbf{x}) > 0$. "At least as extreme as the observed" is in a sense redefined as "at least as extreme as the observed, in the observed direction". If the median of the null distribution of $T(\mathbf{X})$ is 0, then, for $T(\mathbf{x}) > 0$,

$$P_{\theta_0}(T(\mathbf{X}) \geq T(\mathbf{x})|T(\mathbf{x}) > 0) = 2 \cdot P_{\theta_0}(T(\mathbf{X}) \geq T(\mathbf{x})),$$

i.e., twice the unconditional probability that $T(\mathbf{X}) \geq T(\mathbf{x})$. Similarly, for $T(\mathbf{x}) < 0$,

$$P_{\theta_0}(T(\mathbf{X}) \leq T(\mathbf{x})|T(\mathbf{x}) < 0) = 2 \cdot P_{\theta_0}(T(\mathbf{X}) \leq T(\mathbf{x})).$$

Moreover,

$$P_{\theta_0}(T(\mathbf{X}) \geq T(\mathbf{x})) < P_{\theta_0}(T(\mathbf{X}) \leq T(\mathbf{x})) \qquad \text{when} \qquad T(\mathbf{x}) > 0$$

and

$$P_{\theta_0}(T(\mathbf{X}) \geq T(\mathbf{x})) > P_{\theta_0}(T(\mathbf{X}) \leq T(\mathbf{x})) \qquad \text{when} \qquad T(\mathbf{x}) < 0.$$

Consequently, the p-value using this approach can in general be written as

$$\lambda_{TST}(\theta_0, \mathbf{x}) := \min\left(1, \ 2 \cdot P_{\theta_0}(T(\mathbf{X}) \geq T(\mathbf{x})), \ 2 \cdot P_{\theta_0}(T(\mathbf{X}) \leq T(\mathbf{x}))\right).$$

This definition of the p-value is frequently used also in situations where the median of the null distribution of $T(\mathbf{X})$ is not 0, despite the fact that the interpretation of the p-value as being conditioned on whether $T(\mathbf{x}) < 0$ or $T(\mathbf{x}) > 0$ is lost.

At the level α, if $T(\mathbf{x}) > 0$ the test rejects the hypothesis $\theta = \theta_0$ if

$$\lambda_{TST}(\theta_0, \mathbf{x}) = \min\left(1, 2 \cdot P_{\theta_0}(T(\mathbf{X}) \geq T(\mathbf{x}))\right) \leq \alpha.$$

This happens if and only if the one-sided test of $\theta \leq \theta_0$, also based on $T(\mathbf{X})$, rejects its null hypothesis at the $\alpha/2$ level. By the same reasoning, it is seen that the rejection region of a level α twice-the-smaller-tail test always is the union of the rejection regions of two level $\alpha/2$ one-sided tests of $\theta \leq \theta_0$ and $\theta \geq \theta_0$, respectively. The test puts equal weight to the two types of type I errors: false rejection in the two different directions. The corresponding confidence interval is therefore also equal-tailed, in the sense that the non-coverage probability is $\alpha/2$ on both sides of the interval.

Twice-the-smaller-tail p-values are in a sense computed by looking only at one tail of the null distribution. In the alternative approach, `symmetric = TRUE`, we use *strictly two-sided* p-values. Such a p-value is computed using both tails, as follows: $\lambda_{STT}(\theta_0, \mathbf{x}) = P_{\theta_0}\left(|T(\mathbf{X})| \geq |T(\mathbf{x})|\right) = P_{\theta_0}\left(\{\mathbf{X} : T(\mathbf{X}) \leq -|T(\mathbf{x})|\} \cup \{\mathbf{X} : T(\mathbf{X}) \geq |T(\mathbf{x})|\}\right).$

Under this approach, the directional type I error rates will in general not be equal to $\alpha/2$, so that the test might be more prone to falsely reject $H_0(\theta_0)$ in one direction than in another. On the other hand, the rejection region of a strictly two-sided test is typically smaller than its twice-the-smaller-tail counterpart. The coverage probabilities of the corresponding confidence interval $I_\alpha(\mathbf{X}) = (L_\alpha(\mathbf{X}), U_\alpha(\mathbf{X}))$ therefore satisfies the condition that

$$P_\theta(\theta \in I_\alpha(\mathbf{X})) = 1 - \alpha \qquad \text{for all } \alpha \in (0,1),$$

but not the stronger condition

$$P_\theta(\theta < L_\alpha(\mathbf{X})) = P_\theta(\theta > U_\alpha(\mathbf{X})) = \alpha/2 \qquad \text{for all } \alpha \in (0,1).$$

For parameters of discrete distributions, strictly two-sided hypothesis tests and confidence intervals can behave very erratically (Thulin & Zwanzig, 2017). Twice-the-smaller-tail methods are therefore always preferable when working with count data.

It is also worth noting that if the null distribution of $T(\mathbf{X})$ is symmetric about 0,

$$P_{\theta_0}(T(\mathbf{X}) \geq T(\mathbf{x})) = P_{\theta_0}(T(\mathbf{X}) \leq -T(\mathbf{x})).$$

For $T(\mathbf{x}) > 0$, unless $T(\mathbf{X})$ has a discrete distribution,

$$\begin{aligned} \lambda_{TST}(\theta_0, \mathbf{x}) &= 2 \cdot P_{\theta_0}(T(\mathbf{X}) \geq T(\mathbf{x})) \\ &= P_{\theta_0}(T(\mathbf{X}) \geq T(\mathbf{x})) + P_{\theta_0}(T(\mathbf{X}) \leq -T(\mathbf{x})) = \lambda_{STT}(\theta_0, \mathbf{x}), \end{aligned}$$

meaning that the twice-the-smaller-tail and strictly two-sided approaches coincide in this case. The ambiguity related to the definition of two-sided p-values therefore only arises under asymmetric null distributions.

14.4 Deviance tests

Consider a model with $p = n$, having a separate parameter for each observation. This model will have a perfect fit, and among all models, it attains the maximum achievable likelihood. It is known as the *saturated model*. Despite having a perfect fit, it is useless for prediction, interpretation and causality, as it is severely overfitted. It is however useful as a baseline for comparison with other models, i.e., for checking goodness-of-fit: our goal is to find a *reasonable and useful* model with almost as good a fit.

Let $L(\hat{\mu}, y)$ denote the log-likelihood corresponding to the ML-estimate for a model, with estimates $\hat{\theta}_i$. Let $L(y, y)$ denote the log-likelihood for the saturated model, with estimates $\tilde{\theta}_i$. For an exponential dispersion family, i.e., a distribution of the form

$$f(y_i; \theta_i, \phi) = \exp\left([y_i\theta_i - b(\theta_i)]/a(\phi) + c(y_i, \phi)\right),$$

(the binomial and Poisson distributions being examples of this), we have

$$L(y, y) - L(\hat{\mu}, y) = \sum_{i=1}^{n}(y_i\tilde{\theta}_i - b(\tilde{\theta}_i))/a(\phi) - \sum_{i=1}^{n}(y_i\hat{\theta}_i - b(\hat{\theta}_i))/a(\phi).$$

Typically, $a(\phi) = \phi/\omega_i$, in which case this becomes

$$\sum_{i=1}^{n}\omega_i\left(y_i(\tilde{\theta}_i - \hat{\theta}_i) - b(\tilde{\theta}_i) + b(\hat{\theta}_i)\right)/\phi =: \frac{D(y, \hat{\mu})}{2\phi},$$

where the statistic $D(y, \hat{\mu})$ is called the *deviance*.

The deviance is essentially the difference between the log-likelihoods of a model and of the saturated model. The greater the deviance, the poorer the fit. It holds that $D(y, \hat{\mu}) \geq 0$, with $D(y, \hat{\mu}) = 0$ corresponding to a perfect fit.

Deviance is used to test whether two models are equal. Assume that we have two models:

- M_0, which has p_0 parameters, with fitted values $\hat{\mu}_0$,
- M_1, which has $p_1 > p_0$ parameters, with fitted values $\hat{\mu}_1$.

We say that the models are *nested*, because M_0 is a special case of M_1, corresponding to some of the p_1 parameters of M_1 being 0. If both models give a good fit, we prefer M_0 because of its (relative) simplicity. We have $D(y, \hat{\mu}_1) \leq D(y, \hat{\mu}_0)$, since simpler models have larger deviances. Assuming that M_1 holds, we can test whether M_0 holds by using the likelihood ratio-test statistic $D(y, \hat{\mu}_0) - D(y, \hat{\mu}_1)$. If we reject the null hypothesis, M_0 fits the data poorly compared to M_1. Otherwise, the fit of M_1 is not significantly better and we prefer M_0 because of its simplicity.

14.5   Regularised regression

Linear regression is a special case of generalised linear regression. Under the assumption of normality, the least squares estimator is the maximum likelihood estimator in this setting. In what follows, we will therefore discuss how the maximum likelihood estimator is modified when using regularisation, bearing in mind that this also includes the ordinary least squares estimator for linear models.

In a regularised GLM, it is not the likelihood $L(\beta)$ that is maximised, but a regularised function $L(\beta) \cdot p(\lambda, \beta)$, where p is a penalty function that typically forces the resulting estimates to be closer to 0, which leads to a stable solution. The shrinkage parameter λ controls the size of the penalty, and therefore how much the estimates are shrunk toward 0. When $\lambda = 0$, we are back at the standard maximum likelihood estimate.

The most popular penalty terms correspond to common L_q-norms. On a log-scale, the function to be maximised is then

$$\ell(\beta) + \lambda \sum_{i=1}^{p} |\beta_i|^q,$$

where $\ell(\beta)$ is the loglikelihood of β and $\sum_{i=1}^{p} |\beta_i|^q$ is the L_q-norm, with $q \geq 0$. This is equivalent to maximising $\ell(\beta)$ under the constraint that $\sum_{i=1}^{p} |\beta_i|^q \leq \frac{1}{h(\lambda)}$, for some increasing positive function h.

In Bayesian estimation, a *prior distribution* $p(\beta)$ for the parameters β_i is used. The estimates are then computed from the conditional distribution of the β_i given the data, called the *posterior distribution*. Using Bayes' theorem, we find that

$$P(\beta|\mathbf{x}) \propto L(\beta) \cdot p(\beta),$$

i.e., that the posterior distribution is proportional to the likelihood times the prior. The Bayesian *maximum a posteriori estimator* (MAP) is found by maximising the above expression (i.e., finding the mode of the posterior). This is equivalent to the estimates from a regularised frequentist model with penalty function $p(\beta)$, meaning that regularised regression can be motivated both from a frequentist and a Bayesian perspective.

When the L_2 penalty is used, the regularised model is called *ridge regression*, for which we maximise

$$\ell(\beta) + \lambda \sum_{i=1}^{p} \beta_i^2.$$

In a Bayesian context, this corresponds to putting a standard normal prior on the β_i. This method has been invented and reinvented by several authors, from the 1940s onwards, among them Hoerl & Kennard (1970). The β_i can become very small but are never pushed all the way down to 0. The name comes from the fact that in a linear model, the OLS estimate is $\hat{\beta} = (\mathbf{X}^T\mathbf{X})^{-1}\mathbf{X}^T\mathbf{y}$, whereas the ridge estimate is $\hat{\beta} = (\mathbf{X}^T\mathbf{X} + \lambda\mathbf{I})^{-1}\mathbf{X}^T\mathbf{y}$. The $\lambda\mathbf{I}$ is the "ridge".

When the L_1 penalty is used, the regularised model is called the *lasso* (Least Absolute Shrinkage and Selection Operator), for which we maximise

$$\ell(\beta) + \lambda \sum_{i=1}^{p} |\beta_i|.$$

In a Bayesian context, this corresponds to putting a standard Laplace prior on the β_i. For this penalty, as λ increases, more and more β_i become 0, meaning that we can simultaneously perform estimation and variable selection!

References

Andersen P.K., Gill R.D. (1982). Cox's regression model for counting processes: a large sample study. *Annals of Statistics*, 10, 1100-1120.

Agresti, A. (2013). *Categorical Data Analysis*. Wiley.

Barnard, J., Rubin, D.B. (1999). Small sample degrees of freedom with multiple imputation. *Biometrika*, 86, 948-955.

Bates, D., Mächler, M., Bolker, B., Walker, S. (2015). Fitting linear mixed-effects models using lme4. *Journal of Statistical Software*, 67, 1.

Boehmke, B., Greenwell, B. (2019). *Hands-On Machine Learning with R*. CRC Press.

Bollen, K.A. (1989). *Structural Equations with Latent Variables*. Wiley Series in Probability and Mathematical Statistics. Wiley.

Box, G.E. (2013). *An Accidental Statistician*. Wiley.

Box, G.E., Cox, D.R. (1964). An analysis of transformations. *Journal of the Royal Statistical Society: Series B (Methodological)*, 26(2), 211-243.

Breiman, L., Friedman, J., Stone, C.J., Olshen, R.A. (1984). *Classification and Regression Trees*. CRC Press.

Breiman, L. (2001). Random forests. *Machine Learning*, 45(1), 5-32.

Brown, L.D., Cai, T.T., DasGupta, A. (2001). Interval estimation for a binomial proportion. *Statistical Science*, 16(2), 101-117.

Buolamwini, J., Gebru, T. (2018). Gender shades: Intersectional accuracy disparities in commercial gender classification. *Proceedings of Machine Learning Research*, 81, 1-15.

Cameron, A.C., Trivedi, P.K. (1990). Regression-based tests for overdispersion in the Poisson model. *Journal of Econometrics*, 46(3), 347-364.

Casella, G., Berger, R.L. (2002). *Statistical Inference*. Brooks/Cole.

Charytanowicz, M., Niewczas, J., Kulczycki, P., Kowalski, P.A., Lukasik, S. & Zak, S. (2010). A Complete Gradient Clustering Algorithm for Features Analysis of X-ray Images. In: *Information Technologies in Biomedicine*, Ewa Pietka, Jacek Kawa (eds.), Springer-Verlag, Berlin-Heidelberg, 15-24.

Chollet, F., Allaire, J.J. (2022). *Deep Learning with R*. Second edition. Manning.

Cochran, W.G. (1954). Some methods of strengthening the common χ^2 tests. *Biometrics*, 10, 417-451.

Committee on Professional Ethics of the American Statistical Association. (2018). *Ethical Guidelines for Statistical Practice*. https://www.amstat.org/ASA/Your-Career/Ethical-Guidelines-for-Statistical-Practice.aspx

Cook, R.D., & Weisberg, S. (1982). *Residuals and Influence in Regression*. Chapman & Hall.

Cortez, P., Cerdeira, A., Almeida, F., Matos, T., Reis, J. (2009). Modeling wine preferences by data mining from physicochemical properties. *Decision Support Systems*, 47(4), 547-553.

Costello, A.B., Osborne, J. (2005). Best practices in exploratory factor analysis: Four recommendations for getting the most from your analysis. *Practical Assessment, Research, and Evaluation*, 10(1), 7.

Cox, D. R. (1972). Regression models and life-tables. *Journal of the Royal Statistical Society: Series B (Methodological)*, 34(2), 187-202.

Dastin, J. (2018). Amazon scraps secret AI recruiting tool that showed bias against women. Reuters.

Davison, A.C., Hinkley, D.V. (1997). *Bootstrap Methods and their Application*. Cambridge University Press.

Delacre, M., Lakens, D., Leys, C. (2017). Why psychologists should by default use Welch's t-test instead of Student's t-test. *International Review of Social Psychology*, 30(1).

Drymon, M.M. (2008). *Disguised As the Devil: How Lyme Disease Created Witches and Changed History*. Wythe Avenue Press.

Eck, K., Hultman, L. (2007). One-sided violence against civilians in war: Insights from new fatality data. *Journal of Peace Research*, 44(2), 233-246.

Eddelbuettel, D., Balamuta, J.J. (2018). Extending R with C++: a brief introduction to Rcpp. *The American Statistician*, 72(1), 28-36.

Efron, B. (1983). Estimating the error rate of a prediction rule: improvement on cross-validation. *Journal of the American Statistical Association*, 78(382), 316-331.

Elston, D.A., Moss, R., Boulinier, T., Arrowsmith, C., Lambin, X. (2001). Analysis of aggregation, a worked example: numbers of ticks on red grouse chicks. *Parasitology*, 122(05), 563-569.

Fine, J.P., Gray, R.J. (1999). A proportional hazards model for the subdistribution of a competing risk. *Journal of the American statistical association*, 94(446), 496-509.

Fisher, R.A. (1935). *The Design of Experiments*. Oliver & Boyd.

Fleming, G., Bruce, P.C. (2021). *Responsible Data Science: Transparency and Fairness in Algorithms*. Wiley.

Franks, B. (Ed.) (2020). *97 Things About Ethics Everyone in Data Science Should Know*. O'Reilly Media.

Friedman, J.H. (2002). Stochastic Gradient Boosting, *Computational Statistics and Data Analysis*, 38(4), 367-378.

Gao, L.L, Bien, J., Witten, D. (2022). Selective inference for hierarchical clustering. *Journal of the American Statistical Association*, DOI: 10.1080/01621459.2022.2116331.

Groll, A., Tutz, G. (2014). Variable selection for generalized linear mixed models by L1-penalized estimation. *Statistics and Computing*, 24(2), 137-154.

Hall, P. (1992). *The Bootstrap and Edgeworth Expansion*. Springer Science & Business Media.

Hartigan, J.A., Wong, M.A. (1979). Algorithm AS 136: A k-means clustering algorithm. *Journal of the Royal Statistical Society: Series C (Applied Statistics)*, 28(1), 100-108.

Henderson, H.V., Velleman, P.F. (1981). Building multiple regression models interactively. *Biometrics*, 37, 391–411.

Herr, D.G. (1986). On the history of ANOVA in unbalanced, factorial designs: the first 30 years. *The American Statistician*, 40(4), 265-270.

Hoerl, A.E., Kennard, R.W. (1970). Ridge regression: biased estimation for nonorthogonal problems. *Technometrics*, 12(1), 55-67.

Holzinger, K., Swineford, F. (1939). *A Study in Factor Analysis: The Stability of a Bifactor Solution. Supplementary Educational Monograph, no. 48.* University of Chicago Press.

Hu, L.; Bentler, P.M. (1999). Cutoff criteria for fit indexes in covariance structure analysis: conventional criteria versus new alternatives. *Structural Equation Modeling*. 6 (1): 1-55.

Hyndman, R. J., Athanasopoulos, G. (2018). *Forecasting: Principles and Practice*. OTexts.

Imai, K., Keele, L., Yamamoto, T. (2010). Identification, inference, and sensitivity analysis for causal mediation effects. *Statistical Science*, 25(1), 51-71.

James, G., Witten, D., Hastie, T., Tibshirani, R. (2021). *An Introduction to Statistical Learning with Applications in R*. Springer.

Kuznetsova, A., Brockhoff, P. B., Christensen, R. H. (2017). lmerTest package: tests in linear mixed effects models. *Journal of Statistical Software*, 82(13), 1-26.

Liero, H., Zwanzig, S. (2012). *Introduction to the Theory of Statistical Inference*. CRC Press.

Liu, X., Swenson, N.G., Lin, D., Mi, X., Umaña, M.N., Schmid, B., Ma, K. (2016). Linking individual-level functional traits to tree growth in a subtropical forest. *Ecology (Durham)*, 97(9), 2396-2405.

Long, J.D., Teetor, P. (2019). *The R Cookbook*. O'Reilly Media.

Moen, A., Lind, A.L., Thulin, M., Kamali–Moghaddamd, M., Roe, C., Gjerstad, J., Gordh, T. (2016). Inflammatory serum protein profiling of patients with lumbar radicular pain one year after disc herniation. *International Journal of Inflammation*, 2016, Article ID 3874964.

Persson, I., Arnroth, L., Thulin, M. (2019). Multivariate two-sample permutation tests for trials with multiple time-to-event outcomes. *Pharmaceutical Statistics*, 18(4), 476-485.

Petterson, T., Högbladh, S., Öberg, M. (2019). Organized violence, 1989-2018 and peace agreements. *Journal of Peace Research*, 56(4), 589-603.

Picard, R.R., Cook, R.D. (1984). Cross-validation of regression models. *Journal of the American Statistical Association*, 79(387), 575–583.

Prentice R.L., Williams B.J., Peterson A.V. (1981). On the regression analysis of multivariate failure time data. *Biometrika*, 68, 373-379.

Rasch, D., Kubinger, K.D., Moder, K. (2011). The two-sample t test: pre-testing its assumptions does not pay off. *Statistical Papers*, 52(1), 219.

Recht, B., Roelofs, R., Schmidt, L., Shankar, V. (2019). Do ImageNet classifiers generalize to ImageNet?. arXiv:1902.10811.

Rubin, D.B. (1987). *Multiple Imputation for Nonresponse in Surveys*. John Wiley & Sons.

Schoenfeld, D. (1982). Partial residuals for the proportional hazards regression model. *Biometrika*, 69(1), 239-241.

Scrucca, L., Fop, M., Murphy, T.B., Raftery, A.E. (2016). mclust 5: clustering, classification and density estimation using Gaussian finite mixture models. *The R Journal*, 8(1), 289.

Smith, G. (2018). Step away from stepwise. *Journal of Big Data*, 5(1), 32.

Tibshirani, R. (1996). Regression shrinkage and selection via the lasso. *Journal of the Royal Statistical Society: Series B (Methodological)*, 58(1), 267-288.

Tibshirani, R., Walther, G., Hastie, T. (2001). Estimating the number of clusters in a data set via the gap statistic. *Journal of the Royal Statistical Society: Series B (Statistical Methodology)*, 63(2), 411-423.

Thulin, M. (2014a). The cost of using exact confidence intervals for a binomial proportion. *Electronic Journal of Statistics*, 8, 817-840.

Thulin, M. (2014b). *On Confidence Intervals and Two-Sided Hypothesis Testing*. PhD thesis. Department of Mathematics, Uppsala University.

Thulin, M. (2014c). Decision-theoretic justifications for Bayesian hypothesis testing using credible sets. *Journal of Statistical Planning and Inference*, 146, 133-138.

Thulin, M. (2016). Two-sample tests and one-way MANOVA for multivariate biomarker data with nondetects. *Statistics in Medicine*, 35(20), 3623-3644.

Thulin, M., Zwanzig, S. (2017). Exact confidence intervals and hypothesis tests for parameters of discrete distributions. *Bernoulli*, 23(1), 479-502.

Tobin, J. (1958). Estimation of relationships for limited dependent variables. *Econometrica*, 26, 24-36.

Wasserstein, R.L., Lazar, N.A. (2016). The ASA statement on p-values: context, process, and purpose. *The American Statistician*, 70(2), 129-133.

Wei, L.J. (1992). The accelerated failure time model: a useful alternative to the Cox regression model in survival analysis. *Statistics in Medicine*, 11(14-15), 1871-1879.

Wickham, H. (2019). *Advanced R*. CRC Press.

Wickham, H., Bryan, J. (2023). *R Packages*. O'Reilly Media.

Wickham, H., Grolemund, G. (2017). *R for Data Science*. O'Reilly Media.

Wickham, H., Navarro, D., Lin Pedersen, T. (forthcoming). *ggplot2: Elegant Graphics for Data Analysis*. Third edition.

Wilke, C.O. (2019). *Fundamentals of Data Visualization*. O'Reilly Media.

Xie, Y., Allaire, J.J., Grolemund, G. (2018). *R Markdown: the definitive guide*. Chapman & Hall.

Zeileis, A., Hothorn, T., Hornik, K. (2008). Model-based recursive partitioning. *Journal of Computational and Graphical Statistics*, 17(2), 492-514.

Zhang, D., Fan, C., Zhang, J., Zhang, C.-H. (2009). Nonparametric methods for measurements below detection limit. *Statistics in Medicine*, 28, 700–715.

Zhang, Y., Yang, Y. (2015). Cross-validation for selecting a model selection procedure. *Journal of Econometrics*, 187(1), 95-112.

Zou, H., Hastie, T. (2005). Regularization and variable selection via the elastic net. *Journal of the Royal Statistical Society: Series B (Methodological)*, 67(2), 301-320.

Further reading

Below is a list of some highly recommended books that either partially overlap with the content in this book or serve as a natural next step after you finish reading this book. All of these are available for free online.

- *The R Cookbook* (https://rc2e.com/) by Long & Teetor (2019) contains tons of examples of how to perform common tasks in R.
- *R for Data Science* (https://r4ds.had.co.nz/) by Wickham & Grolemund (2017) is similar in scope to Chapters 2-6 of this book, but with less focus on statistics and greater focus on tidyverse functions.
- *Advanced R* (http://adv-r.had.co.nz/) by Wickham (2019) deals with advanced R topics, delving further into object-oriented programming, functions, and increasing the performance of your code.
- *R Packages* (https://r-pkgs.org/) by Wickham and Bryan (2023) describes how to create your own R packages.
- *ggplot2: Elegant Graphics for Data Analysis* (https://ggplot2-book.org/) by Wickham, Navarro & Lin Pedersen is an in-depth treatise of `ggplot2`.
- *Fundamentals of Data Visualization* (https://clauswilke.com/dataviz/) by Wilke (2019) is a software-agnostic text on data visualisation, with tons of useful advice.
- *R Markdown: the definitive guide* (https://bookdown.org/yihui/rmarkdown/) by Xie et al. (2018) describes how to use R Markdown for reports, presentations, dashboards, and more.
- *An Introduction to Statistical Learning with Applications in R* (https://www.statlearning.com/) by James et al. (2021) provides an introduction to methods for regression and classification, with examples in R (but not using `caret`).
- *Hands-On Machine Learning with R* (https://bradleyboehmke.github.io/HOML/) by Boehmke & Greenwell (2019) covers a large number of machine learning methods.
- *Forecasting: principles and practice* (https://otexts.com/fpp2/) by Hyndman & Athanasopoulos, G. (2018) deals with forecasting and time series models in R.
- *Deep Learning with R* (https://livebook.manning.com/book/deep-learning-with-r/) by Chollet & Allaire (2018) delves into neural networks and deep learning, including computer vision and generative models.

Online resources

- A number of reference cards and cheat sheets can be found online. I like the one at: https://cran.r-project.org/doc/contrib/Short-refcard.pdf
- R-bloggers (https://www.r-bloggers.com/) collects blog posts related to R. A great place to discover new tricks and see how others are using R.
- RSeek (http://rseek.org/) provides a custom Google search with the aim of only returning pages related to R.
- Stack Overflow (https://stackoverflow.com/questions/tagged/r) and its sister-site Cross Validated (https://stats.stackexchange.com/) are questions-and-answers sites. They are great places for asking questions, and in addition, they already contain a

ton of useful information about all things R-related. The RStudio Community (https://community.rstudio.com/) is another good option.

- The R Journal (https://journal.r-project.org/) is an open-access peer-reviewed journal containing papers on R, mainly describing new add-on packages and their functionality.

Index

For Product Safety Concerns and Information please contact our EU
representative GPSR@taylorandfrancis.com
Taylor & Francis Verlag GmbH, Kaufingerstraße 24, 80331 München, Germany